RIPARIAN AREAS

Functions and Strategies for Management

Committee on Riparian Zone Functioning and Strategies for Management
Water Science and Technology Board
Board on Environmental Studies and Toxicology
Division on Earth and Life Studies
National Research Council

NATIONAL ACADEMY PRESS
Washington, D.C.

NATIONAL ACADEMY PRESS • 2101 Constitution Avenue, N.W. • Washington, D.C. 20418
NOTICE: The project that is the subject of this report was approved by the Governing Board of the National Research Council, whose members are drawn from the councils of the National Academy of Sciences, the National Academy of Engineering, and the Institute of Medicine. The members of the committee responsible for the report were chosen for their special competences and with regard for appropriate balance.

Support for this project was provided by the U.S. Bureau of Land Management under Contract No. PAA007017, U.S. Bureau of Reclamation under Contract No. 99-FG-81-0-0154, the U.S. Department of Agriculture Natural Resources Conservation Service and Forest Service under Contract No. 68-3A75-0-170, the National Science Foundation under Contract No. DEB-9909095, the U.S. Environmental Protection Agency under Contract No. CR 826316-01-0, and the U.S. Geological Survey under Contract No. 98HQSA0429.

Any opinions, findings, and conclusions or recommendations expressed in this material are those of the authors and do not necessarily reflect the views of the U.S. Government.

International Standard Book Number: 0-309-08295-1
Library of Congress Control Number: 2002105957

Additional copies of this report are available from the National Academy Press, 2101 Constitution Avenue, N.W., Box 285, Washington, DC 20055; (800) 624-6242 or (202) 334-3313 (in the Washington metropolitan area); Internet <http://www.nap.edu>.

Copyright 2002 by the National Academy of Sciences. All rights reserved.

Printed in the United States of America

THE NATIONAL ACADEMIES

National Academy of Sciences
National Academy of Engineering
Institute of Medicine
National Research Council

The **National Academy of Sciences** is a private, nonprofit, self-perpetuating society of distinguished scholars engaged in scientific and engineering research, dedicated to the furtherance of science and technology and to their use for the general welfare. Upon the authority of the charter granted to it by the Congress in 1863, the Academy has a mandate that requires it to advise the federal government on scientific and technical matters. Dr. Bruce M. Alberts is president of the National Academy of Sciences.

The **National Academy of Engineering** was established in 1964, under the charter of the National Academy of Sciences, as a parallel organization of outstanding engineers. It is autonomous in its administration and in the selection of its members, sharing with the National Academy of Sciences the responsibility for advising the federal government. The National Academy of Engineering also sponsors engineering programs aimed at meeting national needs, encourages education and research, and recognizes the superior achievements of engineers. Dr. Wm. A. Wulf is president of the National Academy of Engineering.

The **Institute of Medicine** was established in 1970 by the National Academy of Sciences to secure the services of eminent members of appropriate professions in the examination of policy matters pertaining to the health of the public. The Institute acts under the responsibility given to the National Academy of Sciences by its congressional charter to be an adviser to the federal government and, upon its own initiative, to identify issues of medical care, research, and education. Dr. Harvey V. Fineberg is president of the Institute of Medicine.

The **National Research Council** was organized by the National Academy of Sciences in 1916 to associate the broad community of science and technology with the Academy's purposes of furthering knowledge and advising the federal government. Functioning in accordance with general policies determined by the Academy, the Council has become the principal operating agency of both the National Academy of Sciences and the National Academy of Engineering in providing services to the government, the public, and the scientific and engineering communities. The Council is administered jointly by both Academies and the Institute of Medicine. Dr. Bruce M. Alberts and Dr. Wm. A. Wulf are chairman and vice chairman, respectively, of the National Research Council.

COMMITTEE ON RIPARIAN ZONE FUNCTIONING AND STRATEGIES FOR MANAGEMENT

MARK M. BRINSON, *Chair*, East Carolina University, Greenville, North Carolina
LAWRENCE J. MacDONNELL, *Vice Chair*, Porzak, Browning & Bushong, Boulder, Colorado
DOUGLAS J. AUSTEN, Illinois Department of Natural Resources, Springfield
ROBERT L. BESCHTA, Oregon State University, Corvallis
THEO A. DILLAHA, Virginia Polytechnic Institute and State University, Blacksburg
DEBRA L. DONAHUE, University of Wyoming, Laramie
STANLEY V. GREGORY, Oregon State University, Corvallis
JUDSON W. HARVEY, U.S. Geological Survey, Reston, Virginia
MANUEL C. MOLLES, Jr., University of New Mexico, Albuquerque
ELIZABETH I. ROGERS, White Water Associates, Inc., Amasa, Michigan
JACK A. STANFORD, University of Montana, Polson

NRC Staff

LAURA J. EHLERS, Study Director
ANITA A. HALL, Senior Project Assistant

WATER SCIENCE AND TECHNOLOGY BOARD

RICHARD G. LUTHY, *Chair*, Stanford University, Stanford, California
JOAN B. ROSE, *Vice Chair*, University of South Florida, St. Petersburg
RICHELLE M. ALLEN-KING, Washington State University, Pullman
GREGORY B. BAECHER, University of Maryland, College Park
KENNETH R. BRADBURY, Wisconsin Geological and Natural History Survey, Madison
JAMES CROOK, CH2M Hill, Boston, Massachusetts
EFI FOUFOULA-GEORGIOU, University of Minnesota, Minneapolis
PETER GLEICK, Pacific Institute for Studies in Development, Environment, and Security, Oakland, California
STEVEN P. GLOSS, U.S. Geological Survey, Flagstaff, Arizona
JOHN LETEY, Jr., University of California, Riverside
DIANE M. McKNIGHT, University of Colorado, Boulder
CHRISTINE L. MOE, Emory University, Atlanta, Georgia
RUTHERFORD H. PLATT, University of Massachusetts, Amherst
JERALD L. SCHNOOR, University of Iowa, Iowa City
LEONARD SHABMAN, Virginia Polytechnic Institute and State University, Blacksburg
R. RHODES TRUSSELL, Montgomery Watson, Pasadena, California

Staff

STEPHEN D. PARKER, Director
LAURA J. EHLERS, Senior Staff Officer
JEFFREY W. JACOBS, Senior Staff Officer
WILLIAM S. LOGAN, Senior Staff Officer
MARK C. GIBSON, Staff Officer
M. JEANNE AQUILINO, Administrative Associate
ELLEN A. DE GUZMAN, Research Associate
PATRICIA JONES KERSHAW, Study/Research Associate
ANITA A. HALL, Administrative Assistant
ANIKE L. JOHNSON, Project Assistant
JON Q. SANDERS, Project Assistant

BOARD ON ENVIRONMENTAL STUDIES AND TOXICOLOGY

GORDON ORIANS, *Chair*, University of Washington, Seattle
JOHN DOULL, *Vice Chair*, University of Kansas Medical Center, Kansas City
DAVID ALLEN, University of Texas, Austin
INGRID C. BURKE, Colorado State University, Fort Collins
THOMAS BURKE, Johns Hopkins University, Baltimore, Maryland
WILLIAM L. CHAMEIDES, Georgia Institute of Technology, Atlanta
CHRISTOPHER B. FIELD, Carnegie Institute of Washington, Stanford, California
J. PAUL GILMAN, Celera Genomics, Rockville, Maryland
DANIEL S. GREENBAUM, Health Effects Institute, Cambridge, Massachusetts
BRUCE D. HAMMOCK, University of California, Davis
ROGENE HENDERSON, Lovelace Respiratory Research Institute, Albuquerque, New Mexico
CAROL HENRY, American Chemistry Council, Arlington, Virginia
ROBERT HUGGETT, Michigan State University, East Lansing
JAMES H. JOHNSON, Howard University, Washington, D.C.
JAMES F. KITCHELL, University of Wisconsin, Madison
DANIEL KREWSKI, University of Ottawa, Ottawa, Ontario
JAMES A. MacMAHON, Utah State University, Logan
WILLEM F. PASSCHIER, Health Council of the Netherlands, The Hague
ANN POWERS, Pace University School of Law, White Plains, New York
LOUISE M. RYAN, Harvard University, Boston, Massachusetts
KIRK SMITH, University of California, Berkeley
LISA SPEER, Natural Resources Defense Council, New York, New York

Staff

JAMES J. REISA, Director
DAVID J. POLICANSKY, Associate Director and Senior Program Director for Applied Ecology
RAYMOND A. WASSEL, Senior Program Director for Environmental Sciences and Engineering
KULBIR BAKSHI, Program Director for the Committee on Toxicology
ROBERTA M. WEDGE, Program Director for Risk Analysis
K. JOHN HOLMES, Senior Staff Officer
RUTH E. CROSSGROVE, Managing Editor

Preface

Lands next to water are fundamental to the livelihood of many species of plants and animals, including humans. Birds and other wildlife aggregate in riparian areas, often in great abundance. At the same time, society values riparian areas for production of food and fiber, access to transportation, opportunities for recreation, and natural scenic beauty. This report examines the structure and functioning of riparian areas, how they have been altered by human activity, their legal status, and their potential for management and restoration.

The committee assembled to write this report represents diverse backgrounds, among them various aspects of ecology, hydrology, environmental engineering, and water resources law and policy. Further, committee members come from different geographical areas and have had varied research and management experiences. The group met five times over a period of two years.

This study is an outgrowth of the National Research Council (NRC) report *Wetlands: Characteristics and Boundaries*. The 1995 study recognized that floodplains of rivers in very different climates had similar functions, but that only those in humid climates were wet enough to be designated as wetlands. It became apparent that wetlands were being defined by the presence of a minimum amount of soil saturation necessary to select for plants capable of tolerating oxygen deficiency in the rooting zone—a definition that excludes many riparian areas. Riparian areas on the other hand are primarily defined by their position as those lands bordering streams, rivers, and lakes. (And there is no justifiable reason to exclude shorelines of estuaries and marine coasts.) Although wetlands and riparian areas provide many of the same environmental functions, the differences between their definitions are reflected in vastly different levels of protection.

The current legal and regulatory status of riparian areas is amazingly diverse. There are significant differences in how riparian areas are treated depending on whether they are publicly or privately owned; whether they are under federal, state, or local jurisdiction; and whether the land is agricultural, silvicultural, rangeland, or urban. This is in stark contrast to the situation with wetlands where a federal presence has provided some stability to local practices for more than two decades.

Largely because of the geographic diversity of riparian areas, it is unrealistic to expect that one approach for restoration, or a handful of legal strategies, can resolve the multitude of problems and issues that face riparian areas. In spite of that, many of the recommendations in this report are derived from the fundamental principle that riparian areas are driven by hydrology, and that hydrologic alterations are among the most pernicious impacts.

For a variety of reasons, the committee did not specifically treat economic issues, although several guest speakers provided economic perspectives and analyses. The costs and benefits of managing and protecting riparian areas weighed heavily in our deliberations. This is especially evident in the section on legal and social issues, both inherently economic topics.

Any study of this scope should be comprehensive and regionally balanced. However, much of the imagery generated during riparian discussions at a national level had the western states as a backdrop. Further, the topics of dry-land irrigation, water law, and livestock grazing on public lands are strongly associated with the term "riparian." Although these same topics apply in other regions of the country, they are nowhere as obvious (and in need of attention) than in the American West where so much land is held in the public trust. Where federal policies have caused problems, in most cases federal policies must be invoked to fix them.

The committee would like to thank those who participated in its deliberations. Presentations from sponsoring organizations were made by Joe Williams, John Meagher, and Steve Schmelling from the U.S. Environmental Protection Agency; Gail Mallard and Jonathan Friedman from the U.S. Geological Survey; Mitch Flanagan, Dennis Thompson, and Dave Seery from USDA Natural Resources Conservation Service; Don Prichard from the Bureau of Land Management; and Jerry Christner and Jim Sedell from the U.S. Forest Service. Invited presenters also included Anne Hairston-Strang from the Maryland Department of Natural Resources; Juliet Stromberg and Robert Ohmart from Arizona State University; Patrick McCarthy from The Nature Conservancy; David Kovacic and John Braden from the University of Illinois at Urbana; Joe Colletti from Iowa State University; Mike Dosskey from the U.S. Forest Service National Agroforestry Center; and Chuck Elliot and Dennis Peters from the U.S. Fish and Wildlife Service.

The committee was fortunate to have taken several field trips in conjunction with committee meetings. The following individuals are thanked for their participation in organizing and guiding these trips: Robert Parmenter, Cliff Dahm, and James Gosz from the University of New Mexico; Richard Schultz, Tom Isenhart, and Bill Simpkins from Iowa State University; Jerry Hatfield and Dana Dinnes of the U.S. Agricultural Research Service; Wayne Elmore of the Bureau of Land Management; Susan Holzman of the U.S. Forest Service; and John Anderson, retired.

Committee members are grateful to the leadership provided by Laura Ehlers of the National Research Council in serving as the institutional memory of the committee, organizing committee meetings, and synthesizing, coordinating, and editing the report. Anita Hall was instrumental in creating problem-free arrangements for our meetings.

More formally, the report has been reviewed by individuals chosen for their diverse perspectives and technical expertise, in accordance with procedures approved by the NRC's Report Review Committee. The purpose of this independent review is to provide candid and critical comments that will assist the authors and the NRC in making the published report as sound as possible and to ensure that the report meets institutional standards for objectivity, evidence, and responsiveness to the study charge. The reviews and draft manuscripts remain confidential to protect the integrity of the deliberative process. We thank the following individuals for their participation in the review of this report: Robert Adler, University of Utah College of Law; Paul Barten, University of Massachusetts Amherst; Nancy Grimm, Arizona State University; Clayton Marlow, Montana State University; Robert Naiman, University of Washington; Brian Richter, The Nature Conservancy; Richard Schultz, Iowa State University; and Juliet Stromberg, Arizona State University.

Although the reviewers listed above have provided many constructive comments and suggestions, they were not asked to endorse the conclusions or recommendations, nor did they see the final draft of the report before its release. The review of this report was overseen by Wilford Gardner, University of California at Berkeley. Appointed by the NRC, he was responsible for making certain that an independent examination of this report was carried out in accordance with institutional procedures and that all review comments were carefully considered. Responsibility for the final content of this report rests entirely with the authoring committee and the NRC.

Mark M. Brinson, Chair

Contents

Summary ... 1

1 INTRODUCTION ... 23
Historical Use of Riparian Areas, 24
Definition of "Riparian," 29
Distinguishing Riparian Areas from Other Areas, 34
Scope of the Study, 43
Conclusion, 46
References, 46

2 STRUCTURE AND FUNCTIONING OF RIPARIAN AREAS
ACROSS THE UNITED STATES ... 49
Fluvial Processes and Sediment Dynamics, 49
Hydrologic and Biogeochemical Processes, 58
Regional Climate and Resulting Riparian Vegetation, 77
Riparian Areas as Habitat, 109
Environmental Services of Riparian Areas, 120
Conclusions, 123
References, 127

3 HUMAN ALTERATIONS OF RIPARIAN AREAS ... 144
Hydrologic and Geomorphic Alterations, 145
Agriculture, 161
Industrial, Urban, and Recreational Impacts, 177

Current Status of Riparian Lands in the United States, 199
Conclusions and Recommendations, 207
References, 211

4   EXISTING LEGAL STRATEGIES FOR RIPARIAN
    AREA PROTECTION   225
    Protection of Privately Owned Riparian Areas, 227
    Protection of Federal Lands, 250
    Protection of Water Resources, 266
    Critical Evaluation of the Potentially Most Influential Programs, 276
    Conclusions and Recommendations, 291
    References, 295

5   MANAGEMENT OF RIPARIAN AREAS   299
    Goals of Management, 300
    Tools for Assessing Riparian Areas, 317
    Management Strategies, 352
    Conclusions and Recommendations, 407
    References, 409

APPENDIX A: COMMITTEE MEMBER AND STAFF BIOGRAPHIES   425

Summary

OVERVIEW

The federal Clean Water Act requires that wetlands be protected from degradation because of their multiple, important ecological roles including maintenance of high water quality and provision of habitat for fish and wildlife. For the last 15 years, this protection has slowed the precipitous decline in wetland acreage observed in the United States since European settlement. However, protection of wetlands generally does not encompass riparian areas—the lands bordering waterbodies such as rivers, lakes, and estuaries—even though they often provide many of the same functions as wetlands. Especially in more arid regions of the country, riparian areas support the vast majority of wildlife species, they are the predominant sites of woody vegetation including trees, and they surround what are often the only available surface water supplies. These features have made riparian areas attractive for human development, leading to their alteration on a scale similar to that of wetlands degradation.

Growing recognition of the similarities in functioning of wetlands and riparian areas and the differences in their legal protection led the National Research Council (NRC) in 1999 to undertake a comprehensive study of riparian areas (with sponsorship from seven federal agencies). The goals of the study were to:

- define what the term "riparian" encompasses,
- describe the structure and functioning of natural riparian areas (noting differences across the United States as well as between riparian areas and adjacent waterbodies and uplands),

- document the impacts to riparian areas from humans and assess present-day riparian acreage,
- evaluate methods that assess the condition of riparian areas,
- suggest improved management of riparian areas on forested, agricultural, and developed land (including strategies for complete ecological restoration of riparian areas as well as partial functioning to reflect various objectives), and
- explore the myriad federal, state, and local laws, regulations, policies, and guidance documents affecting riparian areas.

As outlined below, the NRC committee reached several overarching conclusions and recommendations intended to heighten awareness of riparian areas commensurate with their ecological and societal values. More detailed conclusions and recommendations are found in this summary and throughout the report.

Restoration of riparian functions along America's waterbodies should be a national goal. Over the last several decades, federal and state programs have increasingly focused on the need for maintaining or improving water quality, ensuring the sustainability of fish and wildlife species, protecting wetlands, and reducing the impacts of flood events. Because riparian areas perform a disproportionate number of biological and physical functions on a unit area basis, their restoration can have a major influence on achieving the goals of the Clean Water Act, the Endangered Species Act, and flood damage control programs.

Protection should be the goal for riparian areas in the best ecological condition, while restoration is needed for degraded riparian areas. Management of riparian areas should give first priority to protecting those areas in natural or nearly natural condition from future alterations. The restoration of altered or degraded areas could then be prioritized in terms of their relative potential value for providing environmental services and/or the cost effectiveness and likelihood that restoration efforts would succeed. Where degradation has occurred—as it has in many riparian areas throughout the United States—there are vast opportunities for restoring functioning to these areas.

Patience and persistence in riparian management is needed. The current degraded status of many riparian areas throughout the country represents the cumulative, long-term effects of numerous, persistent, and often incremental impacts from a wide variety of land uses and human alterations. Substantial time (years to decades) will be required for improving and restoring the functions of many degraded riparian areas. Commensurate with restoration must be efforts to improve society's understanding of what riparian functions have been lost and what can be recovered.

Although many riparian areas can be restored and managed to provide many of their natural functions, they are not immune to the effects of poor management in adjacent uplands. Upslope management can significantly alter the magnitude and timing of overland flow, the production of sediment, and the quality of water arriving at a downslope riparian area, thereby influencing the capability of riparian areas to fully function. Therefore, upslope practices contributing to riparian degradation must be addressed if riparian areas are to be improved. In other words, riparian area management must be a component of good watershed management.

DEFINING "RIPARIAN"

Riparian areas have received variable levels of attention depending on the field of inquiry. For over 100 years, the term "riparian" has been closely associated with water law. A "riparian" water right generally provides a landowner whose property borders a stream, river, or other body of water the right to use a portion of that water for various purposes. Recognition of the term "riparian" in the basic sciences has been much more recent; since 1970 there has been an explosion of information addressing ecological, hydrologic, biogeochemical, aesthetic, cultural, and social topics related to riparian areas.

Because of the relative newness of our understanding of how riparian areas function, a single precise ecological definition of the term has not yet emerged. However, most scientific definitions of "riparian" share common features, including mention of location in the landscape, hydrology, and sometimes vegetation and soil type. Because the lack of a consistent definition has been identified as a major problem of federal and state programs that might manage and protect these areas, the NRC committee developed the following definition.

> **Riparian areas are transitional between terrestrial and aquatic ecosystems and are distinguished by gradients in biophysical conditions, ecological processes, and biota. They are areas through which surface and subsurface hydrology connect waterbodies with their adjacent uplands. They include those portions of terrestrial ecosystems that significantly influence exchanges of energy and matter with aquatic ecosystems (i.e., a zone of influence). Riparian areas are adjacent to perennial, intermittent, and ephemeral streams, lakes, and estuarine-marine shorelines.**

An important feature of this definition is the concept of riparian areas having gradients in environmental conditions and in functions between uplands and aquatic ecosystems. The shaded zone of influence in Figure ES-1 represents this gradient. Although riparian areas encompass some of the wetlands in a typical landscape setting and also include portions of adjacent aquatic and upland environments, important distinctions between these systems are made (Chapter 1).

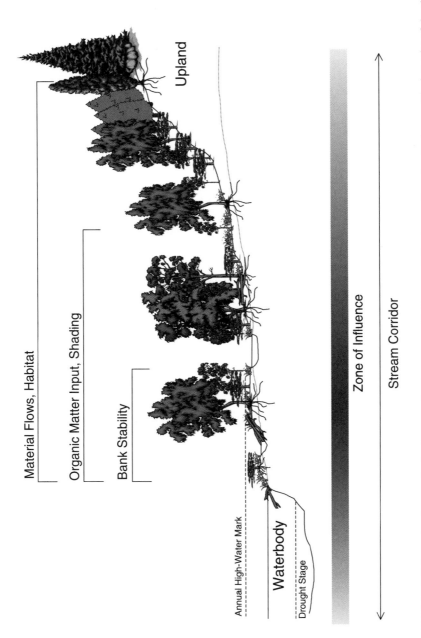

FIGURE ES-1 Schematic of a generic riparian area showing a zone of influence relative to aquatic and upland areas. The intensity of riparian influence is depicted with shading. "Material flows" refers to energy, organic matter, water, sediment, and nutrient flow.

STRUCTURE AND FUNCTIONING OF RIPARIAN AREAS

Riparian areas are the products of water and material interactions in three dimensions—longitudinal, lateral, and vertical. They include portions of the channel system and associated features (gravel bars, islands, wood debris); a vegetated zone of varying successional states influenced by floods, sediment deposition, soil-formation processes, and water availability; and a transitional zone to the uplands of the valley wall—all underlain by an alluvial aquifer.

Fluvial Processes, Sediment Dynamics, and Biogeochemical Interactions

Riparian areas receive water from three main sources besides precipitation: (1) groundwater discharge, (2) overland and shallow subsurface flow from adjacent uplands (hillslope runoff), and (3) flow from the adjacent surface waterbody by multiple pathways. Riparian areas can both gain or lose water as a result of interactions with groundwater. Depending on the setting, groundwater may pass directly through riparian sediments or bypass them entirely through deeper flow paths. Flow across the surface of riparian areas can occur via overbank flow from the channel, hillslope runoff, and rainfall directly onto those saturated areas. The relative importance of overbank flow versus hillslope runoff tends to increase with increasing stream order.

The channel can also supply water to riparian areas via infiltration through channel banks. Indeed, bank storage refers to channel water moving laterally into subsurface riparian areas when river stage is high, and then gradually moving back to the channel when river stage drops. Hyporheic exchange involves flow within the stream channel that enters subsurface sediments and returns to the channel at a downstream location. All of these processes affect water storage, physical and chemical transformations in riparian areas, and the composition and extent of riparian plant communities. Riparian areas that become hydrologically disconnected from their adjacent stream channels (e.g., via levees or channel incision) lose many of their ecological functions.

Varying flow regimes have corresponding sediment dynamics that help shape riparian areas. Although precipitation and runoff promote erosion of uplands and the transport of sediment into stream channels, riparian areas may trap some of these particles. Once sediment-laden water reaches the stream system, the decrease in channel gradient downstream leads first to the deposition of course material (gravel, cobble) in the middle reaches followed by the deposition of fine materials (sand, silt) in the lowest-velocity environments (e.g., coastal segments). Floods tend to rework channel sediments such that riparian areas continually gain and lose substrates—key processes that determine their nature and productivity. Riparian vegetation plays a role in sediment dynamics by providing hydraulic resistance during floods. Subsequent to floods, riparian vegetation becomes established on newly exposed areas of the channel bed, streambanks, and floodplains, providing stability to these areas during subsequent floods.

Soils found in riparian areas have pronounced spatial variability in structure, particle size distribution, and other properties—not only across a riparian area, but also vertically within a given soil profile. This variability is the result of interactions between streamflow patterns and sediment transport in conjunction with variations in local geology, channel morphology, and streamside vegetation. Periods of high flow and the lateral migration of rivers over long periods of time can create a multitude of landforms common to many riparian systems, particularly in unconstrained or relatively wide alluvial valleys. These include meanders, oxbows, natural levees, point bars, secondary channels, floodplains, and terraces.

For waterborne chemicals and particles, the major transformation and transport processes associated with riparian areas include infiltration, deposition, filtration, adsorption, degradation, and assimilation. Infiltration, during which dissolved chemicals and particulates enter the subsurface, is facilitated by macroporous vegetation or litter layers in riparian areas that offer high resistance to overland flow and decrease its velocity (which promotes deposition of sediments formerly suspended in runoff). Filtration of solid particles by riparian vegetation during overland flow traps larger soil particles, aggregates, and particulate organic matter, while adsorption to clay and organic matter in soils intercepts dissolved compounds such as orthophosphorus, heavy metals, and some pesticides. Finally, the plant and microbial assemblages in riparian areas can transform chemicals via many different mechanisms, e.g., denitrification and assimilation. These fate and transport processes occur in subsurface water moving through riparian soils, in slack-water habitats (i.e., shallow and slowly moving sections of channels), and in hyporheic zones. The importance of these mechanisms in controlling water quality depends on the amount of time that hillslope runoff is retained in the riparian area, on whether overland and subsurface flows concentrate and flow through only a portion of the riparian area, and on the stream order of the adjacent channel.

Climate and Resulting Riparian Vegetation

Climate has a strong influence on the structure and functioning of riparian areas, through temperature, precipitation, evapotranspiration, and runoff. Floods play a significant role in determining the composition of riparian vegetation by controlling the germination and successful establishment of seedlings as well as their long-term survival. Rivers in arid regions experience flood discharges that can be orders of magnitude greater than their base flows. This results in significantly greater flow variation and physical disturbance, which promote the growth of tree species and associated plant assemblages that can tolerate these disturbance conditions. The microtopographic variation created by diverse fluvial processes supports a species richness that would not otherwise occur.

Although physical disturbances are a prevalent controlling feature in many riparian areas, soil moisture and the depth to the water table also influence the composition of plant communities. For example, riparian areas in humid climates at the lowest elevations commonly experience soil anoxia brought on by persistent flooding or saturation, which in those areas is a greater factor in controlling riparian species composition than is physical disturbance.

Regional variations in climate have thus resulted in variable riparian plant communities across North America. For example, while there are from 26 to 33 tree genera in riparian areas for regions east of the Great Plains, for regions to the west the number ranges from 9 to 22. Despite wide regional variations in riparian tree genera represented, a core of genera—alders, cottonwoods, and willows—is found in riparian areas across the continent.

Riparian vegetation has many critical functions. It provides friction and resistance to flowing water and to runoff during floods, creates soil macropores by root growth and decay, and stabilizes streambanks via roots. Trees intercept, store, and evaporate a portion of incoming precipitation. Indeed, the amount of water utilized by deep-rooted riparian vegetation has been an issue of significant concern, particularly in arid regions. Riparian plant canopies have an important role in influencing stream temperature and the health of aquatic species. Finally, forested riparian areas contribute wood to streams and lakes, which helps maintain their physical habitat, slows the downstream routing of sediment and organic matter, provides increased hydraulic resistance to flow, and provides a food supply to microorganisms and invertebrates.

Environmental Services of Riparian Areas

The fundamental ecological functions that riparian areas perform fall into three major categories: (1) hydrology and sediment dynamics, (2) biogeochemistry and nutrient cycling, and (3) habitat and food web maintenance. Functions related to hydrology and sediment dynamics include storage of surface water and sediment, which reduces damage from floods downstream from the riparian area. Riparian areas intercept, cycle, and accumulate chemical constituents in shallow subsurface flow to varying degrees, with the societal benefit of removing pollutants from overland flow and shallow groundwater that might otherwise contaminate nearby waterbodies.

Maintaining biodiversity is one of the most important functions of riparian areas and is the basis for many valued fisheries, in addition to bird and other wildlife habitat. The benefits of functioning riparian areas to fish stem directly from the role of vegetation in controlling temperatures, stream structure, and sedimentation. Riparian areas themselves are home to an abundance of animal life, including invertebrates, almost all amphibian species and many reptiles, the majority of bird species (particularly in the semiarid West), and many mammal species with semiaquatic habitats. In addition to being characterized by unique assemblages of plants compared to uplands and wetlands, riparian areas frequently harbor rare plant species.

Except for support of biodiversity, some of the environmental services of riparian areas can be provided by technologies, such as reservoirs for flood control and treatment plants for pollutant removal. However, these substitutions are directed at single functions rather than the multiple functions that riparian areas carry out simultaneously and with little direct costs to society.

Riparian areas perform important hydrologic, geomorphic, and biological functions. They encompass complex above- and below-ground habitats created by the convergence of biophysical processes in the transition zone between aquatic and terrestrial ecosystems. The characteristic geomorphology, plant communities, and associated aquatic and wildlife species of riparian systems are intrinsically linked to the role of water as both an agent of disturbance and a critical requirement of biota.

Riparian areas, in proportion to their area within a watershed, perform more biologically productive functions than do uplands. They provide stream microclimate modification and shade, bank stabilization and modification of sedimentation processes, organic litter and wood to aquatic systems, nutrient retention and cycling, wildlife habitat, and food-web support for a wide range of aquatic and terrestrial organisms. Even though they occupy only a small proportion of the total land base in most watersheds, riparian areas are regional hot spots of biodiversity and exhibit high rates of biological productivity in marked contrast to the larger landscape. This is particularly dramatic in arid regions, as evidenced by the high number of plant and animal species found along watercourses and washes.

HUMAN ALTERATIONS OF RIPARIAN AREAS

Because humans worldwide use more than half of the geographically accessible river runoff, their significant impact on the structure and functioning of riparian areas is not surprising. Effects include changes in the hydrology of rivers and riparian areas, alteration of geomorphic structure, and the removal of riparian vegetation. Drastic declines in the acreage and condition of riparian lands in the United States over the last 100 years are testimony to these effects.

Hydrologic and Geomorphic Alterations

Manipulation of the hydrologic regime via the construction of dams and other structures, interbasin diversion, and irrigation has served to disconnect rivers from their riparian areas. Changes in hydrologic disturbance regimes and patterns of sediment transport include alteration of the timing of downstream flow, attenuation of peakflows, and other effects.

Dams have an immediate upstream effect—the complete loss of riparian structure and functioning due to inundation. Downstream effects include changes in the transport of sediment due to retention behind the dam such that channels below a dam can become increasingly "sediment starved." A second type of downstream alteration is related to the pattern of river flow following dam construction. Large dams can dampen the magnitude of high flows that would occur normally, increase the duration of moderate flows, or even dewater downstream reaches causing substantial declines of riparian forests. Levees are similarly effective at severing hydrologic linkages (i.e., frequency, magnitude, and duration of overbank flows) between a channel and its adjacent riparian areas. The degree of impairment is less for those levees located farther from the stream system (particularly if located outside the meander belt of a river).

Bank stabilizing structures—revetments and rip-rap, gabions, groins, and jetties—have also influenced the characteristics of riparian systems. Rip-rap eliminates microhabitats of plant species that naturally stabilize banks. In addition, because many bank structures reduce the hydraulic roughness (i.e., the frictional resistance to flow) along the channel margins, flow velocities are greater along the bank during high flows, which often precludes the survival of many riparian plant species.

Channelization converts streams into deeper, straighter, and often wider waterbodies to facilitate conveyance of water downstream so that the immediate floodplain area will not flood as long or as deeply, resulting in reduced soil water content. Channelization has the direct effect of destroying riparian vegetation via the use of heavy equipment or by moving the stream channel to a new location where no natural riparian vegetation exists. Indirectly, channelization reduces the survivability of riparian vegetation by lowering the water table and reducing the frequency of overbank flow. The increased flow capacity afforded by channelization compresses the period of water conveyance, making streams "flashier" and increasing erosion rates. Downstream effects include higher flood peaks and greater loading of sediment, nutrients, and contaminants.

Water withdrawals, both from surface waters and groundwater, can have serious deleterious effects on riparian area functioning caused by the lowering of water tables in the vicinity of riparian vegetation. Groundwater pumping for water supply throughout large areas of the West is increasingly common. Because groundwater and surface water are generally connected in floodplains, declines in groundwater level can indirectly be caused by surface water withdrawals or by the regulation of surface water flow by dam construction. Lowering groundwater levels by just one meter beneath riparian areas is sometimes sufficient to induce water stress in riparian trees, especially in the western United States.

Phreatophytic (water-loving) plants historically have been cleared from riparian areas in arid and semiarid climates because they have been viewed as competing with other users of water, particularly irrigated agriculture and municipalities. However, phreatophyte eradication destroys nearly all ecological and geomorphic benefits provided by riparian vegetation, including stabilization of alluvial fill, shading, and provision of wood and microhabitats.

Agriculture

A second major impact to riparian areas has been their conversion to other plant species via land uses such as forestry, row crop agriculture, and livestock grazing. The periodic removal of trees by forestry has the potential to alter the long-term composition and character of riparian forests. Where large portions of the standing timber are harvested or where the period between harvest operations is short, substantial changes to the composition, structure, and function of riparian forests will almost certainly result. The harvest of riparian forests can increase the amount of solar radiation reaching a stream, which can increase water temperatures and affect aquatic primary production. The removal of vegetative cover can impair the ability of riparian areas to retain water, sediment, and nutrients such as nitrogen and phosphorus. In general, the effects of forestry on riparian structure and function are much greater when forests are clear-cut or harvested right up to streambanks and lake shorelines.

Nationwide, traditional agriculture is probably the largest contributor to the decline of riparian areas. Conversion of undeveloped riparian land to agriculture has the potential to decrease infiltration and increase overland flow volumes and peak runoff rates. This results in high erosion rates that inundate riparian vegetation with sediment and limit the filtering functions of the riparian area. Stream channels accommodate the higher flows by increasing their cross-sectional area through widening of the channel or downcutting of the streambed. Tile-drained agricultural areas additionally experience the circumventing of many biological processes that typically occur in riparian areas. Finally, the transport of agricultural chemicals from upslope can negatively impact fauna and flora located in riparian areas and downstream receiving waters.

The primary effects of livestock grazing include the removal and trampling of vegetation, compaction of underlying soils, and dispersal of exotic plant species and pathogens. Grazing can also alter both hydrologic and fire disturbance regimes, accelerate erosion, and reduce plant or animal reproductive success and/or establishment of plants. Long-term cumulative effects of domestic livestock grazing involve changes in the structure, composition, and productivity of plants and animals at community, ecosystem, and landscape scales. Livestock have a disproportionate effect on riparian areas because they tend to concentrate in these areas, which are rich in forage and water. Although native ungulates can inflict similar types of damage to riparian vegetation, their impact is generally much less than that of livestock in areas that support both.

Industrial, Urban, and Recreational Impacts

A variety of mining practices can severely degrade riparian areas. Depending upon the type, size, and location of the mining operation, total hillsides can be excavated and their stream systems moved or buried. Mining spoils are sometimes deposited along stream channels and can destroy riparian vegetation, particularly if they contain toxic metals such as arsenic, cadmium, chromium, copper, lead, mercury, and zinc. When a mining operation exposes large areas of bare ground, substantial increases in overland flow and sediment production can occur during rainfall. Unless a well-designed and operated system of detention ponds is in place, such runoff may greatly increase sediment delivery to nearby riparian areas. Gold mining in valley bottoms has been particularly detrimental in that all riparian vegetation was removed and soils and underlying gravel substrates were mechanically dredged.

Transportation systems have directly and indirectly altered a large number of riparian areas. River transportation has often necessitated the removal of large wood and other obstructions from streambanks. Also significant to riparian systems have been the widespread impacts of channelization, lock construction, and other facets of maintaining these transportation corridors. Road and rail systems have been frequently sited along rivers and lakes, leading to the removal of riparian vegetation from the area occupied by the roadbed, the alteration of topography to provide a roadbed foundation, and local hydrologic modifications to reroute surface water and groundwater. Where sinuous rivers or streams were encountered during highway or railroad construction, portions of the channel were often filled to maintain a straight road alignment at the cost of reduced channel length. Bridges or culverts require the construction of abutments along the bank to provide roadway support. Because the abutments physically constrain the stream, future lateral adjustments by the stream are effectively eliminated. As discussed below for urban development, highway systems and urban roads outside of riparian areas can increase peak overland flow, thus fundamentally altering the hydrologic disturbance regime of adjacent riparian areas.

Urbanization and the accompanying increase in impervious surfaces have profoundly modified watershed hydrology and vegetation, and consequently the structure and functioning of riparian areas. As vegetation is replaced by impervious surfaces (roads, buildings, parking lots), infiltration, groundwater recharge, groundwater contributions to streams, and stream base flows all decrease, while overland flow volumes and peak runoff rates increase. Stream channels respond by increasing their cross-sectional area to accommodate the higher flows. This channel instability triggers a cycle of streambank erosion and habitat degradation in riparian areas similar to that seen with channelization. Above a certain percent imperviousness (approximately 10 to 20 percent), urban stream quality is consistently classified as poor. A secondary effect of urbanization is caused by changes in how overland flow and shallow subsurface flow enter and transverse riparian areas following development. Development promotes the formation of concentrated flows that are less likely to be dispersed within riparian areas, greatly reducing their potential for pollutant removal. For the most part, urbanization and development permanently impair the functioning of riparian areas.

Riparian areas are popular sites for recreational activities that can introduce sediment, nutrients, bacteria, petrochemicals, pesticides, and refuse to adjacent water bodies. Effects on riparian soils include trampling by foot, animal, or vehicle traffic that leads to compaction, destruction of soil biota, and increased erosion. Damage to vegetation can be incidental, as through trampling, or deliberate, as in its removal for the construction of recreational facilities or collection of firewood. Animal life can be affected negatively by recreation in riparian areas in ways that include direct disturbance, modification, or destruction of habitat; pollution; or introduction of pathogens.

The introduction of exotic plant and animal species for various purposes has had a substantial effect on riparian areas. The most common concern about exotic organisms is their displacement of native species and the subsequent alteration of ecosystem properties. For example, saltcedar has replaced cottonwood and other native riparian trees throughout much of the southwestern United States. This situation has been exacerbated by a reduction in flood flows caused by dams and by the lowering of water tables caused by water withdrawal. Other exotic plants that have become abundant in riparian communities include reed canary grass, buckthorns, scotch broom, Chinese privet, and kudzu.

Global climate change could bring about changes to riparian structure and functioning, including the shifting of riparian areas in response to sea-level rise and temperature change. Nonetheless, as significant as climate changes are likely to be, land- and water-use changes have had and will continue to have the greatest effect on riparian areas in the near and medium term.

Current Status of Riparian Lands in the United States

There have been few assessments of national riparian acreage and only a handful of comprehensive studies on the condition of riparian lands. Current estimates of riparian acreage range from 38 million to 121 million acres. Although the available data are highly variable, it is clear that riparian areas constitute a small fraction of total land area in the United States, probably less than 5 percent. Case histories show that in some areas loss of natural riparian vegetation is as much as 95 percent—indicating that riparian areas are some of the most severely altered landscapes in the country.

Information on riparian condition is similarly variable and sparse. Less than half of public riparian areas administered by the Bureau of Land Management (excluding Alaska) are rated as healthy (although this reveals little about the condition of riparian areas in the East, where the percentage of public lands is small). Water-quality impairments to 300,000 miles of streams (10 percent of the total) and to more than 5 million acres of lakes suggest that riparian areas adjacent to impaired streams are suffering similar degradation. Finally, historical trends for wetlands provide clues about trends in riparian lands, given that these areas sometimes overlap. Between 1780 and 1980, every state experienced declines in wetland acreage, with greater than 50 percent loss in 22 states.

The majority of riparian areas in the United States have been converted or degraded. Although landscape studies assessing the status of riparian areas are limited, they reveal that the spatial extent of riparian forests has been substantially reduced, plant communities on floodplains have been converted to other land uses or have been replaced with developments, and the area of both woody and non-woody riparian communities has decreased. The functions of these riparian areas are greatly diminished in comparison to what occurred historically.

There is no comprehensive or methodologically consistent monitoring of trends in riparian areas. It has only been relatively recently that assessments of the areal extent and condition of riparian systems have been undertaken. Unfortunately, these efforts have been limited in scope (covering a small fraction of perennial streams and almost no intermittent and ephemeral streams), they are difficult to compare because of differing methodologies, and they provide only a fragmented view of the nation's riparian systems.

Given the profound lack of information on riparian land status and trends, a comprehensive and rigorous assessment of riparian coverage is greatly needed. A national program to map riparian areas should incorporate broadly available remotely sensed data, such as satellite multispectral data, which could be used to classify and map land cover and land use information in each of the states.

EXISTING LEGAL STRATEGIES
FOR RIPARIAN AREA PROTECTION

Only during the last decade have riparian areas begun to receive legal recognition as places requiring special attention. The degree of protection, the focus, and the spatial coverage of laws and programs are highly variable at federal, state, and local levels. A variety of laws offer mechanisms to help protect some riparian areas or aspects of riparian areas. Few of these laws, however, reflect awareness of riparian areas as unique physical and natural systems in their own right and as landscapes supporting multiple important functions and warranting special management and protection. Rather, protection of riparian areas is an indirect consequence of other objectives, such as water-quality protection or habitat management.

Five approaches have been used to protect riparian areas, depending on whether the land is publicly or privately owned. First, certain federal laws require the evaluation of adverse effects that would be caused by federal actions, along with consideration of less environmentally damaging alternatives. Such an approach is not specific to riparian areas, nor does it require their protection, but it does ensure attention to their environmental values if they would be potentially affected by a proposed federal action. A second approach is to place special limits on activities in riparian areas on public lands. For example, in the Pacific Northwest, logging and other activities are restricted in riparian reserves that have been established on federal lands in order to protect salmon. A third approach is to regulate activities in privately owned riparian areas. Examples are found in statewide programs such as the Massachusetts Rivers Protection Act and the New Hampshire Comprehensive Shoreland Protection Act. Fourth, incentives such as cost-sharing, low-cost loans, or tax reductions may be used to encourage stewardship on private riparian areas. At the national level, several Farm Bill programs provide incentives for moving intensive agricultural practices away from streams by installing riparian buffers. Fifth, privately owned riparian lands can be purchased—either in fee or by easement—for public management.

For the regulatory and nonregulatory approaches used by the states to address protection of privately owned riparian areas, a significant limitation is that their success is measured by the number of practices implemented and rarely by actual environmental improvements. Indirect metrics of success (such as "miles of riparian buffer installed") are typical of state and federal conservation programs rather than measured improvements in water quality. Because of these uncertain metrics, and because many restoration programs are relatively new, it is difficult to know whether the federal, state, and local programs have been or will be effective in restoring structure and functioning to riparian areas on privately owned land. Interest seems to be growing in the use of conservation easements and other incentives to induce landowners to hold riparian areas as buffers, natural areas, or open space, as well as in the purchase of riparian lands for greenways or wildlife areas. The Total Maximum Daily Load (TMDL) Program (stemming from the Clean Water Act) is expected to have a significant impact on riparian areas because many of the TMDL implementation plans being developed call for restoration of riparian areas as a required management measure to achieve needed reductions in nonpoint source pollution.

The use and management of public lands and resources are governed by both federal and state laws. The specific federal laws that apply depend on which system (e.g., national forest, Bureau of Land Management land, wild and scenic river) the land is included within and what resources are at issue. Each managing agency affords some consideration to riparian areas and resources, whether by regulation or in an internal manual or policy handbook. Few specific provisions for riparian areas have been established in legislation or in executive orders, but agencies have considerable latitude to decide how and to what extent their planning and management activities will account for these areas. One result is that individual districts or units within agencies may vary in their interpretation and implementation of riparian measures established administratively. Thus, while different and additional constraints apply to management of federal riparian lands as compared to privately owned riparian lands, the constraints are not uniform from agency to agency, nor are they even uniformly interpreted and applied within agencies. For the most part, they have been established principally by administrative action, not by legislation, and thus are subject to administrative change. Riparian areas on federal lands are seldom managed as natural systems, though they may receive management attention or protection when they support resources of concern (such as wildlife or fisheries) and are threatened by certain land uses (such as livestock grazing or mining). Federal statutes contain very little guidance for land managers who face conflicts between riparian area protection and permissible land uses. Only if a federal agency proposes an activity in or affecting a riparian area that would jeopardize threatened or endangered species or violate water-quality requirements is riparian area protection clearly required.

There are opportunities for protection of riparian areas by extending water rights to instream uses. Current water law in the western states follows the doctrine of prior appropriation, whereby the first to take control of and actively develop and use a water resource holds a protectable legal right to the water against all other claimants. In the eastern states, water rights are afforded to those landowners adjacent to a water body (riparian doctrine). Though neither of these systems protects the water needs of streams or their riparian areas, there have been attempts to amend state laws to acknowledge instream water use and afford it water rights.

Management guidelines and regulations differ drastically among forest, range, agricultural, residential, and urban lands on private lands. No state has a general land-use law or framework to coordinate management of the landscape for multiple uses (e.g., forestry, grazing, agriculture, mining, urban development). Although many states have been willing to regulate or manage timber harvesting on private lands in riparian areas, they have not been nearly as willing to restrict other agricultural activities, except in some areas with demonstrated water-quality problems. Instead, the preference has been to induce change in farming practices through incentives provided by programs such as the Conservation Reserve Program, the Conservation Reserve Enhancement Program, and the Water Quality Incentive Program.

States should consider designating riparian buffer zones adjacent to waterbodies within which certain activities would be excluded and others would be managed. The broad importance of protecting riparian areas for water quality and fish and wildlife benefits calls for state-level programs of land-use regulation that treat all riparian landowners equally, such as the Massachusetts Riverfront Protection Act. At the very least, states should consider establishing such buffers for sensitive areas (as has been done for the Chesapeake Bay). In the absence of a statewide program, local governments should be encouraged to develop riparian buffer zones.

Few, if any, federal statutes refer expressly to riparian area values and as a consequence generally do not require or ensure protection of riparian areas. Even the National Wild and Scenic Rivers Act refers only to certain riparian values or resources; it does not consider riparian areas as natural systems, nor does it require integrated river corridor management. Moreover, statutes governing federal land management do not direct agencies to give priority to riparian area protection when conflicts among permissible land uses arise. This absence of a national riparian mandate stands in stark contrast to the existence of a federal wetlands law.

Public lands should be managed to protect and restore functioning riparian areas. Federal land management agencies should promulgate regulations requiring that the values and services of riparian areas (habitat-related, hydrologic, water quality, aesthetic, recreational) under their jurisdiction be restored and protected. At a minimum, agencies should assess the condition of riparian areas, develop and implement restoration plans where necessary, exclude incompatible uses, and manage other uses to ensure their compatibility with riparian area protection. Ideally, Congress should enact legislation that recognizes the myriad values of riparian areas and direct federal land management and regulatory agencies to give priority to protecting those values.

Instream flow laws can help protect riparian areas if river and stream flows are managed to mimic the natural hydrograph. Water allocation has historically favored human claims to water over using it for environmental needs. Recently, the needs of natural systems have been addressed in some cases by preserving minimum stream flows. Because riparian functioning is dependent on the full range of variation in the hydrologic regime, the reintroduction or maintenance of such flow regimes (in addition to minimum stream flow) is essential for restoring and sustaining, respectively, healthy riparian systems.

MANAGEMENT OF RIPARIAN AREAS

Strategies that reflect a spectrum of goals are needed for improving the ecological functions and the sustainability and productivity of existing riparian areas. Protection (preservation or maintenance) of intact riparian areas is vital because these areas represent valuable reference sites for understanding the goals and efficacy of other restoration efforts, and they are important sources of genetic material for the reintroduction of native biota into degraded areas. Measures to protect intact areas are often relatively easy to implement, have a high likelihood of being successful, and are less expensive than the restoration of degraded systems.

Restoration refers to the process of repairing the condition and functioning of degraded riparian areas. Ecological restoration in particular has the stated goal of regaining predisturbance characteristics. There are many riparian systems where ecological restoration is achievable. However, there are also situations where permanent changes in hydrologic disturbance regimes (e.g., dams), natural processes (e.g., global climate change), channel and floodplain morphology (e.g., channel incision), and other impacts (e.g., extirpation of species, biotic invasions) may preclude a recovery to the composition and functions that previously existed. Nevertheless, even in such situations, there are often opportunities to effect significant ecological improvement of riparian areas and to restore, at least in part, many of the functions they formerly performed.

Assessment Tools

For decision-makers to be effective in managing riparian areas, they need information on the status and condition of these areas. A variety of assessment tools are available for this purpose, although most have been developed for application to wetlands. Nonetheless, if further refined, these tools can be instrumental in prioritizing restoration activities and in more efficiently allocating resources toward restoration projects. Although there are no nationally recognized protocols for assessing the ecological condition of riparian areas, several methods and approaches are available, ranging from landscape-level to site-specific, from rapid and qualitative to research-level and model-based, and from those designed to answer ecological questions to those oriented toward socioeconomic issues.

Three reference-based methods may be particularly useful—Proper Functioning Condition (PFC), the Hydrogeomorphic Approach (HGM), and the Index of Biological Integrity (IBI). All are oriented toward evaluating the condition of ecosystems by comparing the project site with conditions expected in the absence of human activities or in least-disturbed sites. Once methods are developed in the form of guidebooks (for HGM) or indices (in the case of IBI), their application is relatively straightforward. PFC is the most rapid assessment method in that it is conducted in the field and the results are "immediately" known. While PFC is qualitative and dependent on the knowledge and judgment of a team of experts, HGM and IBI are based on quantitative data gathered and analyzed from unaltered to degraded sites prior to assessor involvement (although the collection and analysis of such data require considerable expertise). Unlike PFC, neither HGM nor IBI was developed primarily for riparian areas, and both would require modification in their approaches to data collection and analysis.

The concepts underlying most assessment tools currently used for wetlands and aquatic ecosystems are transferable to riparian areas, suggesting that these tools can be modified to assess the condition of riparian areas. In some cases, this would require an expansion from the aquatic portion to the floodplain and terrestrial parts of riparian areas. In cases where wetlands are major components of riparian areas, modifications would be minimal.

Proper Functioning Condition provides a good first-generation framework for riparian assessment. This method, which can be rapidly applied, may have its greatest utility in quickly identifying riparian areas that have been significantly degraded. However, there is currently no consideration of riparian biology in PFC because the assessment principally evaluates physical factors. Independent testing and evaluation are critical to ensure PFC's accuracy, usability, and credibility across the diverse suite of riparian areas in any given region.

The Hydrogeomorphic Approach and the Index of Biological Integrity hold considerable potential for assessing the condition of riparian areas. HGM (originally developed for wetlands) provides data useful not only for the assessment of condition, but also for the overall design of regional or watershed-scale restoration efforts. Most IBI assessments have been limited to aquatic ecosystems. Both HGM and IBI should be revised for use in riparian areas, for example by developing indices of integrity for riparian plant communities.

Management Strategies

The range of possible restoration activities in riparian areas is broad, spanning from simple activities at a single site to large-scale projects. In many cases, relatively easy things can be done to improve the condition of riparian areas, such as planting vegetation, removing small flood-control structures, or reducing or removing a stressor such as grazing or forestry. Where the objective of restoration is to improve the entire river system, more holistic watershed approaches will be necessary, and management strategies such as removing impediments to the natural hydrologic regime may be required. Restoration can be either passive (halting those activities that are causing degradation) or active (management to accelerate the development of self-sustaining and ecologically healthy systems) or both. Successful restoration requires local knowledge of hydrology and ecology including the range of natural variability, disturbance regimes, soils and landforms, and vegetation.

Reestablishing the Hydrologic Regime

Where natural hydrologic regimes (and corresponding sediment transport regimes) have been significantly altered by dams, levees, locks, low-water diversion channels, or off-stream storage ponds, perhaps the most important restoration need is to reestablish or restore these disturbance regimes to the extent possible. In the majority of cases to date, restoration involving changes in flow regime such as dam re-regulation has targeted fish populations and has not considered riparian objectives. Restoring the natural flow regime should focus on reestablishing the magnitude, frequency, and duration of peak flows needed to reconnect and periodically reconfigure channel and floodplain habitats. Fortunately, research is now demonstrating the essential functions performed by periodic flooding in shaping river channels, building floodplains, creating backwater sloughs, and supporting riparian vegetation. As a result, dam operations are changing in some locations to allow at least some controlled flooding. Given the current level of water resources infrastructure, dammed rivers will probably never have flow releases that fully replicate pre-dam flow regimes, and upstream portions of dammed rivers may never be restored. However, in many areas there may be major opportunities for altering flow release patterns so that they are increasingly "friendly" to the hydrologic needs of downstream riparian areas. There is also considerable potential for restoring riparian areas by altering levees, dikes, and other structures designed to impede the movement of water away from a channel.

Strategies that focus on returning the hydrologic regime to a more natural state have the greatest potential for restoring riparian functioning. Riparian vegetation has evolved with and adapted to the patterns of changing flows associated with stream and river environments. Altering dam operations, removing levees, and otherwise re-creating a more natural flow regime and associated sediment dynamics are of fundamental importance for recovering riparian vegetation and the functions that it provides. Geomorphic restoration alone cannot bring about the complexity that would result from a fully functioning river corridor with a free-flowing exchange of sediment and organic matter between the channel and riparian area.

Dam operations should be modified where possible to help restore downstream riparian areas. There is an increasing need to send greater flows down long segments of rivers to improve riparian plant communities. The effects on downstream riparian areas of manipulating dam discharges should be monitored to help identify those practices that show potential for aiding riparian restoration. In most cases it may not be possible to reinstate pre-dam conditions, but it may be possible to create a smaller, more natural stream that mimics many characteristics of the historical one.

Future structural development on floodplains should occur as far away from streams, rivers, and other waterbodies as possible to help reduce its impact on riparian areas. Structural developments typically have significant and persistent effects on the size, character, and functions of many riparian areas. Thus, preventing unnecessary structural development in near-stream areas should be a high priority at local, regional, and national levels. In addition, acquisition of land through conservation easements can be used to retain currently undeveloped land within floodplains in a more natural state.

Vegetation Management

Because of the fundamental importance of vegetation to the ecological functioning of riparian areas, where such vegetation has been degraded or removed, its recovery is a necessary part of any restoration effort. In many instances, recovery of riparian vegetation can be attained simply by discontinuing those land- and water-use practices that caused degradation (passive restoration). Attempts to actively restore altered systems without first removing or reversing the cause of decline are not likely to achieve functional riparian vegetation.

With regard to historically harvested riparian forests, there may be opportunities to combine passive and active restoration approaches. The protection of riparian vegetation from future harvest would be a passive approach. Active restoration approaches include planting native trees, encouraging more rapid development of late-successional stages through intermediate harvests, and augmenting large wood in streams to meet ecological goals.

A common restoration tactic in riparian areas of the Pacific Northwest has been to reintroduce large wood into streams. Unfortunately, this has frequently been done without consideration of restoring the riparian forest over the long term. Wood has even been added to streams that never contained appreciable amounts historically (e.g., meadows). Before large wood is introduced into streams, it must be determined whether wood is appropriate for creating the habitat conditions and ecological processes associated with restoration goals.

For overgrazed riparian areas, the passive restoration approach is simply to exclude domestic livestock from riparian areas via fencing, herd management, or other approaches. Excluding cattle from riparian areas is the most effective tool for restoring and maintaining water quality and hydrologic function, vegetative cover and composition, and native species habitats. Once ecological and hydrologic functions are restored, grazing in some cases could have minor impacts if well managed.

Where cattle are not excluded from riparian areas degraded by livestock grazing, conditions are unlikely to improve without changes in grazing management. Changing the season of use, reducing the stocking rate or grazing period, resting the area from livestock use for several seasons, and/or implementing a different grazing system can lead to improvements in riparian condition and functioning. Although grazing strategies other than full exclusion may promote restoration, they are likely to proceed more slowly and run a greater risk of failure.

In riparian areas that support agricultural crops, the long-term loss of native plants and the widespread occurrence of exotic plants increase the difficulty of accomplishing restoration goals, such that active management of riparian areas (using constructed buffer zones) is likely to be needed. Buffer zones are a valuable conservation practice with many important water-quality functions. Under proper conditions, these buffers are highly effective in removing a variety of pollutants from overland and shallow subsurface flow. They are most effective for water-quality improvement when hillslope runoff passes through the riparian zone slowly and uniformly and along lower-order streams where more of the flow transverses riparian areas before reaching the stream channel. Riparian buffer zones should be viewed as a secondary practice that assists in-field and upland conservation practices and "polishes" the hillslope runoff from an upland area.

Even when riparian buffer zones are marginally effective for pollutant removal, they are still valuable because of the numerous habitat, flood control, groundwater recharge, and other environmental services they provide. Unless new evaluation procedures are developed that consider both the water quality and ecological functions of riparian areas, it is unlikely that riparian zone size (width and length) and composition (vegetation types, other features) will be determined in a way that optimizes their potential for environmental protection.

Riparian areas associated with forested, grazed, and agricultural lands provide some of society's best opportunities for restoring habitat connectivity across the landscape. Management of riparian vegetation in ways that optimize their value as habitat for plants and animals will require planning and action at both site-specific and landscape scales. In addition, more integrated management that uses a functional approach and seeks to optimize habitats for a variety of native species is needed. Much riparian management currently suffers from focusing on a single species or taxon.

Management of Other Activities

Many of the restoration strategies discussed above involve land- and water-use changes that will require a new understanding of why riparian areas are important. Through improved educational programs, the ecological importance and intrinsic human values associated with these lands may be better balanced against the competing wants and needs of a modern society.

Although recreational use provides an excellent opportunity to foster stewardship of riparian areas, most recreational development in riparian areas lacks sound ecological assessment and planning. Future management should combine careful design using a landscape perspective, limitations on certain uses that are incompatible with preservation or rehabilitation of riparian areas, and involvement of the local community and other stakeholders. The goal of managing recreational activities in riparian areas is to perpetuate natural functions (e.g., wildlife habitat) while still allowing human use and enjoyment of these areas.

More formal education on riparian areas needs to reach broad and diverse audiences if it is to succeed in effecting positive change in riparian management. It should include traditional educational institutions and reach out directly to policy makers, natural resources personnel, government officials, developers, landowners, and the public at large. Natural resources professionals need to expand their perspectives beyond their formal background and training. The public's aesthetic appreciation of waterbodies is already high. This appreciation should be harnessed to further public stewardship of riparian areas.

CONCLUSIONS

Riparian areas provide essential life functions such as maintaining streamflows, cycling nutrients, filtering chemicals and other pollutants, trapping and redistributing sediments, absorbing and detaining floodwaters, maintaining fish and wildlife habitats, and supporting the food web for a wide range of biota. The future success of at least five national policy objectives—protection of water quality, protection of wetlands, protection of threatened and endangered species, reduction of flood damage, and beneficial management of federal public lands—depends on the restoration of riparian areas.

1

Introduction

Riparian areas are commonly thought of as those lands bordering streams, rivers, and lakes. This association with adjacent waters is an intrinsic part of the structure and functioning of riparian areas. In an ecologically healthy landscape, streams and their riparian areas form an inseparable unit—the stream corridor. The stream corridor encompasses not only the active river channel, but also the exposed bars and areas of ponded water near the channel, as well as the floodplain surfaces above and outside the channel banks. A river channel that has become disconnected from its riparian area no longer stores water and accumulates sediment, thus losing many of its ecological functions.

Ecologically healthy stream corridors and lakeshores are more than just sediment and water, channels and floodplains. They include assemblages of riparian plant communities and wildlife that depend upon the natural hydrologic regimes representative of a particular landscape. In the absence of human alteration, riparian plant communities support numerous functions including bank stabilization through root strength, sediment deposition on floodplains during periods of overbank flow, interstitial flow through the sediments, and large wood supply, which has a substantial influence on channel complexity and instream habitat features. Ecologically intact riparian areas naturally retain and recycle nutrients, modify local microclimates, and sustain broadly based food webs that help support a diverse assemblage of fish and wildlife. Like the loss of floodplain connectivity caused by altered channels and flow regimes, the removal of streambank vegetation has a large ecological impact—affecting aesthetics, recreational opportunities, and other characteristics of these areas that humans value.

HISTORICAL USE OF RIPARIAN AREAS

Prior to the settlement of the United States by Europeans and others, Native Americans utilized riparian areas for a number of purposes. Lakes and waterways, bounded by riparian plant communities and landforms, provided important transportation corridors. Riparian areas were natural producers of berries, seeds, roots, herbs, and other plant parts useful to these early societies. A plethora of wildlife species commonly found within riparian areas complemented the fisheries resources of the adjacent streams and lakes. And because of their proximity to water, riparian systems became synonymous with availability of water for human consumption as well as with relief from the hot and dry conditions common to many portions of the western United States.

With the advent of European settlement, initially in the eastern United States and subsequently across the Midwest and West, rivers and riparian systems were heavily utilized and significantly altered, a trend that continues today. Major rivers continued to serve as transportation corridors, and streamside forests provided fuel for steam-powered riverboats. The use of waterways for transportation provided an impetus for both clearing large wood from channels and reducing the potential recruitment of large wood into stream channels by harvesting streamside trees (Maser and Sedell, 1994). Furthermore, floodplain soils were extremely fertile, and thus vast areas of riparian forests were cleared for farming. In the Midwest, the ditching and draining of extensive floodplains and other low-lying areas for agricultural production ensured loss of many riparian systems. Riparian trees, because of their size, quality of wood, and closeness to a river or stream where they could easily be floated to a downstream sawmill, were highly valued, which greatly increased their likelihood of being harvested.

Thousands of miles of the nation's highways and railroads have been constructed along waterways (Rose, 1976; Jensen, 1993; Lewty, 1995), creating significant impacts to riparian systems including removal of riparian vegetation; "hardening" of streambanks with concrete, rip-rap, or other means; realignment of channels; and increased sediment production. In the western United States, the construction of dams and other water control structures for power generation and irrigation diversions followed by the subsequent alteration to downstream hydrologic regimes have additionally influenced the extent, quality, and functioning of many riparian systems (Reisner, 1987). In other instances, concern about the water use of streamside vegetation in the southwestern United States during the 1960s led to the initiation of programs directed at the removal of "phreatophytic" (water-loving) vegetation along watercourses (Culler, 1970). Historic livestock production has also impaired riparian function, with western riparian areas being repeatedly overgrazed during most of the nineteenth and twentieth centuries. In recent decades, the rates of urbanization and recreational development along waterways have accelerated and greatly altered many of the nation's riparian areas. By any ledger—physical, biological, or social—the impacts and extent of

Timber rafts were floated down the Mississippi from Prescott Wisconsin to sawmills and markets to the south in 1885. The rafts were made up of cribs—16- by 32-foot sets of logs roped together. SOURCE: Neuzil (2001).

change to riparian systems across the country have been extensive, diverse, and persistent.

America's rivers, streams, and lakes and their attendant riparian systems have been utilized for centuries, generally with limited knowledge about the environmental consequences of such actions on either current or subsequent generations. Many of the impacts to riparian systems have been directly or indirectly related to policies of proactive resource development that have dominated the history of this nation. An expanding population base coupled with an increasing standard of living has ensured a high and increasing demand upon the productivity not only of riparian areas, but also of all the nation's natural resources. Continued population growth and increasing resource demands remain a dominant force in the national agenda. Even though large areas (e.g., national parks and national forests) have been set aside and policies developed to protect some of their natural resources, protection of other portions of the American landscape has been less stringent, less organized, and not always implemented. Riparian areas are characteristic of this latter situation.

Variable Riparian Area Protection

One of the areas where protection of natural resources has been a central theme nationally is forest law and policy (Committee of Scientists, 1999). Yet, even forestlands provide an example of how riparian areas have only recently assumed more explicit recognition and the variable nature of their protection. When Congress authorized the establishment of forest reserves in the Creative Act of 1891, the predominant reason for the legislation was to meet the request of municipalities and irrigation districts for watershed protection. In the Organic Act of 1897, the first listed purpose for establishing forest reserves (which subsequently became national forests) was that of "securing favorable conditions of water flows," which has generally been interpreted as protecting upland watersheds from significant adverse effects. However, maintaining hydrologic connectivity and the integrity of riparian systems would seemingly be an important part of securing and sustaining favorable flow conditions. The Weeks Act of 1911 also emphasized watershed protection.

In the latter half of the twentieth century, legislation advocated multiple uses of national forests, but "without impairment of the productivity of the land" (Multiple-Use Sustained-Yield Act of 1960). Although in the National Forest Management Act (NFMA) of 1976, the production of timber and other resources were indicated as important multiple uses of federal forestlands, Congress emphasized that "the Forest Service has both a responsibility and an opportunity to be a leader in assuring that the Nation maintains a natural resource conservation posture that will meet the requirements of our people in perpetuity." Congress further indicated a need to ensure that "protection is provided for streams, streambanks, shorelines, lakes, wetlands, or other bodies of water from detrimental changes in water temperatures, blockages of water courses, and deposit of sediment." Although riparian areas are not explicitly identified in NFMA, it has been increasingly accepted in agency practice that water resources protection cannot occur without the protection of riparian systems.

Although the protection afforded many riparian areas has increased in recent years on national forests and grasslands, lands administered by the Bureau of Land Management (BLM), state lands, and industrial forestland and private woodlands, loss and degradation of riparian systems throughout the much of United States continue. If the ecological, economic, or cultural assets of riparian systems are not more fully identified, understood, and appreciated by society, there will likely be little incentive or desire by individuals, communities, or the nation to protect and maintain these areas.

Growing Recognition of the Term "Riparian"

The term "riparian," which has long been used in the United States, has generally been limited to policy and regulations associated with water law. A "riparian" water right generally provides a landowner whose property borders a stream, river, or other body of water the right to use a portion of that water for irrigation, human consumption, or other purposes. In the eastern United States, "riparian rights" were used as a basis for allocating water, while in the western United States, where water is relatively scarce, the doctrine of "prior appropriation" was typically used. In both situations, the vast majority of water rights were conferred long before the ecological importance of riparian areas and their dependence upon flow regimes were documented and understood. It has only been during the last 30 years that there has been a dramatic increase in the number of published studies that address riparian issues. Thus, only recently have scientists, from a variety of disciplines, begun to explore and better understand the widespread importance of these systems for a range of ecological functions and human values.

To illustrate the increase in recent scientific literature about riparian areas, three databases were investigated. The first was the Library of Congress where a keyword search for all "riparian" citations, by decade, was undertaken for the period 1900 through 1999. Of the 210 documents identified by this keyword search, approximately 37 percent addressed ecological topics, 52 percent addressed water law and policy issues, and 10 percent were indeterminate in their emphasis. Because some of the citations represent symposium proceedings and conferences, a relatively large number of riparian and riparian-related publications could be represented by any given citation. Hence, the actual number of publications contained within the Library of Congress database is much larger than the overall numbers would indicate. Interestingly, of the U.S. riparian citations identified as having an ecological emphasis, only 5 percent were published prior to 1970 (Figure 1-1).

The second database queried was one developed jointly by the U.S. Forest Service (USFS) and the University of Washington (http://www.lib.washington.edu/Forest/). This database contains over 11,000 references to journal articles, government documents, monographs, conference proceedings, and other papers associated with streams, rivers, streamside vegetation, water quality, and other riparian-related topics. This database has a western bias because most national forests and rangelands are located in the western states, and many ecological studies associated with western streams and rivers have occurred in recent decades. Even so, the database contains citations from every state and a number of foreign countries (see Figure 1-2). As with documents listed by the Library of Congress, only a small portion (less than 10 percent) of the references included in the USFS/University of Washington database were published prior to 1970. Of those references that explicitly utilized the term "riparian" as a keyword identifier (approximately a fifth of the total), fewer than 5 percent were published before 1970.

A third compilation of riparian and wetland publications for the western United States (Koehler and Thomas, 2000) reveals similar temporal trends in the

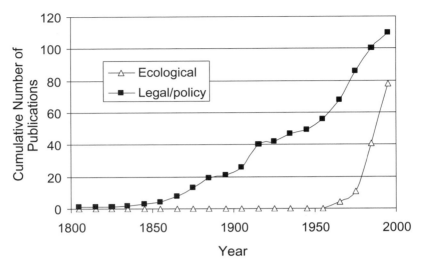

FIGURE 1-1 Cumulative number of citations, by date of publication, obtained from a search of the Library of Congress catalog using the term "riparian" for a keyword, December 2000.

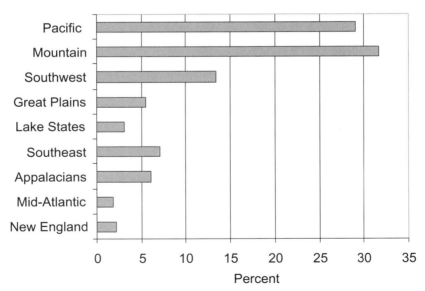

FIGURE 1-2 Percentage of riparian and riparian-related publications, by region, from the U.S. Forest Service and University of Washington (College of Forest Resources) database as of December 2000.

number of publications (Figure 1-3). Furthermore, it was only in 1985 that the "First North American Riparian Conference" was held in the United States (Johnson et al., 1985).

Trends revealed by these databases are consistent with the relatively recent mushrooming of scientific information that specifically addresses riparian and riparian-related topics. Although much fundamental biological, geomorphic, and hydrologic research undoubtedly preceded this recent period of riparian research and provided a useful context for understanding many riparian issues, the last few decades have shown an exceptional trend toward increased research productivity on a wide variety of riparian topics. These trends imply that many current natural resource managers, city councilors, state and federal politicians, and the general public acquired their educational backgrounds during a period when riparian issues, functions, and values were likely never mentioned or discussed, much less emphasized. Even in today's era of general environmental awareness and concerns, educational programs have been slow to incorporate subject matter that addresses the importance of riparian functions and values.

DEFINITION OF "RIPARIAN"

The lack of a consistent definition for "riparian" has been identified as a major problem of federal and state programs that might manage and protect these areas (Steiner et al., 1994). As discussed in detail below, riparian areas generally

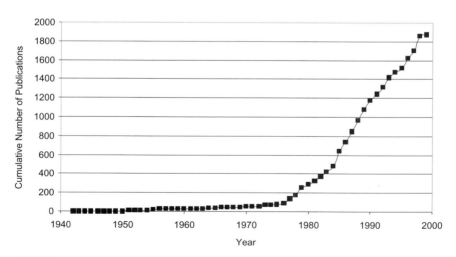

FIGURE 1-3 Cumulative number of riparian/wetland publications, by date of publication, for the western United States. SOURCE: Koehler and Thomas (2000).

do not satisfy regulatory and other definitions of "wetland," and thus are not encompassed by regulatory programs for wetland protection. A goal of this report is to develop a working definition of "riparian" that can be used to define those areas that require protection and explain the need for such protection.

Webster's Ninth New Collegiate Dictionary defines riparian as "relating to or living or located on the bank of a natural watercourse (as a river) or sometimes of a lake or a tidewater." The terms "streamside areas," "streambanks," and "bottomlands" are frequently used interchangeably with "riparian areas." As one might expect, the simple dictionary definition has been expanded or altered innumerable times by scientists and others, frequently for specific purposes or to reflect certain disciplinary preferences. However, an informal survey of definitions from a wide variety of sources—some of which are compiled in Table 1-1—revealed some general traits that most definitions of "riparian" have in common.

Reference to location is the most frequent characteristic of definitions of "riparian." Riparian areas are invariably defined as being directly adjacent to a waterbody, typically a stream. Definitions vary to the extent that they include all stream types, from perennial to ephemeral (see Box 1-1). Some are restricted to fresh waters, while others incorporate marine and estuarine waters as well. Although typically thought of in relation to streams and rivers, many "riparian" definitions (such as the dictionary definition above) include more static hydrologic regimes that incorporate lakes, estuaries, and other waters in addition to streams. Finally, expansive definitions include manmade waters, such as reservoirs and drainage ditches.

Hydrology is the primary emphasis of most definitions of wetlands and is also used to define riparian areas. Indeed, their proximity to water foreshadows the importance of hydrology in some definitions of riparian areas. However, not all definitions include hydrology, and those that do share little common language. The only statement universally found or strongly implied in various definitions is that riparian areas are wetter than adjacent uplands. More detailed hydrologic descriptions mention the extent and frequency of wetness, the width of wetted area (Freeman and Dick-Peddie, 1970), the role of flooding (Naiman et al., 1993), and interactions with the saturated zone. Although some have suggested definitions as precise as the 100-year floodplain (Lewis, 1996), such definitions have not received wide acceptance.

Regulatory and reference definitions for wetlands include vegetation and soil conditions. Although these factors may be somewhat less recognized for their importance to riparian areas, nonetheless many definitions of "riparian" include them. Soil characteristics and vegetation in riparian areas are frequently noted as being different or distinguishable from adjacent upland areas, particularly in semiarid and arid regions of the country. Invariably, the soils and vegetation of riparian areas are noted as being adapted to distinct hydrologic regimes such as elevated water tables, relatively high levels of soil moisture, or frequent flooding.

TABLE 1-1 Federal Agency Definitions of "Riparian"

Agency	Definition
Bureau of Land Management (1999)	A riparian area is an area of land directly influenced by permanent water. It has visible vegetation or physical characteristics reflective of permanent water influence. Lake shores and stream banks are typical riparian areas. Excluded are such sites as ephemeral streams or washes that do not exhibit the presence of vegetation dependent upon free water in the soil.
U.S. Fish and Wildlife Service (1998)	Riparian areas are plant communities contiguous to and affected by surface and sub-surface hydrologic features of perennial or intermittent lotic and lentic water bodies (rivers, streams, lakes, or drainage ways). Riparian areas have one or both of the following characteristics: (1) distinctively different vegetative species than adjacent areas, and (2) species similar to adjacent areas but exhibiting more vigorous or robust growth forms. Riparian areas are usually transitional between wetlands and upland.
U.S. Forest Service (2000)	Riparian areas are geographically delineated areas, with distinctive resource values and characteristics, that are comprised of the aquatic and riparian ecosystems, floodplains, and wetlands. They include all areas within a horizontal distance of 100 feet from the edge of perennial streams or other water bodies....A riparian ecosystem is a transition between the aquatic ecosystem and the adjacent terrestrial ecosystem and is identified by soil characteristics and distinctive vegetation communities that require free and unbound water.
U.S. Forest Service Region 9 (Parrott et al., 1997)	Riparian areas are composed of aquatic ecosystems, riparian ecosystems and wetlands. They have three dimensions: longitudinal extending up and down streams and along the shores; lateral to the estimated boundary of land with direct land-water interactions; and vertical from below the water table to above the canopy of mature site-potential trees.
U.S. Department of Agriculture NRCS (1991)	Riparian areas are ecosystems that occur along watercourses and water bodies. They are distinctly different from the surrounding lands because of unique soil and vegetation characteristics that are strongly influenced by free or unbound water in the soil. Riparian ecosystems occupy the transitional area between the terrestrial and aquatic ecosystems. Typical examples would include floodplains, streambanks, and lakeshores.
U.S. EPA and NOAA Coastal Zone Management Act (EPA, 1993)	Riparian areas are vegetated ecosystems along a water body through which energy, materials and water pass. Riparian areas characteristically have a high water table and are subject to periodic flooding and influence from the adjacent waterbody. These systems encompass wetlands, uplands, or some combinations of these two land forms. They will not in all cases have all the characteristics necessary for them to be classified as wetlands.
Forest Ecosystem Management Assessment Team (FEMAT, 1993)	Riparian Reserves are portions of watersheds where riparian-dependent resources receive primary emphasis and where special standards and guidelines apply to attain Aquatic Conservation Strategy objectives. Riparian Reserves include those portions of a watershed required for maintaining hydrologic, geomorphic, and ecologic processes that directly affect standing and flowing waterbodies such as lakes and ponds, wetlands, and streams.

> **BOX 1-1**
> **Perennial, Intermittent, and Ephemeral Streams**
>
> Although there are no universally accepted definitions for perennial, intermittent, or ephemeral stream types, most definitions include or imply the following characteristics (Hewlett, 1982; Art, 1993; Comín and Williams, 1994; Nevada Division of Water Planning, 1999). Perennial reaches of streams receive substantial groundwater inputs and generally flow continuously throughout the year. Their flows can vary widely from year to year and may dry up during severe droughts, although groundwater is generally near the surface. Perennial streams are found in both humid and arid regions, although in arid regions, the point of initiation for perennial reaches generally occurs further downstream. Intermittent stream reaches typically flow for several weeks or months each year when precipitation and associated groundwater inputs are relatively high. The timing of the flow and drying of intermittent streams is broadly predictable on a seasonal basis. Though sometimes associated with arid and semiarid climates, intermittent streams are well represented in humid regions. Ephemeral portions of streams flow only in direct response to precipitation. Thus, their flow is as unpredictable as the rainfall events that drive them. Because the channel of ephemeral streams is generally well above the water table, these streams flow for only a few hours or days following a storm of sufficient magnitude to produce overland flow. Many of the dry washes or arroyos of the more arid regions of North America may be classified as ephemeral streams.

Descriptions of other biota, particularly assemblages that are unique within a landscape, are also sometimes included in definitions of riparian.

Finally, one of the most prevalent characteristics of definitions is the concept of riparian areas as gradients. Occupying the space between land and water, these areas are characterized by multiple transitions in soil, biota, and hydrology. Some scientists have described riparian areas as "ecotones" or interfaces between terrestrial and aquatic ecosystems (Gregory, 1997), while others have embraced riparian ecosystems as landscape units comprising an array of zones that extend from aquatic to upland environments (Brinson et al., 1981). In either case, riparian areas clearly are characterized by gradients in environmental conditions, ecological processes, and species that make it difficult to assign them discrete boundaries (Naiman and Décamps, 1990).

It should be noted that for management and regulatory purposes, riparian areas are frequently given distinct spatial boundaries in order to achieve specific goals and are thereby called "riparian zones" or "riparian management areas." Such management designations incorporate inherent trade-offs between proportions of riparian functions included within and outside of the boundaries of the zone.

The working definition developed in this report is broad in the sense that it encompasses all the characteristics mentioned above, including reference to location, hydrology, vegetation, soils, and the concept of gradients:

Riparian areas are transitional between terrestrial and aquatic ecosystems and are distinguished by gradients in biophysical conditions, ecological processes, and biota. They are areas through which surface and subsurface hydrology connect waterbodies with their adjacent uplands. They include those portions of terrestrial ecosystems that significantly influence exchanges of energy and matter with aquatic ecosystems (i.e., a zone of influence). Riparian areas are adjacent to perennial, intermittent, and ephemeral streams, lakes, and estuarine–marine shorelines.

This definition is consistent with other definitions developed by interdisciplinary groups of scientists with expertise in riparian issues. For example, Ilhardt et al. (2000) describe riparian areas as "three-dimensional ecotones of interaction that include terrestrial and aquatic ecosystems, that extend down into the groundwater, up above the canopy, outward across the floodplain, up the near slopes that drain to the water, laterally into the terrestrial ecosystem, and along the water course at a variable width." Lowrance et al. (1985) defines riparian areas as "a complex assemblage of plants and other organisms in an environment adjacent to water. Without definite boundaries, it may include streambanks, floodplain, and wetlands . . . forming a transitional zone between upland and aquatic habitat. Mainly linear in shape and extent, they are characterized by laterally flowing water that rises and falls at least once within a growing season." Even programs with disparate goals have developed similar definitions of "riparian."

Over the last 15 years, several federal agencies have developed considerably narrower definitions of "riparian" for application to their programs, as summarized in Table 1-1. Most definitions reflect the particular goals of individual agencies, including mandates to protect, manage, or restore riparian areas (e.g., BLM and USFS) and to map riparian areas, a responsibility of the U.S. Fish and Wildlife Service (FWS).

It is useful to define two terms sometimes used interchangeably with "riparian." Almost all rivers have *floodplains*—aggraded reaches composed of complex bed sediments (alluvia) where flood waters spread out laterally. Clearly part of riparian areas, floodplains are dynamic structures composed of the channel system and adjacent depositional levees, interfluvial bars and low-lying, depositional shelves, often with ridge and swale topography reflecting backfilling of ancient river channels. A *river* or *stream corridor* generally refers to riparian areas *and* their adjacent waterbodies as a unit defined longitudinally from headwaters to the ocean. Figure 1-4 is a schematic of a stream corridor that shows the many interconnections between its different components. Because floodplains are porous and contain aquifers that are closely linked to and controlled by the channel system, waterbodies and their riparian areas are linked longitudinally, vertically, and horizontally—not just by the movement of water and sediments, but also by the movement of biota (Stanford and Ward, 1993).

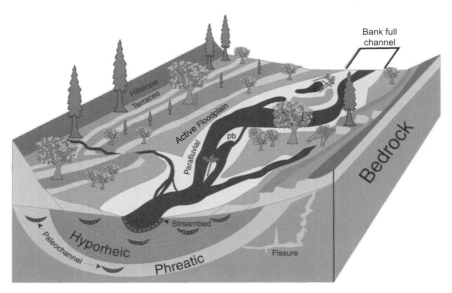

FIGURE 1-4 Idealized view of an alluvial river corridor showing the longitudinal (upstream–downstream), vertical (interstitial), and lateral (floodplain) dimensions that interact both hydrologically and ecologically. The riparian area includes the channel system, its floodplain, and the transition zone into the uplands. An alluvial aquifer underlies the channel and includes both a hyporheic zone and a deeper phreatic zone of groundwater. The phreatic zone contains groundwater that has had no contact or mixing from surface sources for very long time periods, often hundreds of years or longer. In contrast, in the hyporheic zone river water moves rapidly (on the order of days) through surficial alluvia characterized by very high hydraulic conductance. Other terms (parafluvial and bank-full) and interactions are described in detail in Chapter 2. pb = point bar. SOURCE: Adapted from Stanford (1998).

DISTINGUISHING RIPARIAN AREAS FROM OTHER AREAS

Another way of characterizing riparian areas is to identify what they are not and to contrast them with other adjacent land and water units. The definition in the preceding section describes riparian areas as "zones of influence" between aquatic and terrestrial areas. As such, riparian areas encompass some or all of the wetlands in a typical landscape setting, but they also include portions of adjacent aquatic and upland environments. Figure 1-5(A) shows a "generic" riparian zone of influence in detail; Figures 1-5(B–E) illustrate this zone of influence in small stream, river, lake, and estuarine-marine settings and approximate the scale at which the zone applies.

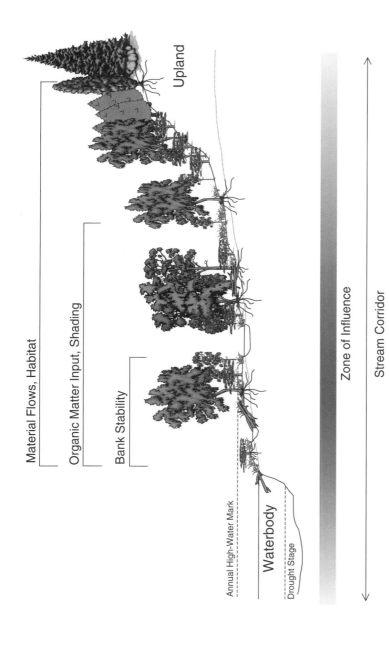

FIGURE 1-5(A) Schematic of a generic riparian area (A) showing a zone of influence relative to aquatic and upland areas. Four common riparian settings are illustrated: (B) a small stream, (C) a large river, (D) a lake, and (E) an estuarine–marine setting. Horizontal scales differ and are provided for perspective. The intensity of riparian influence is depicted with shading. "Material flows" refers to energy, organic matter, water, sediment, and nutrient flow. SOURCE: Adapted from Ilhardt et al. (2000).

Small Stream

10 m

upland

upland

Zone of Influence

FIGURE 1-5(B)

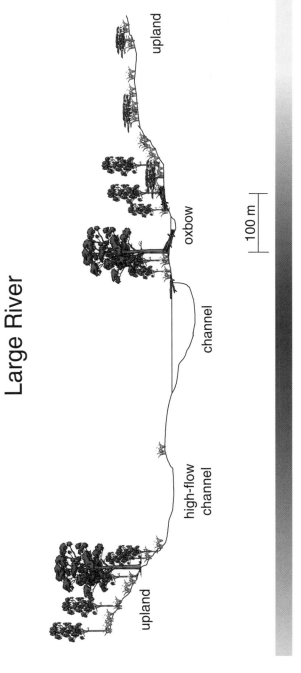

FIGURE 1-5(C)

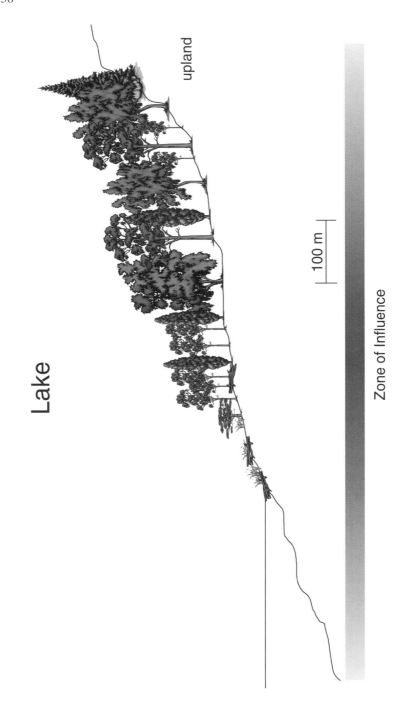

FIGURE 1-5(D)

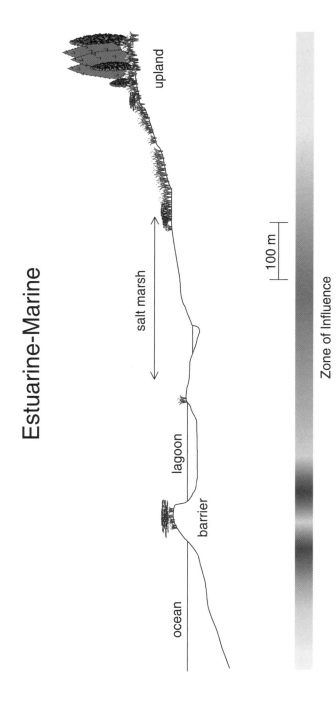

FIGURE 1-5(E)

Riparian Areas Versus Aquatic Ecosystems

Riparian areas are differentiated from aquatic ecosystems, wetlands, and uplands based on the same characteristics—hydrology, soils, and vegetation—used to define them. With regard to vegetation, riparian areas are normally dominated by woody plants (e.g., trees), grasses, and emergent herbaceous plant cover, in contrast to aquatic ecosystems where these plant types are absent. Rather, aquatic plants commonly found in North America include bullrushes and arrowhead in the shallower waters, water lilies in still deeper water with little current, and submerged aquatic plants (e.g., pondweed, milfoil, coontail, waterweed, and bladderwort), some of which tolerate moderate current velocities.

Although riparian areas and aquatic ecosystems obviously differ in their moisture regimes and water depth, it is difficult to identify a boundary that can be used consistently to separate them. In fact, definitions of riparian may include intermittent to permanent aquatic environments, particularly in headwater streams where the vegetated portions of riparian areas provide the aquatic portion with organic matter, shade, fluvial structure, and biotic exchanges. [Some definitions of wetland ecosystems (and deepwater habitats) extend to two meters of water depth (Cowardin et al., 1979) and even to six meters in marine environments (Scott and Jones, 1995)].

In the transition from headwater to larger streams, there is a tendency for flooding to be more protracted and less abrupt because of the cumulative effects to the hydrograph of multiple tributaries originating from regions differing in the quantity and timing of flow. Floodplains become more expansive in a downstream direction and, in some cases, wetter (Rheinhardt et al., 1998). Ultimately riparian areas of rivers grade into coastal estuaries where tidally driven hydrologic regimes and salinity combine to support estuarine riparian areas.

In anticipation of practical reasons for identifying an aquatic boundary, especially for large rivers, lakes, and estuaries, the following perspective is offered. The aquatic boundary of riparian areas could be established where permanent water begins. For bodies of water that have relatively constant elevations, such as estuarine shorelines influenced by slowly changing sea level, a boundary such as mean high tide or mean low tide is relatively easy to identify. For waterbodies with relatively large fluctuations such as streams, "permanence" is a relative term. Just as it is conventional to assign return periods or frequencies to flood elevations for streams, the same approach could be used to establish a return period for a permanence level during drought—e.g., the low level that would occur at a frequency of every ten years. For ephemeral and intermittent streams, this low stage lies below the channel, and thus the riparian area using this criterion would include the entire streambed and underlying alluvium. (This approach to delineating the zone of influence is consistent with the fact that forest canopies shade the entire stream width of many intermittent streams.) For perennial streams and lakes, a zone of influence could include aquatic portions shaded by riparian vegetation or even a distance equivalent to the height of a canopy tree. For the shoreline of some lakes (e.g., the Great Lakes), levels fluctuate over a multi-year cycle, causing the boundary of the riparian area to migrate back and forth.

Riparian Areas Versus Wetlands

Two definitions of wetlands provide a useful starting point for differentiating them from riparian areas. One is the jurisdictional, or legal, definition used as the basis for delineating wetlands in the U.S. Army Corps of Engineers regulatory program (WES, 1987):

> The term "wetlands" means those areas that are inundated or saturated by surface or groundwater at a frequency and duration sufficient to support, and that under normal circumstances do support, a prevalence of vegetation typically adapted to life in saturated soil conditions. Wetlands generally include swamps, marshes, bogs, and similar areas.

This definition highlights three factors—water, vegetation, and soils—used to characterize wetlands. In practice, a number of primary and secondary field indicators are used to identify wetlands, as described in the Corps' manual. These indicators constitute criteria that are related to one or more of the three factors. They are prone to vary geographically, which results in the need for regionalization of the manual through supplementary materials, as discussed in NRC (1995). The NRC report found the wetland definition and the delineation approach to be scientifically sound even though they were developed specifically for regulatory purposes with inevitable policy overtones that govern their application.

NRC (1995) provides a reference definition of "wetland" to which specific regulatory definitions can be compared. Although the NRC's definition is consistent with the three criteria mentioned above, it recognizes hydrologic conditions as paramount because they give rise to the diagnostic features of hydric soils and hydrophytic vegetation. NRC (1995) defines "wetland" as follows:

> A wetland is an ecosystem that depends on constant or recurrent, shallow inundation or saturation at or near the surface of the substrate. The minimum essential characteristics of a wetland are recurrent, sustained inundation or saturation at or near the surface and the presence of physical, chemical, and biological features reflective of recurrent, sustained inundation or saturation. Common diagnostic features of wetlands are hydric soils and hydrophytic vegetation. These features will be present except where specific physicochemical, biotic, or anthropogenic factors have removed them or prevented their development.

In both definitions, the emphasis is on minimum conditions that characterize a wetland–upland boundary (i.e., "sufficient to support" and "minimal essential characteristics") rather than on describing the full wetness gradient. With emphasis on the dry end of the gradient, a point in the landscape is identified between wetter (definitely wetland) and drier (definitely upland) conditions.

In contrast to most wetland definitions, a balanced characterization for riparian areas should treat the riparian area as a whole and not place undue emphasis on either its driest or its wettest margin. Moreover, fundamental characteristics of riparian areas are their zonal nature and their position in the landscape. Except for the FWS definition that refers to wetlands as "lands transitional between terrestrial and aquatic systems" (Cowardin et al., 1979), most wetland definitions do not incorporate concepts of zonation and landscape position. To omit these characteristics of riparian areas, however, would be to disregard two of their diagnostic characteristics.

Hence, compared with wetlands, riparian areas are both more expansive from some perspectives and more restrictive from others. They are more expansive because they may include not only portions of wetlands, but also nonvegetated portions of point bars. At the dry end, they encompass terrestrial areas that do not necessarily require inundation and saturation near the surface, as do wetlands. Riparian areas are more restrictive than wetlands because they are confined to specific geomorphic settings of streams, lakes, and estuarine-marine environments. Extensive peatlands or flatwood wetlands, for example, would not be considered riparian areas because most of them lack landscape attributes of linear configuration, distinct zonation, and adjacency to waterbodies.

Riparian Areas Versus Uplands

Unlike uplands that receive precipitation as their principal or only source of water, moisture in riparian areas may be supplied from both adjacent uplands and aquatic ecosystems. Water enters riparian areas from uplands in the form of groundwater discharge, shallow subsurface flow, and overland flow (see Chapter 2). From the aquatic side, water is supplied by overbank flow, infiltration through adjacent channel banks (bank storage), and by hyporheic flow from alluvium upstream. Unlike uplands, floods are important agents of geomorphic change in riparian areas, eroding soils and sediments from some parts of the riparian area while depositing them in others to create new geomorphic surfaces. The lateral transport of materials to riparian areas from outside the floodplain constitutes a fundamental difference between riparian areas and uplands.

These hydrologic conditions are clearly reflected in the distinctive vegetation of riparian and upland areas. First, the flooding regimes of riparian areas select for disturbance-tolerant species, and they may also constrain colonization of riparian areas by flood-sensitive species. Second, because riparian areas exist where depth to the water table is relatively shallow, specific assemblages of shrubs and trees withdraw water directly from the saturated area (Robinson, 1958), a phenomenon not generally possible in uplands. In low-elevation settings of arid landscapes, riparian areas may be the only place where cottonwoods and willows are present, and the only place where mesquite grows vigorously (Stromberg et al., 1996). These areas may extend hundreds of meters from stream channels on terraces where surface connections are entirely absent but shallow groundwater is available to plants. Riparian forests persist in arid regions where there is insufficient precipitation to support upland forests (Brinson, 1990), thus creating a distinct boundary. In humid climates, water availability is not as crucial, and the boundary between uplands and riparian areas is not always easy to distinguish. Nonetheless, there are distinct assemblages of trees in humid riparian areas, including some species of willow, alder, cypress, tupelo, and sycamore (Hupp and Osterkamp, 1985). In drainage basins with relatively fine sediments, riparian tree species are more likely to be distinguished by their ability to withstand the stresses of sediment anoxia than a requirement for a dependable source of groundwater (Wharton et al., 1982; Friedman and Auble, 2000).

Steep slopes, such as rock escarpments or abrupt sediment banks, present a more ambiguous case because there may be no active hydrologic connection with the adjacent stream, river, or lake for the majority of the riparian area. Nevertheless, riparian functions remain strong, with pronounced flows of biotic and abiotic materials between aquatic and terrestrial ecosystems. Such "upland" riparian areas provide important habitat for a variety of species that respond to topography, water-influenced microclimates, and the presence of a natural movement corridor, among other factors. In areas with steep slopes, distinguishing between riparian and the non-riparian areas should be based on function. This can be justified because the zone of influence by moisture or disturbance alone may be quite narrow and insufficient to account for other influences. Consistent with the working definition of "riparian" developed in this report, which includes areas that "significantly influence exchanges of energy and matter with aquatic ecosystems," the upper boundary of such riparian areas can be functionally identified by the potential for trees to contribute portions of their wood to a stream channel, should they fall in that direction.

SCOPE OF THE STUDY

This study is an outgrowth of the National Research Council's (NRC) study on the Characterization of Wetlands (NRC, 1995). The major intent of that work was to identify both the strengths and weaknesses of current regulatory practice regarding wetlands. Beginning in the mid-1970s, the regulation of wetlands as "waters of the United States" under Section 404 of the Clean Water Act has elicited strong opposition by property rights advocates, developers, and landowners. Because portions of riparian areas contain "waters" and, as explained later, carry out many of the same functions as wetlands, questions on how riparian areas should be managed are often very similar to those for wetlands. A brief background of how the wetland study came about provides historic context to the present study.

In order for a landowner to fill a wetland, the owner must request and receive a permit from the U.S. Army Corps of Engineers (Corps), with oversight authority provided by the U.S. Environmental Protection Agency (EPA). The first step is to identify the location of the wetland to be affected and its boundaries, for which technical manuals were developed to assist regulators in consistently applying delineation procedures in the field. Until 1989, the Corps and EPA had different manuals for this task. In order to provide more uniformity in the regulatory program, a new and revised manual—the "1989 Manual"—was produced by an interagency team of scientists (Federal Interagency Committee for Wetland Delineation, 1989).

In part as a result of its inconsistent application, landowners and developers perceived the 1989 manual as extending the reach of wetlands to include lands that previously had not been designated as wetlands. Further, opponents of the new manual complained that it had been adopted without public comment and that its application constituted a "taking of property" (without just compensation) (Kusler, 1992). In response, the White House released in August 1991 a document—1991 proposed revisions (56 Fed. Reg. 40,446; 10,991)—much different from the approach that the scientists had taken. Field testing found the 1991 proposed revisions not only to be technically wanting, but to propose a much narrower definition of wetlands. Unlike the 1989 Manual, the 1991 proposed revisions would substantially reduce the surface area of wetlands that would fall under jurisdiction of Section 404 of the Clean Water Act. The groundswell of complaints and other objections brought the issue of wetlands definition under scrutiny by Congress. Faced with the need to respond to public outcries from both sides, Congress ordered the Corps to revert to using a 1987 manual (WES, 1987) and mandated that the NRC study the situation and provide recommendations.

That NRC committee reviewed the science of wetland identification and delineation, identified the functions and values of wetlands, and examined the variation among wetland types. During this process, it became obvious that some areas of the landscape, especially riparian areas, were not always wet enough to be encompassed by any of the technical manuals or by the reference definition for wetlands. In other words, they did not have sufficient wetness to develop hydric soils and to support hydrophytic vegetation—key criteria (along with hydrology) necessary to define wetlands. These marginally dry areas were prevalent in arid and semiarid climates where wetlands are exceedingly rare, but were often in the same landscape position as floodplain wetlands of more humid climates. To a large degree, riparian areas perform many of the same functions as areas that are jurisdictional wetlands in more humid climates, such as water storage and conveyance, nutrient and sediment removal, and plant and animal habitat maintenance.

The committee considered the protection and maintenance of riparian areas important toward meeting the goals of the Clean Water Act. However, because the study commissioned by Congress was restricted to wetlands, any attempt to include riparian areas "would unreasonably broaden the definition of wetland and undermine the specificity of criteria and indicators that have developed around wetland delineation" (NRC, 1995). The following was one of the recommendations of the committee: "If national policy extends to protection of riparian zones pursuant to the goals of the Clean Water Act, regulation must be achieved through legislation that recognizes the special attributes of these landscape features, and not by attempting to define them as wetlands." The purpose of the present study is to recognize and identify the attributes of riparian areas and make recommendations for managing and maintaining these attributes.

Chapter 2 describes riparian structure and how riparian areas affect water quality, provide habitat for fish and wildlife, and serve as corridors for species movement, among many other functions. The report explores these and other aspects of functioning, especially as they vary between climatic extremes of North America. Because of the importance of riparian areas in arid and semiarid regions of the nation, special attention is given to these locations. While riparian areas are also a fundamental part of more humid landscapes, they may not be as easily separated from their adjacent uplands as in arid regions. Nevertheless, principles that govern the functioning of riparian areas are broadly transferable across a wide variety of landscapes.

Knowledge of the status of riparian lands in the United States, both in quantity and quality, is fundamental to any management program that sets goals for improvement. This information should include whether the resource is increasing or decreasing over time, the geographic distribution of these changes, and, ideally, data on the condition of existing riparian areas. But even without detailed inventories on riparian resources nationwide, local knowledge of riparian condition provides clues to the types of impacts they have experienced. Dam and levee construction, cattle grazing, conversion to agricultural production, and withdrawal of water for domestic consumption, power generation, and irrigation are just a few examples. Chapter 3 comprehensively describes the general impacts of water resources development and other human activities on the condition and functioning of riparian areas. The current status of riparian lands in the United States is assessed in terms of overall increases or decreases in acreage, habitat condition, and other important trends.

In spite of an increasing wealth of scientific information, riparian policy issues are complicated by a long and complex history of changing land use and ownership, a proliferation of legal statutes that influence their development for various purposes, and a relatively recent overlay of environmental regulations. The complexity is a result, in part, of development of legislation over time with little attention to removing redundancies and avoiding conflicts. Because policy differs geographically at national and state levels, and jurisdictionally among agencies, disputes are sometimes resolved in the courts. These issues are analyzed in Chapter 4, which describes the regulatory landscape surrounding riparian areas. Chapter 4 focuses on the differences between eastern and western regions and between public and private lands, as well as on the variability observed from state to state. It considers policy goals reflected in current laws and regulatory and nonregulatory programs for riparian areas and makes recommendations regarding the protection of these areas in the future. Programs of federal and state agencies particularly relevant to or dependent on riparian areas are highlighted.

Finally, Chapter 5 considers the management of riparian areas, which is confounded by many present-day conflicts regarding the functioning of riparian areas and by multiple desired uses of riparian land and water. The report describes our current scientific understanding of these conflicts and the additional scientific information needed to resolve them. Several management strategies, including dam re-operation and other hydrologic manipulations, design of riparian management systems for agriculture, grazing and forestry practices, and education, are explored for restoring, enhancing, and preserving riparian areas across the United States.

CONCLUSION

A large body of scientific information on riparian areas has developed over the last several decades documenting their importance as elements of regional landscapes. Early publications on riparian topics focused predominantly on water law. However, since 1970 there has been an explosion of information addressing ecological and hydrologic topics, processes and functions, aquatic and terrestrial wildlife habitats and food web support, aesthetics, production of goods and services, cultural and social values, and other topics related to riparian areas. As discussed in greater detail in Chapters 4 and 5, new scientific information about the structure, functioning, and importance of riparian areas should be included forthrightly in education at all levels and given full consideration in environmental regulatory and policy processes including decision-making.

REFERENCES

Art, H. W. 1993. The dictionary of ecology and environmental science. New York: Henry Holt and Co.

Brinson, M. M., B. L. Swift, R. C. Plantico, and J. S. Barclay. 1981. Riparian ecosystems: their ecology and status. FWS/OBS–81/17. Kearneysville, WV: U.S. Fish and Wildlife Service.

Brinson, M. M. 1990. Riverine forests. Pp. 87–141 In: Forested wetlands. A. E. Lugo, M. M. Brinson, and S. Brown (eds.). Amsterdam, The Netherlands: Elsevier.

Bureau of Land Management (BLM). 1999. Draft environmental impact statement for riparian and aquatic habitat management in the Las Cruces field office, New Mexico.

Comín, F. A., and W. D. Williams. 1994. Parched continents: our common future? In: Limnology now: a paradigm of planetary problems. R. Maralef (ed.). New York: Elsevier Science B. V.

Committee of Scientists. 1999. Sustaining the people's lands: Recommendations for stewardship of the National Forests and Grasslands into the next century. Washington, DC: U.S. Department of Agriculture. 193 pp.

Cowardin, L. M., V. Carter, F. C. Golet, and E. T. LaRoe. 1979. Classification of wetlands and deepwater habitats of the United States. FWS/OBS-79/31. Washington, DC: U.S. Fish and Wildlife Service.

Culler, R. C. 1970. Objectives, methods, and environment—Gila River phreatophyte project, Graham County, Arizona. Washington, DC: U.S. Govt. Printing Office. 25 pp.

Environmental Protection Agency (EPA). 1993. Guidance specifying management measures for sources of nonpoint pollution in coastal waters. Washington, DC: EPA.

Federal Interagency Committee for Wetland Delineation. 1989. Federal manual for identifying and delineating jurisdictional wetlands. Washington, DC: U.S. Army Corps of Engineers, U.S. Environmental Protection Agency, U.S. Fish and Wildlife Service, and USDA Soil Conservation Service. Cooperative technical publication. 76 pp. plus appendixes.

Fish and Wildlife Service (FWS). 1998. A system for mapping riparian areas in the western U.S. Washington, DC: U.S. Fish and Wildlife Service.

Forest Ecosystem Management Assessment Team (FEMAT). 1993. Forest ecosystem management: an ecological, economic, and social assessment. Washington, DC: U.S. Department of Agriculture.

Freeman, and Dick-Peddie. 1970. Woody riparian vegetation in the Black and Sacramento Mountain Ranges, Southern New Mexico. The Southwestern Naturalist 15(2):145–164.

Friedman, J. M., and G. T. Auble. 2000. Floods, flood control, and bottomland vegetation. Pp. 219–237 In: Inland flood hazards: human, riparian, and aquatic communities. E. Wohl (ed.). Cambridge: University Press.

Gregory, S. V. 1997. Riparian management in the 21st century. Pp. 69–85 In: Creating a forestry for the 21st century: the science of ecosystem management. Kohm, K. A., and Franklin, J. F. (eds.). Washington, DC, and Covelo, CA: Island Press.

Hewlett, J. D. 1982. Principles of forest hydrology. Athens, GA: University of Georgia Press.

Hupp, C. R., and W. R. Osterkamp. 1985. Bottomland vegetation distribution along Passage Creek, Virginia, in relation to fluvial landforms. Ecology 66:670–681.

Ilhardt, B. L., E. S. Verry, and B. J. Palik. 2000. Defining riparian areas. Pg. 29 In: Riparian management in forests of the continental eastern United States. Verry, E. S., J. W. Hornbeck, and C. A. Dolloff (eds.). New York: Lewis Publishers.

Jensen, O. 1993. The American heritage history of railroads in America. New York: American Heritage/Wing Books. 314 pp.

Johnson, R. R., C. D. Ziebell, D. R. Patton, P. F. Ffolliott, and R. H. Hamre (eds.). 1985. Riparian systems and their management: reconciling conflicting uses. General Technical Report RM-120, Fort Collins, CO: USDA Forest Service. 523 pp.

Koehler, D. A., and A. E. Thomas. 2000. Managing for enhancement of riparian and wetland areas of the western United States: an annotated bibliography. General Technical Report RMRS-GTR-54. Ogden, UT: USDA Forest Service. 369 pp.

Kusler, J. 1992. Wetlands delineation: an issue of science or politics. Environment 34(7-11):29–37.

Lewis, W. M. 1996. Defining the riparian zone: lessons from the regulation of wetlands. In: At the water's edge: the science of riparian forestry. Conference Proceedings June 19-20, 1995. Minnesota Extension Service. BU-6637-S.

Lewty, P. J. 1995. Across the Columbia Plain: railroad expansion in the interior Northwest, 1885–1893. Pullman, WA: Washington State University Press. 326 pp.

Lowrance, R. R., R. Leonard, and J. Sheridan. 1985. Managing riparian ecosystems to control nonpoint pollution. J. Soil Water Conserv. 40:87–91.

Maser, C., and J. R. Sedell. 1994. From the forest to the sea: the ecology of wood in streams, rivers, estuaries, and oceans. Delray Beach, FL: St. Lucie Press. 200 pp.

Naiman, R. J., and H. Décamps (eds.). 1990. The ecology and management of aquatic-terrestrial ecotones. Man and the Biosphere Series, Volume 4. Paris, France: UNESCO.

Naiman, R. J., H. Décamps, and M. Pollock. 1993. The role of riparian area corridors in maintaining regional diversity. Ecol. Appl. 3:209–212.

National Research Council (NRC). 1995. Wetlands: characteristics and boundaries. Washington, DC: National Academy Press.

Neuzil, M. 2001. Views on the Mississippi: the photographs of Henry Peter Bosse. Minneapolis, MN: University of Minnesota Press.

Nevada Division of Water Planning. 1999. Water words dictionary, 8th edition. Carson City, NV: Nevada Department of Conservation and Natural Resources.

Parrott, H. A., D. A. Marions, and R. D. Perkinson. 1997. A four-level hierarchy for organizing stream resources information. Pp. 41–54 In: Proceedings, Headwater hydrology symposium. Missoula, MT: American Water Resources Association.

Reisner, M. 1987. Cadillac desert: the American West and its disappearing water. New York: Penguin Books. 582 pp.

Rheinhardt, R. D., M. C. Rheinhardt, M. M. Brinson, and K. Faser. 1998. Forested wetlands of low order streams in the inner coastal plain of North Carolina, USA. Wetlands 18:365–378.

Robinson, T. W. 1958. Phreatophytes. U.S. Geological Survey Water Supply Paper 1423. 84 pp.

Rose, A. C. 1976. Historic American roads: from frontier trails to superhighways. New York: Crown Publishers. 118 pp.

Scott, D. A., and T. A. Jones. 1995. Classification and inventory of wetlands: a global overview. Pp. 3–16 In: Classification and Inventory of the World's Wetlands. C. M. Finlayson and A. G. van der Valk (eds.). Dordrecht, The Netherlands: Kluwer Academic Publishers.

Stanford, J. A. 1998. Rivers in the landscape: introduction to the special issue on riparian and groundwater ecology. Freshwater Biology 40(3):402–406.

Stanford, J. A. and J. V. Ward. 1993. An ecosystem perspective of alluvial rivers: connectivity and the hyporheic corridor. J. N. Am. Benthol. Soc. 12(1):48–60.

Steiner, F., S. Pieart, E. Cook, J. Rich, and V. Coltman. 1994. State wetlands and riparian area protection programs. Environmental Management 18(2):183–201.

Stromberg, J. C., R. Tiller, and B. Richter. 1996. Effects of groundwater decline on riparian vegetation of semiarid regions: the San Pedro, Arizona. Ecological Applications 6:113–131.

U.S. Department of Agriculture Forest Service (USFS). 2000. Forest Service Manual, Title 2500, Watershed and Air Management. Section 2526.05. Washington, DC: USDA Forest Service.

U.S. Department of Agriculture Natural Resources Conservation Service. 1991. General Manual, 190-GM, part 411. Washington, DC: USDA NRCS.

WES Environmental Laboratory. 1987. U.S. Army Corps of Engineers wetlands delineation manual. U.S. Army Engineer Waterway Experiment Station Technical Report Y-87-1.

Wharton, C. H., Kitchens, W. M., and Sipe, T. W. 1982. The ecology of bottomland hardwood swamps of the Southeast: a community profile. FWS/OBS-81/37. Washington, DC: U.S. Fish and Wildlife Service. 133 pp.

2

Structure and Functioning of Riparian Areas Across the United States

The interaction of climate with the earth's surface has created a variety of landscapes drained by networks of streams, rivers, lakes, and wetlands. Riparian areas are found adjacent to essentially all of these waterbodies except where human disturbance has intervened. Although riparian areas differ considerably in their structure and function from site to site, there are patterns in the attributes of riparian areas and how they are distributed across the landscape. While a single characteristic (such as the presence of bedrock) may strongly influence the size, characteristics, and functions of a given riparian area, generally the interaction of many climatic, hydrologic, geomorphic, and biological factors shape riparian environments. For example, differences in climate dictate the seasonality of the hydrologic cycle and determine the timing and intensity of flooding. Watershed features such the slope of the land, size of the watershed, storage capacity of the soil, and supplies of groundwater and sediment interact with climate to modulate or amplify these effects. Within the riparian area itself, further sources of variation can be found in channel morphology, sediment dynamics, and floodplain structure. Ultimately, all these factors influence species composition of riparian biota. This chapter focuses on the structure and functions of riparian areas, with an emphasis on those bordering streams and rivers rather than lakes and estuarine–marine waterbodies. Riverine riparian areas, because of their great collective length, comprise the vast majority of riparian areas in the United States.

FLUVIAL PROCESSES AND SEDIMENT DYNAMICS

Streams and rivers, which flow longitudinally downstream from higher elevations, can be classified by their size and the number of tributaries that flow into them. As shown in Figure 2-1, headwater streams are classified as *first order*, with order number increasing in a downstream direction. Headwater networks of very small streams accumulate rainfall, overland flow, snowmelt, or aquifer discharge, sending variable amounts of water downstream to increasingly larger channels.

The water budget of all streams and rivers is determined by climate and by other watershed attributes such as topography, soil type, bedrock substrata, groundwater discharge, and vegetation. Natural flow patterns—unregulated by dams and water diversion—will vary with the dynamics of water delivery and cycling, unless the source is a spring fed by a deep (phreatic) aquifer that has very little surface connection (Gibert et al., 1994; Vervier, 1990). According to Poff et al. (1997), the flow regime of a river can be distinguished by several major components, including magnitude, frequency, duration, timing, and rate of change, as described in Box 2-1. River flows are often described using one or more of these components. Thus, for example the bank-full flow, which defines the bank-

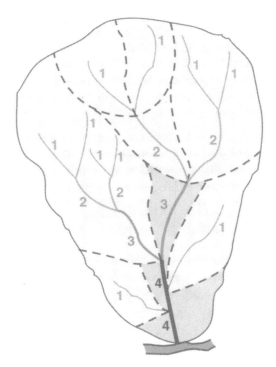

FIGURE 2-1 Stream orders for a watershed that includes first- to fourth-order streams. Ephemeral streams are not shown on this diagram. SOURCE: Reprinted, with permission, Strahler (1952). © 1952 by The Geological Society of America.

> **BOX 2-1**
> **Components of Flow—the "Master" Variable**
>
> Because streamflow is strongly correlated with critical physical and biological characteristics of rivers, such as water temperature, sediment transport, channel morphology and habitat diversity, it represents a "master variable" that influences the functions of associated riparian areas.
>
> *Flow magnitude* represents the amount of water moving past a given location per unit time. It can influence rates of solute, suspended sediment, and bedload sediment transport, and thus is a critical variable with regard to the creation of alluvial landforms (e.g., point bars, floodplains streambanks, and channel sinuosity). As discussed later in this chapter, high flows are needed for some species to create local zones of erosion/deposition for seedling establishment.
>
> *Flow frequency* refers to how often a flow of a given magnitude is equaled or exceeded over some time interval. Flow frequency, in combination with flow magnitude, indicates the amount of energy a stream has to do work (e.g., sediment transport, channel adjustments, etc.).
>
> *Flow duration* represents the period of time associated with a specific flow magnitude. From the perspective of riparian plant communities and floodplain functions, flow duration represents the length of time that overbank flows occur or that soils remain saturated from high flows. Flow duration is often a crucial variable for many riparian plants that have adapted their physiology to accommodate extended periods of high moisture levels.
>
> *Flow timing* generally refers to the seasonality of a given flow. For example, the timing of most snowmelt runoff for many western streams and rivers occurs in late spring and early summer. Fish and other organisms have adapted their life history strategies to the timing of these flow periods. Superimposed upon the long-term water and sediment budget of the watershed, flow timing determines the relative wetness or dryness of the adjacent riparian area and is therefore a primary structuring process.
>
> The *rate of change* in streamflow or water levels represents how quickly a flow changes from one magnitude to another. Streams and rivers that derive their flow from snowmelt are generally considered less "flashy" than those that respond to large amounts of rainfall. Rate of change can influence water sediment transport rates and riparian plant communities. For example, seedlings of deciduous woody species may need a relatively low rate of change during snowmelt recession flows for them to successfully establish.

full channel, is the discharge of the 1.5- to 3-year return period storm (Dingman, 1984). Floods are of larger discharge and generally occur less frequently than bank-full events. Floods move and sort sediments and other materials, forming the physical structures that compose the riparian areas of rivers. Big floods, which are relatively rare, often create a physical template that is continually reworked and modified by lower flows. Hence, diverse alluvial landforms, such as gravel bars, floodplains, islands, terraces, and the channel network, are created by flow-mediated movement of sediments and are in a constant state of change (Ward, 1998; Ward et al., 2000).

The size and character of streamside riparian areas is directly related to water delivery to and flux through the watershed. One pattern is the tendency for riparian areas to be expansive next to big, larger-order rivers, which in part reflects multiple hydrologic sources (e.g., seasonal overbank flows from the river, flood-related flows in secondary channels, and groundwater discharge, all discussed in detail in a later section). Periods of high flow, particularly in unconstrained or relatively wide alluvial valleys, can create a multitude of landforms (e.g., streambanks, floodplains, and terraces) that are common to many riparian systems. However, such basic patterns are often too simplistic to be widely useful in predicting structure and function of riparian areas across many landscapes. For example, the lower Columbia River (9th-order) upstream of Portland, Oregon, is constrained by resistant bedrock, resulting in narrow floodplains and minimal riparian areas (see Figure 2-2 for the difference between constrained and uncon-

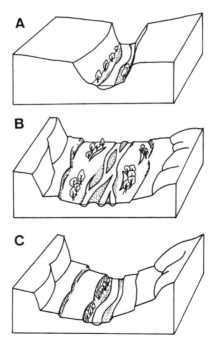

FIGURE 2-2 Geomorphology of a stream corridor in (A) a constrained reach, (B) an unconstrained aggrading reach, and (C) an unconstrained degrading reach. SOURCE: Reprinted, with permission, from Dahm et al. (1998). © 1998 by Blackwell Science Ltd.

strained river reaches). Another generalization is that where fluvial systems encounter relatively wide valleys and low channel gradients, they typically develop a system of meanders and floodplains that represent both sediment and water discharge regimes in dynamic balance with valley and channel gradients, channel morphology, and riparian vegetation. In general, these sinuous channels occur within a definable meander belt, simply defined by a linear boundary that connects the outer margins of the existing channel meanders. More accurate geomorphic delineation of the meander belt would be based on the actual margin of the floodplain, boundaries of historical channel meanders (as shown for the Willamette River in Figure 2-3), or boundaries of past recorded inundation extent.

Because river flow naturally is erosive and water is the universal solvent, particulate materials (sediment, rocks, trees) and dissolved materials (salts, organic compounds) are exported downstream in proportion to stream power and

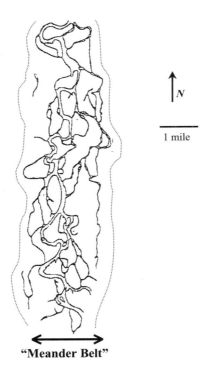

"Meander Belt"

FIGURE 2-3 Meander belt of the Willamette River based on the channel meanders in a reach from Eugene to Harrisburg in 1850. SOURCE: Modified from Williamson et al. (1995).

deposited in a wide array of fluvial landforms within the river channel and floodplain. A generalized three-dimensional view of riparian areas includes portions of the channel and associated features (gravel bars, islands, large wood); a parafluvial zone (which corresponds to the bank-full width) that experiences the seasonal range of flow variation; a vegetated area of varying successional states influenced by floods, sediment deposition, and water availability; and a transitional zone to the uplands (see Figure 1-4). Furthermore, these features are generally underlain by an alluvial aquifer that can have a major influence on riparian processes, particularly where bed sediments are deep (Stanford, 1998). The longitudinal, lateral, and vertical pathways through which water and materials are conveyed through riparian areas are discussed in detail throughout this chapter.

Erosion and Deposition

The processes of erosion, transport, and deposition continually disturb and reshape the riparian environment. Materials from upstream sources such as erosion zones along hillslopes and riparian terraces or landslides are sorted by flowing water and transported downstream until the physics and energetics of the transport process dictate deposition either in the channel or on the floodplain of the river. As shown in Figure 2-4, flow-mediated erosion of sediment occurs in the steep gradients of lower-order segments, deposition of course material (gravel,

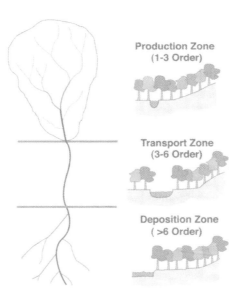

FIGURE 2-4 The geomorphic zones of a fluvial system. SOURCE: Reprinted, with permission, from Schultz et al. (2000). © 2000 by American Society of Agronomy.

cobble) occurs in the middle reaches related to aggradation of the river valley and associated loss of flow velocity (energy dissipation), and deposition of fine materials (sand, silts) occurs in the lowest-velocity environments that characteristically occur in the high-order segments of the Piedmont or coastal plain. Hence, as the size of the stream increases, the size of the floodplain generally increases.

Although these broad patterns in sediment transport explain a trend of downstream fining in the grain size of bed sediment in many river systems (Schumm, 1960), in reality sediments of all sizes are sorted along every channel or floodplain within the river corridor. Flowing water sorts the sediment between different areas of the channel with different capacities for maintaining sediment in suspension. Coarse sediments are suspended and deposited only in the highest-energy environments of the river channel, i.e., areas with relatively high velocity. Fine-grained sediments, in contrast, are generally restricted to the lowest-energy backwaters of the active channel or to the floodplain.

Although bank-full flows maintain channels, floods account for much of the major work in reshaping channels and floodplains. Increased production of sediments from terrestrial sources and acceleration of bank erosion during floods can release large amounts of fine- and coarse-textured sediment into a channel over a short period of time, which are then deposited in downstream channels or on floodplains. Floods also cause substantial realignment of channels because of reoccupation of secondary or abandoned channels by newly released sediment (Beschta et al., 1987a). When significant amounts of coarse sediment become available locally, rapid adjustments to the morphology of the channel can occur. These effects may be transmitted in both the upstream direction (backwater effects, including upstream bed material storage and an altered channel morphology; channel incision and gully head cuts) and the downstream direction (higher levels of sediment transport with the potential for increased channel instabilities).

The net result of fluvial processes over decades to millennia is a slow modification and reworking of the channel and floodplain physical template such that sediment routing must be viewed as a constantly changing feature of all alluvial rivers. Distinct features such as cutbanks, meander scrolls, and point bars migrate over time. Within the meander belt, the deposition of sediments on vegetated floodplains occurs periodically over time and the exact character, dimensions, and location of meanders may incrementally shift as a result of the migration of gravel point bars and the erosion of cutbanks. Secondary channels become plugged, creating backwater sloughs or oxbow lakes. Slugs of sediment derived from an episodic landslide may take many years to move down the river corridor, influencing riparian areas to different extents as they pass through specific segments. Large tree boles eroded from riparian areas substantially increase the variation in sediment transport and deposition, and thus also the variety of habitat types available for biota (Naiman et al., 2001).

Cycling of sediments back and forth between the main channel and the channel's banks and floodplain is an important component of sediment transport in rivers (Meade et al., 1990). Floodplains expand laterally and "grow upward" due to the long-term deposition of fine sediment during recurring overbank flows. Seasonal high flows continually disturb new areas. Newly deposited sediments on floodplains undergo biogeochemical changes (i.e., diagenesis) that will over time transform a flood-deposited sediment into a riparian soil. The overall result is the creation of a complex patchwork of riparian areas, each with a slightly different microenvironment of sediment grain size and nutrient and water availability, and each at a different stage of development since the last disturbance (Amoros et al., 1987; Malanson, 1993).

Because of the dynamic flow and sediment transport regimes often associated with riparian areas, their soils reflect a high degree of unevenness in particle sizes, soil depth, and the amount of associated compounds such as organic matter. Highly variable water levels typically result in morphological soil features such as mottling, gleying, oxidation/reduction, and others. However, in instances where floodplains have been slowly built up via the incremental deposition of fine silt layers over many centuries, soil characteristics across extensive areas may be relatively uniform. Floodplain soils have been some of the most productive areas in the nation for agricultural production due to their high levels of nutrients and organic matter.

Flow Modification Within Riparian Areas

Although the energy from water moving down a channel can be used to do work (e.g., scour banks and transport sediment), the vast majority is used to overcome the frictional resistance provided by a channel's bed and banks and is eventually dissipated as heat. Thus, streamside riparian areas are responsible for the dissipation of energy associated with flowing water. The flow resistance, or roughness, of a stream reach, caused by the physical configuration of its channel, streambanks, and floodplains as well as by the riparian plant communities, can be described by a roughness coefficient, such as Manning's n (Leopold et al., 1964). Cowan (1956) identified several major channel conditions that affect roughness: bed material, degree of surface irregularity, variations in channel cross section, relative effects of obstructions, degree of meandering, and effects of vegetation. Importantly, vegetation can directly or indirectly affect all these conditions, with the possible exception of bed material, thus indicating it often has a major influence on channel roughness and on how channels dissipate stream energy during periods of high flow. Herbaceous riparian vegetation increases local friction on streambanks by creating flexible and three-dimensional barriers to flow. Riparian graminoids (grasses, sedges, rushes) and shrubs are particularly effective at trapping sediments during high flows and helping to maintain stable streambanks. For forest floodplains, roughness increases directly with the density and size of trees (Li and Shen, 1973; Petryk and Bosmajian, 1975). Large wood provided to streams and rivers from riparian forests can also have a significant effect on channel roughness via the occurrence of debris jams and other accumulations that alter flow patterns (Abbe and Montgomery, 1996; Montgomery et al., 1996; Piegay and Gurnell, 1997).

At high flow, streambanks, floodplains, and their associated vegetation provide resistance to flowing water, thus locally altering patterns of scour, sediment transport, and deposition (Sedell and Beschta, 1991). For example, low velocity zones have been observed to develop when floods pass through riparian forests, creating sites for the retention of sediment and organic matter and refuges for aquatic organisms (Swanson et al., 1998). Floodplain vegetation is especially effective at providing protection from scour, which is why well-vegetated floodplains typically are areas of long-term sediment accumulation.

During periods of low flow, woody species have a much less significant effect on flow roughness because of the smaller surface area exposed to surface flow, such that flow resistance tends to be controlled more by the morphology of the channel. In contrast, aquatic macrophytes and graminoids can greatly influence the resistance provided during low-flow periods (Kauffman and Krueger, 1984). Finally, the uptake and transpiration of water by riparian and upslope vegetation during low-flow periods can alter discharge, thereby influencing aquatic habitat (Rothacher, 1970; Troendle, 1983; Cheng, 1989; Keppeler and Ziemer, 1990; Hicks et al., 1991).

Lacustrine Riparian Areas

Unlike the riparian areas of stream and river (lotic) environments, riparian areas bordering lakes differ significantly in the energy sources that drive physical mixing (Wetzel, 2001). In the shallow littoral environments of lakeshores, mixing is generally driven by temperature gradients and storm-generated waves. An important contrast with lotic environments is the type and frequency of water-level changes at lakeshores. Seiches, for example, can cause substantial changes in water level over periods of days to weeks at the shores of large lakes without the kinds of erosive forces of floods that affect channel floodplains. Lakeshores also tend to have much larger water-level changes over longer-term (interannual) cycles, as determined by interannual variation in climate and the regional water balance. Large reservoirs and other river impoundments used for water storage may exhibit nonseasonal fluctuations in water level, with hydrographs varying erratically under the control of hydropower production or irrigation supply. Consequently, riparian areas around reservoirs are highly variable and often are composed of non-native, invasive species because they have little long-term continuity in water supply and occur in areas of the landscape that have no legacy of native plant colonization (e.g., Nilsson et al., 1997).

Despite major differences in flow velocities and extent of water-level changes, the shallow littoral environment and riparian areas adjacent to lakeshores have much in common with riparian areas bordering streams. As in streams, a broad range of sediment types and textures is often available, nutrients are often ample and primary productivity is high, water exchange between the surface and subsurface is conducive to high rates of biogeochemical cycling in sediments, and secondary productivity in these environments is typically high (Wetzel, 2001). In the case of large lakes with inlets from rivers, alluvial deltas may develop by sediment deposition in the river–lacustrine confluence. Often, river deltas in lakes and reservoirs facilitate robust riparian areas in a manner similar to the islands and low terraces that occur in alluvial rivers. Deltaic riparian areas can be large landforms up to many square miles in size. Few studies have been done in such environments (e.g., Stanford and Hauer, 1992).

In summary, riparian areas are characterized by a spatial and temporal mosaic of conditions reflecting variability in sediment type and particle size distribution, timing of water sources and water quality, and time since disturbance by floods. Seasonal dynamics in flow and sediment transport constitute the foundation of riparian structure and thus influence the resulting colonization by riparian species and the many functions performed by these areas. Moisture availability and anoxia in riparian soil are additional factors that closely follow the distribution of grain sizes determined by fluvial processes. In many channels, the natural variability of flow has been regulated and sediment inputs have been curtailed downstream of dams and water diversions. As discussed in Chapter 3, the influence of humans in regulating river flow has had overwhelming effects on ecological processes in rivers and riparian areas, because of the disruption of flow seasonality, sediment dynamics, and moisture availability.

HYDROLOGIC AND BIOGEOCHEMICAL PROCESSES

Hydrologic Pathways in Riparian Areas

Riparian areas receive water from three main sources: (1) groundwater discharge, (2) overland and shallow subsurface flow from adjacent uplands with additional input from direct precipitation, and (3) flow from the adjacent surface water body. The major losses of water from riparian areas include groundwater recharge and evapotranspiration. Plate 2-1 illustrates these major water flow paths for a streamside riparian area. Both the quality (in terms of dissolved and particulate constituents) and the timing of water from these sources vary considerably. For example, the discharge of deep groundwater is on the order of centuries, while overbank flows and intense rainstorms can change flows within minutes.

Groundwater Sources

Winter et al. (1998) outlines some of the basic interrelationships between groundwater and surface water in streams and lakes and shows how interactions vary as a result of differences in climate, topography, and surficial geology.

Streams whose downstream flows increase as a result of groundwater discharge are referred to as *gaining* streams. In contrast, flow in the channel decreases in the downstream direction in *losing* streams that recharge the groundwater system. Because of variability in water sources and hydrogeologic properties of aquifers, it is typical for streams to simultaneously experience discharge in one reach while experiencing recharge in others. For example, steep mountain streams gain water by groundwater discharge in their upper reaches and then lose water as they flow out of constricted mountain valleys onto alluvial fans.

Lakes and wetlands share some of the same relationships with groundwater as do streams. Lakes and wetlands commonly discharge and recharge simultaneously in different parts of the system and experience flow reversals seasonally (Figure 2-5). As in streams and rivers, movement of water between groundwater and surface water is influenced by the nature of the substrata and the water elevation in the lake compared with water levels and gradients in groundwater of the adjacent aquifer (Sebestyen and Schneider, 2001). Water moves from areas of high elevation to areas of low elevation, sometimes involving streams or rivers at inlets or outlets to the lake.

Because the majority of riparian areas are associated with stream and river channels, this discussion focuses on interactions between groundwater and river channels rather than lakes. From a relatively large-scale perspective (miles or greater), the direction of groundwater flow in the vicinity of rivers is typically associated with patterns of floodplain and channel topography. As a result, flow pathways are seldom entirely parallel or entirely perpendicular to the main channel but instead occur diagonally toward the channel in a downstream direction. The major controls on orientation of groundwater flow paths are hydraulic properties of aquifer materials, regional gradient, and sinuosity of channel (Larkin and Sharp, 1992). Groundwater that tends to flow parallel to a channel is referred to as *underflow* (Larkin and Sharp, 1992); in contrast, groundwater flow perpendicular to and toward the channel is referred to as *baseflow* (Hall, 1968) (see Figure 2-6).

At much smaller spatial scales, i.e., feet to tens of feet, interactions between groundwater and riparian areas are influenced primarily by heterogeneities of riparian and channel sediments, which have a critical effect on local direction and flow rate of groundwater. In some settings, baseflow passes directly through riparian sediments, while in others, baseflow may bypass riparian sediments by flowing through coarse material underneath and discharging vertically from directly beneath the stream bed (Phillips et al., 1993). This short-circuiting of the root zone can have important implications for the extent of certain transformation processes that occur in riparian areas. As discussed later, the variation in the specific flow paths characteristic of riparian areas may explain why some buffers are not as effective as others.

An often-overlooked aspect of groundwater–riparian–channel interactions is that groundwater discharge is not equivalent along all parts of a channel. Instead, certain channel subreaches tend to collect a significant proportion of all ground-

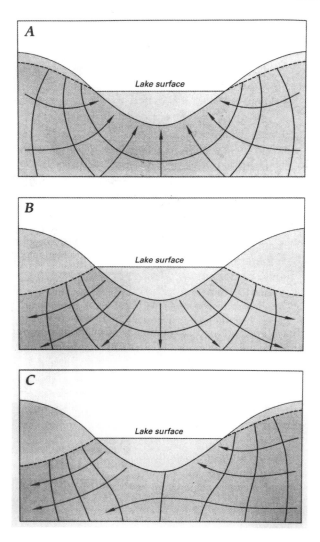

FIGURE 2-5 Lakes, like streams, can receive groundwater inflow (A), lose water as seepage to groundwater (B), or both (C). SOURCE: Winter et al. (1998).

water entering a stream in a given reach. Groundwater discharge points tend to occur at the upstream ends of pools, the upstream side of meanders, anywhere along the channel thalweg (deepest area of central channel), and within side channels or alcoves in streams and rivers (Harvey and Bencala, 1993). An uncommon example of a groundwater discharge point is a visible area of groundwater seepage on the channel bank above the level of the channel.

The importance of groundwater discharge points along channels is twofold. First, areas along channels that collect groundwater discharge tend to favor establishment of rich riparian vegetation, especially in dry climates where water avail-

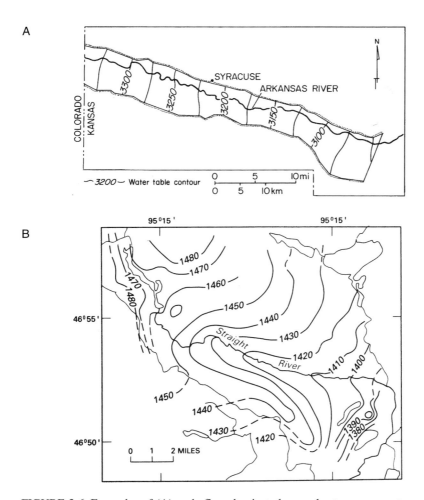

FIGURE 2-6 Examples of (A) underflow-dominated groundwater movement parallel to the channel and (B) baseflow-dominated groundwater movement perpendicular to the channel. Groundwater moves in the direction of decreasing water table contours. SOURCES: (A) Reprinted, with permission, from Larkin and Sharp (1992). © 1992 by Geological Society of America. (B) Winter et al. (1998).

ability is a major factor limiting the establishment and maintenance of riparian vegetation. Second, groundwater discharge points tend to have cooler water locally in the channel during summer and warmer water in winter compared with other channel areas. As such, they may be important in creating thermal refugia for aquatic organisms at particularly sensitive times in their life cycles.

Hillslope Sources

Hillslope sources of water to riparian areas all begin as precipitation falling on the landscape. There are numerous pathways for water to travel from the hillslope to riparian areas. For example, *overland flow* begins as precipitation that exceeds the percolation capacity of the soil. Precipitation can also travel downslope via *shallow subsurface flow* (a portion of which may emerge onto the hillslope surface before reaching the channel). Precipitation that falls directly onto saturated areas of the hillslope further augments these pathways. All of these flow mechanisms are referred to collectively as *hillslope runoff*. Figure 2-7 contrasts hillslope *runoff* in three situations differing in climate and soil development. Humid forested landscapes with deep permeable soils have deep percolation and groundwater flow to riparian areas, in addition to shallow flow on the lower hillslope during intense storms (Figure 2-7A). In areas where local geology includes soil layers of low permeability, drainage is often restricted to shallow permeable soil (Figure 2-7B). In arid areas, intense precipitation onto hillslopes with sparse xerophytic vegetation and impervious soils creates a situation where overland flow is often the dominant pathway of drainage (Figure 2-7C).

Topography and hydraulic properties of sediments influence the degree to which riparian areas store hillslope runoff or transmit it to the channel. If comprised of coarse sediments, the riparian area can usually store large quantities of hillslope runoff and release it to the channel by groundwater discharge. Riparian sediments that are relatively fine and are lower in permeability than other soils of the watershed generally cannot store large quantities of water quickly enough, leading to rapid expansion of saturated areas. During intense storms, shallow subsurface flow that cannot move fast enough laterally emerges as *return flow* onto the surface of the riparian area. This flow is further augmented by rain falling directly on saturated areas. The expanding areas of saturation in the riparian area that result from hillslope runoff are referred to as variable source areas (Hewlett and Hibbert, 1967) or partial contributing areas (Dunne and Black, 1970). Such areas become saturated during the early part of intense storms, expanding further if rainfall continues. Depending on storm intensity and duration, the concave upward areas of valley bottoms and hillslope hollows tend to become saturated because they collect storm water faster than they deliver it to channels.

As is obvious from the preceding discussion, hillslope runoff is a highly variable process in space and time that depends on a variety of factors, particu-

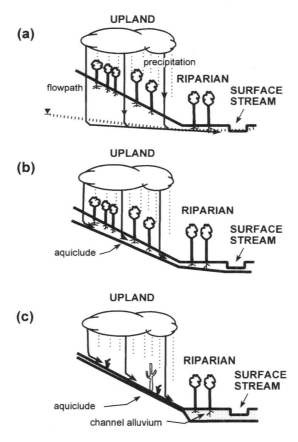

FIGURE 2-7 Hydrologic linkage of upland ecosystems with stream corridor ecosystems in (A) forested landscapes with deep, well-drained soils where water percolates well beyond the rooting zone, (B) forested landscapes with shallow soils where underflow intercepts the rooting zone, and (C) arid and semiarid landscapes where soils of low permeability force overland flow. SOURCE: Reprinted, with permission, from Fisher et al. (1998). © 1998 by Springer-Verlag.

larly topography, sediment hydraulic properties, and antecedent groundwater levels and soil moisture. Weather is also important through its effect on the intensity and duration of precipitation, temperature, and solar radiation patterns. These factors in turn determine the types and amount of vegetation present in riparian areas and thus the extent of evapotranspiration. The interplay between these factors and their effect on hillslope runoff are summarized in Figure 2-8.

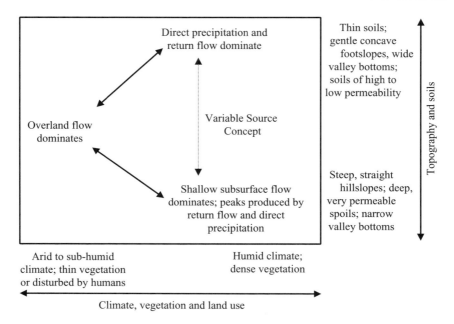

FIGURE 2-8 Dominant processes of hillslope runoff in response to rainfall. SOURCE: Adapted from Dunne (1978).

Channel Sources

A final source of water to riparian areas is from the channel itself via overbank flow, bank storage, or hyporheic flow. The extent and duration of overbank flow, or flooding, of riparian areas is very much dependent on intensity and duration of precipitation, basin area and topography, soil and aquifer type, and morphology of the river channel and floodplain. In general, when channels reach flood stage during storms and floods, riparian areas temporarily store excess water that cannot be quickly conveyed downstream. The overall effect of this water storage is delay and attenuation of the flood peak in downstream areas (Moench et al., 1974; Bhowmilk et al., 1980). Figure 2-9 shows the diversity of flooding environments for a 50-year flood on the Rhône River floodplain system. Floodwaters that overtop channel banks over a period of many decades and centuries and the sediments they carry are essential to the creation and maintenance of floodplain landforms mentioned earlier, such as levees and ridges. Rapid flow across riparian areas can rehydrate abandoned channels that directly flow back to the channel. Floodwaters that become trapped in topographic depressions, such as isolated sloughs, subsequently recharge the groundwater system; deposition of fine sediment, such as fine silt and clay, occurs during these long

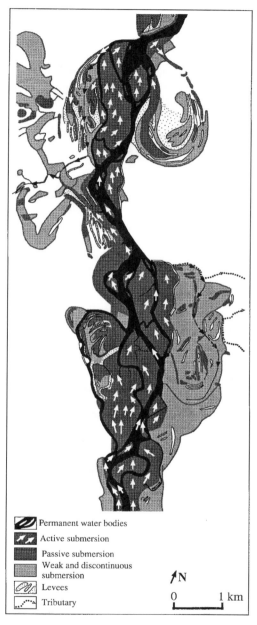

FIGURE 2-9 Inundation patterns of the Rhône River floodplain system during a 50-year flood event, demonstrating the complexity of overbank flooding. SOURCE: Reprinted, with permission, from Tockner et al. (2000). © 2000 by John Wiley and Sons, Inc.

periods of standing water. Hupp (2000) found that deposition rates are relatively high in southeastern coastal plain rivers because of the frequent return intervals of overbank flow and the relatively high sediment loads of the rivers. Riparian plant communities in Oregon have been shown to be associated with floodplain surfaces that receive relatively frequent overbank flows of at least once every five years (Chapin et al., 2000). As shown schematically in Figure 2-10, the relative importance of overbank flow versus hillslope runoff typically increases with increasing stream order.

Subsurface movement of water from the channel into the groundwater aquifer beneath the floodplain is sometimes an important source of water to riparian areas, particularly in ephemeral and intermittent streams. In perennial streams, this usually involves bidirectional exchange back and forth between the surface channel and groundwater beneath the floodplain. Two types of bidirectional interactions deserve special mention. The first is *bank storage*, which involves channel water moving laterally into subsurface riparian areas when river stage is high, and then gradually moving back to the channel when river stage drops (Pinder and Sauer, 1971) (Figure 2-11A). Bank storage in riparian areas can affect water storage, chemical transformations in streams and rivers, surface water temperature, and the composition and extent of riparian plant communities.

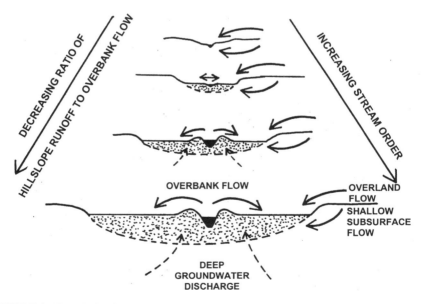

FIGURE 2-10 Relative importance of hillslope runoff versus overbank flow. SOURCE: Reprinted, with permission, from Brinson (1993). © 1993 by Dr. Douglas A. Wilcox, Editor-in-Chief, Wetlands.

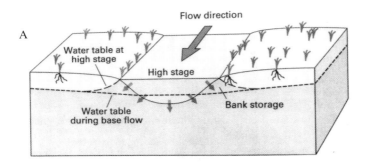
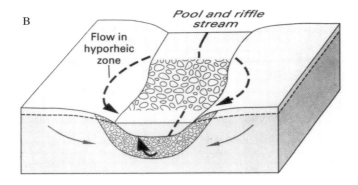
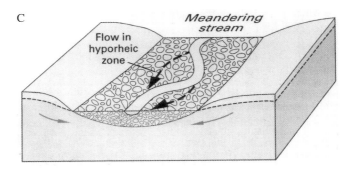

FIGURE 2-11 Water exchange between channel and riparian areas caused by (A) change in stream stage followed by bank storage, (B) streambed topography routing streamflow temporarily through subsurface (hyporheic) flow paths, and (C) hyporheic flows through bends of meandering stream. SOURCE: Winter et al. (1998).

Squillace (1996), for example, demonstrated that bank storage in the Cedar River (Iowa) sequestered (and possibly led to degradation of) pesticides such as atrazine that had previously been transported to the channel by spring runoff from agricultural fields. After the peak spring flows in the Cedar River, pesticides were slowly discharged back to the river over a period of weeks to months, accomplishing a dilution of pesticide reaching the river.

The other type of bidirectional interaction is *hyporheic exchange*, which is the temporary routing of water through gravel bars and the alluvium surrounding stream channels (i.e., the hyporheic zone—see Figure 1-4). The extent of the hyporheic zone is defined operationally using solute tracers, e.g., as the depth in the sediment where tracer concentrations indicate that 10 percent or more of the water is derived from the channel (Triska et al., 1989). Channel flow enters the hyporheic zone due to uneven pressure gradients over a rough streambed, or due to pooling of water at higher elevations behind flow obstructions such as riffles (Harvey and Bencala, 1993; Wroblickly et al., 1998). As shown in Figure 2-11(B) and (C), channel flow that is routed through hyporheic zones generally returns to the channel within a relatively short distance downstream. Passage of stream water through hyporheic flow paths increases oxygen concentrations in the subsurface, creating specialized habitats for burrowing organisms beneath streamside riparian areas (Jones and Mulholland, 2000) and optimal conditions for bull trout and salmon eggs where hyporheic flow returns to the channel. Certain biogeochemical reactions are also enhanced in hyporheic zones, affecting the transport and transformation of nutrients, metals, and organic compounds (Jones and Mulholland, 2000).

Role of Transpiration

Trees, shrubs, and herbaceous plants contribute significantly to water cycling and material movement in riparian areas. The most direct effect of plants on water flow and storage in riparian areas is transpiration. In smaller channels of headwater basins, riparian evapotranspiration accounts for a percentage of the groundwater that would otherwise be discharged to the channel (Daniel, 1976). Early studies along the Gila River in Arizona estimated that evapotranspiration removes 12.3 percent of the water from the system, while evaporation from the river surface and wet sand bars removed 2.5 percent (Gatewood et al., 1950). In a study of approximately 50 basins in the Appalachian Valley and Piedmont areas of the Mid-Atlantic region, riparian transpiration removed approximately 10 percent of recharged groundwater prior to discharge to streams (Rutledge and Mesko, 1996). Transpiration in relatively small headwater streams is significant enough to cause a diel cycle in streamflow, with decreased streamflows during the day and increased streamflows at night. Harvey et al. (1991) found that the size of the hyporheic zone in riparian areas expanded during the day and contracted at night, in accordance with riparian transpiration, providing circumstantial evidence that in some areas, streamflow may be a significant source of water for riparian transpiration.

An important need in western riparian areas is to improve our understanding of the water-use requirements of riparian vegetation. This involves quantifying the relative importance of various water sources for transpiration, including recently recharged precipitation, groundwater, and surface water. Modern tools are being applied to this problem, including the use of water-stable isotopes in tree sap as tracers to identify the source of water for transpiration (Flanagan et al., 1992; Dawson and Ehleringer, 1993). Early results indicate that riparian plants use different components of water in the alluvium; sacaton grass, for example, uses recently recharged precipitation from the unsaturated zone (Moran and Heilman, 2000). In contrast, mesquite uses a mixture of groundwater and unsaturated zone water depending on tree size, cottonwood uses mostly groundwater, and willow uses only groundwater. When combined with sap-flow measurements, water-stable isotopes determine how water use by riparian trees changes with forest age, groundwater levels, and climatic fluctuations.

Unlike evapotranspiration, direct evaporation from stream reaches is usually small in comparison with groundwater discharge or flow inputs from upstream. However, evaporation can make a measurable contribution to the water budgets of lakes.

* * *

As evident from the preceding discussion, hydrologic fluxes through riparian areas are highly variable in both space and time. Time scales range from minutes to hours (hyporheic flow and transpiration), days to months (storm and seasonal snowmelt response), and years to decades (climatic effects on recharge and baseflow discharge). As a result of that complexity, it is entirely possible that a single riparian area could function some of the time as a pathway for groundwater discharge, at other times as a hyporheic zone, and at still other times as a zone of bank storage. Any assessment of the hydrology of riparian areas therefore depends not only on physical attributes of the channel, watershed, and climate, but also on spatial and temporal boundaries of the particular problem. For example, determining only the net groundwater exchange in a stream or lake (i.e., the difference between discharge and recharge) is not always adequate for characterizing groundwater interactions that affect riparian areas. In many situations, both groundwater discharge from the watershed and recharge from the surface water body are important, necessitating more thorough investigations of water fluxes using multiple approaches (Krabbenhoft et al., 1990; Harvey and Bencala, 1993; Hunt et al., 1996; Choi and Harvey, 2000). Thus, there is no universally acceptable approach to characterizing the water balance of riparian areas, and many studies employ significant simplifications, assumptions, or other qualifications. Examples of riparian water balances developed for various purposes include those by Goodrich et al. (2000), MacNish et al. (2000), and Rutledge and Mesko (1996).

Biogeochemical Interactions Between Riparian Areas and the Surrounding Landscape

Along with water flow through riparian areas comes the transport and transformation of chemicals and particulate matter—key factors that affect the ecology of rivers and lakes. These processes have been most intensely studied within stream channels, although the role of riparian areas and groundwater in influencing adjacent aquatic systems is increasingly being explored. Although this section relies primarily on research conducted in riverine settings, broad concepts connecting physical and ecological factors have also been developed for lakes and wetlands (see Labaugh et al., 1996; Kratz et al., 1997; Carpenter et al., 1998; Wetzel and Søndergaard, 1998; and Wetzel, 1999).

Instream Processes

Nutrients are cycled within streams and rivers, moving back and forth between inorganic forms and the living tissue of biota. Dissolved inorganic nutrients and nutrients associated with fine particulate organic or inorganic matter move with the flowing water, while nutrients in biotic compartments, such as microbes, periphyton, aquatic plants, and riparian trees, spend much longer in one place within the stream corridor. The cycling of nutrients between transported and fixed components is the basis of the Nutrient Spiraling Concept, which refers to the sequences of movement and temporary retention that occur during downstream transport (Newbold et al., 1982). The concept has been tested in the field through experimentation using dissolved tracers such as salts to track water flow and isotopic forms of important nutrients such as nitrogen and phosphorus (Figure 2-12). Recent work suggests that variations in hydrologic processes, not in biological and geochemical processes, are the most significant cause of site-to-site variation in the cycling of dissolved and fine particulate materials in streams (Findlay, 1995).

Consistent with the Nutrient Spiraling Concept is the River Continuum Concept (Vannote et al., 1980), which sets forth a classification of ecological processes in streams and rivers that is firmly grounded in the principles of fluvial geomorphology and hydrology. The River Continuum Concept places ecological processes, such as productivity, respiration, and food web structure, in an interdependent upstream–downstream sequence.

A limitation of the Nutrient Spiraling and River Continuum Concepts is that they emphasize longitudinal transport and changes in dominant processes according to channel hydrology. Thus, they refer primarily to the wetted channel and bottom sediment, rather than to riparian areas. The Flood Pulse Concept (Junk et

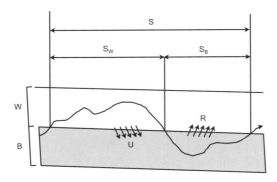

FIGURE 2-12 Two-compartment nutrient spiraling model. The spiraling length S is the average distance a nutrient atom, such as phosphorus, travels downstream during one cycle. A cycle begins with the availability of the nutrient atom in the water column, and includes its distance of transport in the water (S_W) until its uptake (U) and assimilation by the biota, and whatever additional distance the atom travels downstream within the biota (S_B) until that atom is eventually re-mineralized and released. W = water, B = biota, R = release. SOURCE: Modified from Newbold et al. (1982).

al., 1989; Bayley, 1991), on the other hand, emphasizes the role of lateral exchanges between the channel and floodplain (and thus through riparian areas) as an integral driver of ecosystem processes in river corridors. As shown in Figure 2-13, biogeochemical interactions between riparian areas and channels are prob-

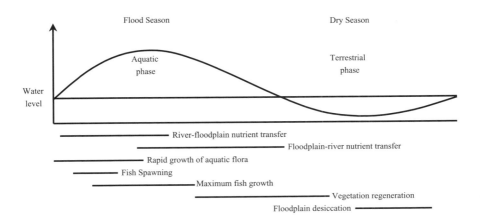

FIGURE 2-13 The influence of the flood pulse within the river–floodplain complex. SOURCE: Reprinted, with permission, from Bard and Wilby (1999). © 1999 by Routledge Publishers.

ably most intense during storms ("river–floodplain nutrient transfer"). Material fluxes of water, sediment, nutrients, and organic matter are maximized at flood stage. For example, floods deposit new sediments with associated nutrients on the riparian surfaces, creating the conditions that favor high primary productivity of riparian vegetation. Flood pulsing is also an efficient means of moving carbon and nutrients out of the forest and into the main channel. The transformation and transport processes that carry out the lateral exchange of material are discussed in detail below.

Fate and Transport Processes in Riparian Areas

Riparian vegetation indirectly influences biogeochemical cycling through transpiration and other effects on water flow. However, it has more direct effects, including uptake or excretion of solutes by roots as well as symbiotic associations with bacteria or fungi that stimulate important biogeochemical reactions. Several characteristics of riparian areas make them important sites for subsurface transformations of nutrients and other chemicals. Riparian soils (defined topographically as valley bottom areas that tend to become saturated during storms) possess greater soil N concentrations, higher-quality particulate organic carbon (as measured by C:N ratio), and greater overall microbial activity than do ridge and slope area soils (Garten et al., 1994). In addition, riparian soil water and dissolved organic carbon are flushed to streams much more quickly than hillslope soil water (Boyer et al., 1997).

Numerous studies have investigated the role of grassed and forested riparian areas in controlling the transport of sediment, nutrients, pesticides, metals, microorganisms, and other pollutants to receiving waters, using buffers both within and upslope of riparian areas. Although the results of these studies are highly variable in terms of "pollutant removal" or "trapping," they have greatly improved our understanding of the mechanisms controlling transport and fate in riparian areas, especially for sediment and nutrients. The major physical, chemical, and biological fate and transport processes associated with riparian areas include infiltration, deposition, filtration, adsorption, degradation, and assimilation. Figures 2-14(A) and 2-14(B) show the important fate and transport processes for nitrogen (which is generally dissolved) and phosphorus (which is generally bound to sediment), respectively, in riparian areas.

Infiltration (also referred to as percolation) is a primary transport process during which water and dissolved chemicals and particulates enter the subsurface. Infiltration is important because it decreases the volume of overland flow, thus reducing the aboveground transport of chemicals and particulates. Once in the soil profile, pollutants are often removed or degraded by a variety of physical, chemical, and biological processes. Infiltration is one of the more easily quantifiable mechanisms affecting the performance of riparian areas that are to be utilized for removing chemicals. Thus, many constructed riparian buffers are de-

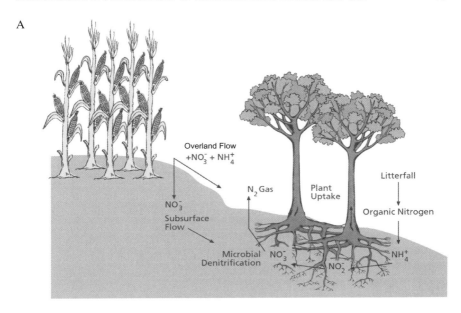

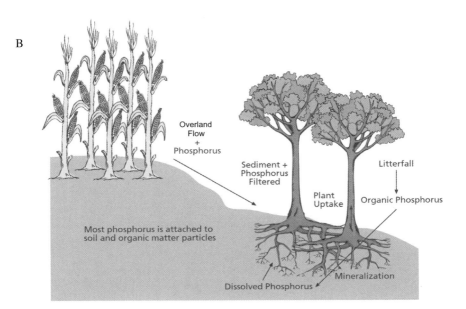

FIGURE 2-14 Fate and transport processes for (A) nitrogen and (B) phosphorus in riparian areas.

signed to maximize the infiltration of precipitation from a storm (Midwest Plan Service, 1985). (This approach necessitates large land requirements, especially if other removal processes are ignored.)

The dense herbaceous vegetation or litter layers of riparian areas offer high resistance to overland flow and decrease its velocity immediately upslope and within riparian areas. This reduction in velocity can promote *deposition* of sediments formerly suspended in hillslope runoff. Sediment-bound chemicals may be removed from overland flow during the deposition process. However, it should be noted that some trapped chemicals, such as organic nutrients, may be released into overland flow in dissolved form. Several mathematical models exist to describe deposition of sediment and sediment-bound chemicals in riparian areas (Hayes and Hairston, 1983; Lee et al., 1989; Inamdar et al., 1999).

Filtration of solid particles by vegetation and litter during overland flow and *adsorption* of dissolved chemicals and microorganisms to soil and plant surfaces are not well understood. Filtration is more significant in trapping larger soil particles, aggregates, and particulate organic matter, while adsorption to clay and organic matter in soils is more effective in trapping dissolved compounds with positive charges, such as orthophosphorus, heavy metals, and some pesticides. Adsorption of chemicals to the soil surface during overland flow is probably not very significant because of the short contact time and because adsorption sites are likely filled with previously adsorbed molecules (Dillaha et al., 1989).

Because the soils of riparian areas are generally enriched with root biomass and organic matter and have diverse soil microbiology, they support a myriad of biological processes that can transform chemicals dissolved in the subsurface. For example, the enhanced biological activity of riparian areas includes the microbial *degradation* of organic pesticides (USDA, 2000) and petroleum products (Brock and Madigan, 1991, p. 654). Both plants and soil microorganisms can *assimilate* large amounts of dissolved subsurface chemicals, particularly nutrients. Plant uptake can lead to either short- or long-term nutrient removal, depending on whether nutrients are stored in woody biomass that is retained at the end of the growing season or lost as leaves and twigs that return to the soil surface (Lowrance et al., 1995; Correll, 1997). The recycling of nutrients through plant uptake and release via decomposition contributes to keeping nutrients in the riparian area rather than releasing them to an adjacent waterbody. To maintain active nutrient assimilation, plant biomass must be removed, as is sometimes done in managed riparian areas where fast-growing trees are harvested for lumber and grasses and herbs are harvested for forage and biomass (Schultz et al., 1995). In addition, whether assimilation by vegetation occurs from the unsaturated zone is critical in determining whether there is an effect on flowpath chemistry (the saturated zone). Nonetheless, assimilation by plants is not a widely exploited mechanism of nutrient removal because the process has not been well described analytically.

Denitrification is a transformation process in which soil microorganisms take up dissolved nitrate from subsurface water and convert it to nitrogen gas. It is believed to be highly effective in removing nitrate from subsurface flow where conditions are favorable, that is, where nitrate-laden groundwater flows through areas that are both enriched with organic matter and anaerobic. Denitrification may also occur under aerobic or unsaturated soil conditions in localized soil micro-sites with high concentrations of particulate organic carbon (Parkin, 1987). Given their ample organic matter and diverse microbiology, it is not surprising that riparian areas support denitrification (Groffman et al., 1992; Addy et al., 1999), with riparian forests reported to remove 30–40 kg N ha^{-1} yr^{-1} under suitable conditions (Lowrance et al., 1995). Denitrification also occurs at seeps, where groundwater comes to the surface through soil horizons that are enriched with organic matter, as well as in wetlands, shallow groundwater, and other areas where substantial amounts of organic matter exist under saturated conditions. Schade et al. (2001) demonstrated the importance of denitrification over plant assimilation as a mechanism for removing nitrate from shallow groundwater. In this case, organic carbon produced by riparian shrubs acted as an energy source to drive denitrification. Despite the popularity of managing riparian areas to enhance denitrification, there is some doubt as to their importance in removing nitrate in certain agricultural settings. For example, Bohlke and Denver (1995) showed that on the Delmarva Peninsula, contact of groundwater flow paths with geochemically reducing sediments at the base of the shallow aquifer, prior to discharge through riparian areas, was more important in accounting for denitrification.

Sites of Fate and Transport Processes. Because the transformation processes described above require contact between chemical-laden water and either riparian vegetation or microbes in soils and sediment, their extent is obviously limited in instances where groundwater passes below the biologically active riparian area. Such bypassing can occur when riparian soils have low hydraulic conductivity (compared with a sand or loam) because of a relatively large proportion of fine sediment such as clay, silt, or humified organic matter. The extent of bypassing, and thus the ability of riparian areas to support chemical transformation such as denitrification, depends on many factors such as antecedent moisture, soil texture, underlying aquifer or bedrock geology, and human-induced landscape and channel conditions (Gilliam et al., 1997). Transformation mechanisms that occur in the unsaturated zone, such as some plant assimilation of nutrients, will have little effect on groundwater chemistry.

Although the preceding processes have been described as occurring in the subsurface water moving through riparian soils, they also occur in slackwater habitats (i.e., shallow and slowly moving sections of surface water channels) and hyporheic zones. For example, nitrate is removed from flowing water via denitrification within hyporheic zones (Hinkle et al., 2001). A significant proportion of nutrients (both dissolved and particulate) and inorganic and organic components are associated with sediments of the hyporheic zone as opposed to the active river channel (Pinay et al., 1992, Vervier et al., 1992, 1993). Periphyton, benthic algae, leaf packs, microbes, and some of their consumers are also present in varying degrees within the sediment.

Contact Time. Longer contact times are required for the transformation of some nutrients and other dissolved substances in overland and shallow subsurface flow. Thus, the extent to which these mechanisms occur is dependent on the amount of time that runoff and associated chemicals are retained in the riparian area, which is in turn largely a function of hydrology. Overall pollutant removal occurs to the greatest extent when overland flow and shallow subsurface flow are distributed uniformly across the riparian area. When overland and shallow subsurface flows concentrate and flow through only a portion of the riparian area (which is dependent partly on local topography), the areas with concentrated flow have shorter detention times, and their transformation mechanisms may be overwhelmed. Activities such as agriculture, silviculture, mining, and urbanization tend to concentrate flows (via gullies, channels, and subsurface tiles—see Chapter 3), such that only a small fraction of the riparian area's chemical and particulate trapping potential is realized (Dillaha et al., 1989).

Role of Stream Order. The importance of biogeochemical transformations, in terms of preventing pollutants from reaching adjacent waters, diminishes as one goes from ephemeral and first- and second-order streams to larger, higher-order streams. A greater portion of the flow passes through riparian areas along low-order streams before reaching the channel network, making their riparian areas more instrumental in removing pollutants from runoff. In contrast, most of the flow in high-order streams comes from low-order stream channels, and only a small portion of the flow in high-order streams actually crosses the riparian areas associated with the high-order stream segment. This suggests that if water-quality protection is a primary objective, priority should be given to restoration of functional riparian areas along ephemeral and first- and second-order streams over larger, higher-order streams (similar to the conclusion reached by Brinson (1993) for wetlands used for water-quality protection). As shown in Figure 2-15, first- and second-order stream channels comprise the vast majority of all stream kilometers in a given watershed (Leopold et al., 1964). It should be noted that the role of stream order is less clear for transformation via hyporheic exchange and the passage of water into and out of the riparian area during longitudinal flow.

Finally, it must be remembered that chemical transformation/removal is just one function of riparian areas (although frequently the primary target of management actions in agricultural areas—see Chapter 5). Even when specific riparian areas are only marginally effective for pollutant removal, they are still essential for wildlife habitat, flood control, and many other environmental services as described below.

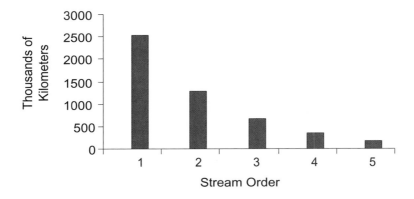

FIGURE 2-15 Relationship between stream order and stream length. Reprinted, with permission, from Leopold et al. (1964). © 1964 by W.H. Freeman and Company.

REGIONAL CLIMATE AND RESULTING RIPARIAN VEGETATION

The streamflow and associated fluvial and sediment processes that shape riparian areas are the products of regional patterns of topography and geomorphology, climate, and runoff. All these factors combine to create the observed distribution of riparian vegetation across the United States.

Climate

Climate has a strong influence on the structure and functioning of riparian areas, mainly through temperature and precipitation. These in turn strongly influence two other factors—evapotranspiration (ET) and runoff. ET refers to the surface water that moves from the liquid phase to water vapor through transpiration by plants and evaporation. (Potential evaporation (PET) is the evapotranspiration that would occur if water were not limiting.) Runoff is water that escapes both evapotranspiration and transport to deeper aquifers, finding its way to streams via overland and shallow subsurface pathways. Regional variation in these parameters contributes to the diversity of riparian vegetation observed across the United States.

Incoming solar radiation from sunlight influences air temperature, precipitation, and the subsequent apportioning of precipitation into evapotranspiration, subsurface recharge, and watershed runoff. Within the contiguous 48 states, solar radiation exhibits significant seasonal variability. Short days, low sun angles, and significant cloud cover result in low solar radiation inputs during winter months. In contrast, long days, high sun angles, and less cloud effects during the summertime combine to more than double the solar radiation of the winter months for many areas. Although the seasonal pattern of incoming solar radiation to the atmosphere varies little from year to year, clouds, which are highly variable, reduce the actual amount and timing at any particular point on the earth. Finally, aspect and slope further influence the amount of energy available at a particular location: during clear sky conditions, south-facing slopes receive more incoming solar radiation per unit area than do north-facing slopes (Reifsnyder and Lull, 1965).

Annual precipitation is highest in the eastern United States, particularly along the Gulf of Mexico coast, with up to 155 cm in Tallahassee, Florida (Figure 2-16). In general, precipitation is 90 cm or more to the east of a line extending from the mouth of the Mississippi River to the western shores of the Great Lakes and is distributed evenly throughout the year. (South Florida, with a pronounced winter dry season and summer wet season, is an exception.) The major moisture source is the Gulf of Mexico, even as far north as New England, but east of the Appalachian Mountains, the Atlantic Ocean is also significant. The lowest annual rainfall in the eastern United States usually exceeds 50 percent of the long-term mean; thus, severe droughts are uncommon. Compared to more arid regions, the eastern United States is characterized by gradual gradients in hydrophytic vegetation away from streams.

West of the Mississippi, the Gulf of Mexico is a less significant source of moisture, and north–south mountain ranges reduce the supply of Pacific moisture reaching the region, especially in the valleys. The western half of the Great Plains, the Great Basin, and the Southwest are mostly semiarid or arid. Non-coastal areas of the Southwest have with a strong summer rainfall maximum. The summer precipitation often comes in heavy thundershowers, and runoff generally travels as overland flow to nearby streams with little infiltration into the soil. Thus, little precipitation becomes available to recharge groundwater. The Sonoran Desert has a significant but smaller winter rainfall maximum arising from storms of Pacific origin. Because of the scarcity of rainfall, desert scrub ecosystems are distributed as far north as the Canadian border (49° N) and beyond, even at elevations of 2,000 m. In these climates, the gradient of vegetation with distance from streams is steep, and the well-developed riparian vegetation is usually striking in contrast to the surrounding desert landscape.

Finally, the Pacific Coast has a Mediterranean or modified Mediterranean climate, with a strong precipitation maximum during winter when cool temperatures limit plant growth. This seasonal peculiarity results in a strong moisture deficit for plants during the summer growing season. Along the coast, annual rainfall ranges from about 25 cm in the south to well over 100 cm in the north, and it can exceed twice that amount in coastal mountains and along the western slopes of the Sierra Nevada–Cascade range. The Pacific Coast and the Southwest have much greater variation in annual rainfall than does the eastern United States.

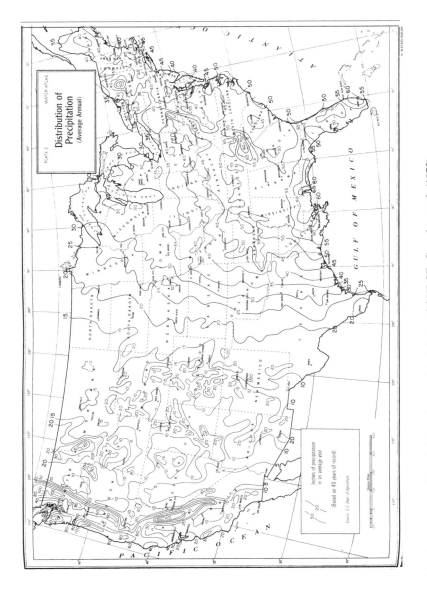

FIGURE 2-16 Average annual precipitation in the United States. SOURCE: Geraghty et al. (1973).

Except in the extreme northern part of the Pacific Coast, the lowest annual rainfall is often considerably less than 50 percent of the long-term average.

Across the United States, mean annual temperature generally decreases from south to north and with increasing elevation. Temperature variability, especially the range between daily and seasonal highs and lows, increases with distance from the ocean, especially the Pacific. Both temperature and precipitation vary with latitude, elevation, and proximity to large waterbodies or mountains.

These regional patterns of solar radiation, precipitation, and temperature, combined with elevation and other local topographic features that affect runoff, result in a highly complex pattern of "life zones" (as illustrated in Plate 2-2). Because of higher variability in precipitation and temperature in the West, well over 75 percent of the possible life zones are found west of the Mississippi River.

Runoff

Runoff describes the flow of water from the terrestrial landscape to surface-water bodies (e.g., stream channels, ponds, and lakes). It comprises all of the water that moves quickly to channels without being evaporated or stored for significant periods in soils and groundwater. Long-term runoff patterns from watersheds, which are reflected in streamflow, play an important role in shaping riparian systems. Streamflow amounts from a given watershed often exhibit general patterns from year to year as a result of topography, soil type, geology, vegetative cover, and the watershed's climate. Figure 2-17 shows the mean annual runoff in the United States.

The variation in runoff between watersheds can be explained by the intensity of precipitation, season of maximum precipitation relative to evapotranspiration, drainage area, land slope, soil and geologic characteristics, and vegetation type (Gregory and Walling, 1973). These factors determine the extent of infiltration versus overland flow, the proportion of soil water that is evaporated, the flow paths of precipitation across or through soils, and the proportion of soil water that is recharged to groundwater. Together, they affect the timing and rate of water delivery to channels, as well as the total runoff.

To evaluate the relative importance of precipitation and evapotranspiration to runoff amounts, these two variables are sometimes combined into a single factor called excess precipitation, defined as the difference between precipitation and evapotranspiration. Another common way to portray this concept of excess precipitation is the difference between precipitation and potential evaporation, or P–PET (Plate 2-3). In general, the percentage of precipitation that occurs as runoff is highest where excess precipitation is greatest, e.g., in the Northeast, Northwest, and Upper Midwest, and in high mountain areas. In those areas, the proportion of incoming precipitation that becomes runoff is typically greater than 50 percent. Excess precipitation is more moderate in the Southeast and Midwest. In the Great Plains, Great Basin, and Southwest, excess precipitation is at or near

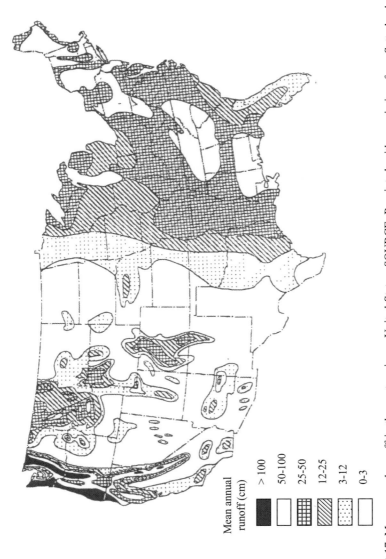

FIGURE 2-17 Mean annual runoff in the conterminous United States. SOURCE: Reprinted, with permission, from Satterlund and Adams (1992). © 1992 by John Wiley & Sons, Inc. NOTE: The white sections in the far Pacific Northwest and East fall into the mean annual runoff category of 50 to 100 cm, while those white sections located in the Great Basin, the arid Southwest, and the Plains states fall into the category of 0 to 3 cm.

zero and P–PET is negative, and consequently runoff percentages are typically well below 20 percent.

Seasonal variability in runoff reflects the timing and type of precipitation, the evapotranspiration rates throughout the year, and the intensity and duration of storms. Winter and early-spring storms in the East are longer in duration than in the West and occur when there is greater antecedent soil moisture, causing highest runoff in late winter or early spring despite fairly even distribution of precipitation throughout the year. Accumulation of snowfall and its subsequent melting are important factors influencing runoff patterns in cooler regions and where streams in the West have their headwaters in mountain ranges. Summer thunderstorms are dominant contributors to runoff in western mountains, the arid Southwest, and the Great Plains because winter precipitation is often low and because sparsely vegetated surfaces contribute a greater percentage of precipitation to overland flow. Together, these climatic factors determine patterns of peak runoff that become progressively later in the year from east to west across the United States. High-flow periods also tend to be shorter in duration the farther west one goes, as the dependence on summer thunderstorms increases. These general patterns of runoff variability do not include the Pacific Coast, where the timing of runoff is more similar to that of the Northeast.

The interannual variability of runoff, which in part determines the disturbance regimes characteristic of riparian areas, is highest in the Southwest (>100 percent), less in the Great Plains, Great Basin, and Midwest (>50 percent), and lowest in the Northwest, Rocky Mountains, and eastern United States (<30 percent) (Figure 2-18). Although in any given year and watershed the pattern of runoff can be relatively unique, over periods of multiple years, runoff patterns for specific watersheds tend to converge where lithology, soils, elevations, vegetation, and climatic inputs are similar.

Topographic and Geomorphic Patterns

In addition to climate and runoff features, there are important topographic and geomorphic patterns that lead to the observed variability in riparian areas across the United States. As shown in Figure 2-19, there are distinct types of riparian areas depending on the shape and steepness of the terrain, the composition of the underlying geologic materials, and their relative position within the stream network. For example, headwater streams tend to be associated with relatively steep watersheds and often have streams or rivers that are laterally constrained by hillslopes or geologic formations. Constrained streams have limited capability to adjust to changes in flow and sediment delivery rates. However, in a down-valley direction, the deposition of alluvial sediments along valley bottoms and lowlands decreases hillslope and geologic constraints on a stream's lateral movement. Hence, unconstrained streams tend to have a greater sinuosity and are

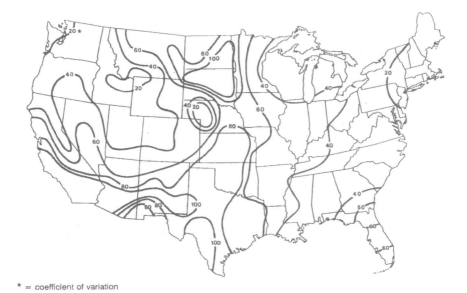

* = coefficient of variation

FIGURE 2-18 Interannual variability of runoff, as measured by coefficients of variation for runoff from the conterminous United States. SOURCE: Reprinted, with permission, from Patrick (1995). © 1995 by John Wiley & Sons, Inc.

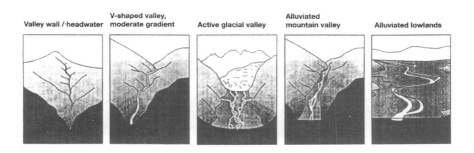

FIGURE 2-19 Generalized pattern of stream systems occurring across a range of terrain types. SOURCE: Reprinted, with permission, from Naiman et al. (1992). © 1992 by John Wiley & Sons, Inc.

more capable of adjusting to varying patterns of flow and sediment transport from upstream sources.

Position in the stream network plays a major role in riparian area structure because riparian areas are highly responsive both to the hydroperiod of the adjacent waterbody (e.g., depth, frequency, duration, etc.) and the source of the water (marine versus freshwater)—characteristics which differ among headwater streams, larger-order rivers, lakes, and estuaries. In many cases, these forces can be more important than climate in determining riparian area structure. For example, salt-influenced tidal marshes are more structurally and functionally distinct from floodplain forests within the same climate zone than are two salt-influenced tidal marshes, one in an arid climate and the other in a humid climate.

There have been various attempts at characterizing regions of distinct topography, hydrology, and geomorphology in order to provide an improved perspective regarding the regional structure of riparian areas. For example, Winter et al. (1998) and Winter (2001) define a "landscape unit" (called a basic building block of all landscapes) as simply an area of uplands adjacent to an area of lowlands, with the two areas being separated by an area of steeper terrain. The landscape unit's hydrology is determined by hillslope gradients, soil permeability, dimensions and permeability of the geologic framework, and atmospheric-water exchange, which is controlled by climate. Conceivably, all of the more complex hydrologic landscapes that are evident in the United States are variations or multiples of these fundamental landscape units. Some examples include (1) mountainous terrain, with narrow lowlands and uplands separated by steep valley sides, (2) a basin and range landscape, with very wide lowlands separated from much narrower uplands by steep valley sides, and (3) plateaus and high plains consisting of narrow lowlands separated from very broad uplands by valley sides of various slopes (Winter et al., 1998).

The purpose of the hydrologic landscape unit classification is to characterize pathways and rates of water movement through landscapes. For example, if a landscape has low land slope and low-permeability soils, overland flow will be slow and recharge to groundwater will be limited. In contrast, if the soils are permeable, overland flow may be limited but subsurface flow and groundwater recharge will be high. The key variables used are the distribution of landscape relief (maximum minus minimum elevation), average slope, slope distribution (percent flatland and percent upland or lowland), geologic texture and permeability, and available atmospheric water exchange (annual average precipitation-potential evapotranspiration). Wolock (2001) used statistical analyses of the existing nationwide datasets (averaged over approximately 200-km^2 watersheds) to classify landscapes across the United States, with the results clustering into 20 hydrologic landscape units (shown in Plate 2-4). Unlike with the life zone map (Plate 2-2), regional differences in hydrologic landscape units are significant across the entire country primarily because of the high variability in surface and subsurface properties.

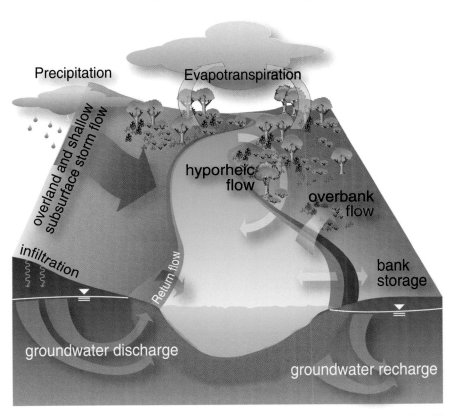

PLATE 2-1 Major pathways of water movement through riparian areas emphasizing (1) groundwater flow, (2) overland flow and shallow subsurface flow from adjacent uplands, and (3) instream water sources such as overbank flow, bank storage and hyporheic exchange.

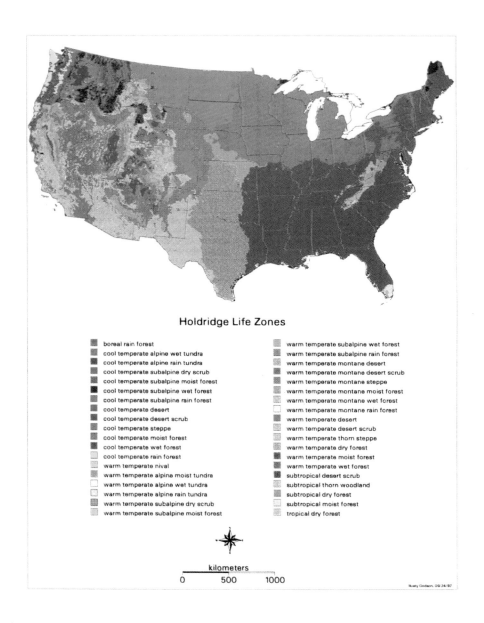

PLATE 2-2 Life zone map for the United States based on enhanced VEMAP climate data. SOURCE: Reprinted, with permission, from Lugo et al. (1999). © 1999 by Blackwell Science Ltd.

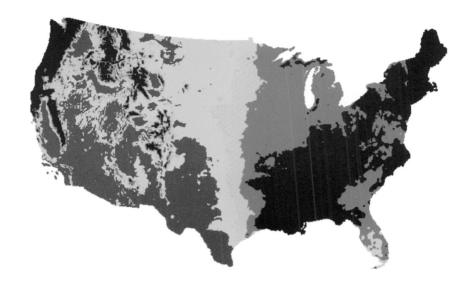

Mean annual precipitation minus potential evapotranspiration (mm/year)

- -300 and lower
- -299 to -100
- -99 to 100
- 101 to 300
- 301 and higher

PLATE 2-3 Average annual precipitation minus potential evapotranspiration requirements (P–PET). SOURCE: Reprinted, with permission, from Daley et al. (1994). © 1994 by American Meteorological Society.

Explanation

1 – Variably wet plains having highly permeable surface and highly permeable subsurface
2 – Wet plains having highly permeable surface and moderately permeable subsurface
3 – Variably wet plains having poorly permeable surface and highly permeable subsurface
4 – Wet plains having moderately permeable surface and moderately permeable subsurface
5 – Arid plains having moderately permeable surface and moderately to highly permeable subsurface
6 – Wet plains having poorly permeable surface and poorly permeable subsurface
7 – Wet plains having moderately permeable surface and poorly permeable subsurface
8 – Semiarid plains having poorly permeable surface and poorly permeable subsurface
9 – Wet plateaus having poorly permeable surface and highly permeable subsurface
10 – Semiarid plateaus having moderately to poorly permeable surface and highly permeable subsurface
11 – Wet plateaus having poorly permeable surface and poorly permeable subsurface
12 – Semiarid plateaus having highly to moderately permeable surface and poorly permeable subsurface
13 – Semiarid plateaus having poorly permeable surface and poorly permeable subsurface
14 – Arid playas having highly to moderately permeable surface and moderately permeable subsurface
15 – Arid mountains having poorly to moderately permeable surface and moderately permeable subsurface
16 – Wet mountains having moderately permeable surface and poorly permeable subsurface
17 – Semiarid mountains having moderately permeable surface and poorly permeable subsurface
18 – Variably wet mountains having moderately permeable surface and poorly permeable subsurface
19 – Very wet mountains having moderately permeable surface and poorly to moderately permeable subsurface
20 – Wet, extreme-relief mountains having moderately permeable surface and poorly permeable subsurface

PLATE 2-4 Hydrologic landscape units of the United States. SOURCE: Wolock (2001).

Disturbance Regimes

A variety of natural, physical disturbances that span scales ranging from tectonic activity to localized erosion help shape riparian areas. Geologic uplift that increases or decreases channel slope can cause, respectively, greater channel incision or greater overbank flow frequencies (Burnett and Schumm, 1983). Even in tectonically stable areas, however, the primary influences on the structure of riparian plant communities are fluvial processes—in particular, floods and the associated transport of sediment within streams. Floods and overbank flows occur when stream discharge exceeds channel capacity. Where snowmelt is an important contributor, rain-on-snow events produce exceptional floods. At lower elevations, high-intensity thunderstorms are of great importance in generating floods, particularly in the Southwest and Great Basin. Floods along the Gulf of Mexico and the southern Atlantic Coast are often associated with landfall of hurricanes and tropical depressions.

As shown in Figure 2-20, fluvial disturbances play a significant role in determining the composition of riparian vegetation by controlling the germination and successful establishment of seedlings as well as their long-term survival. The recruitment of woody riparian species in particular is dependent on interannual variability in flooding, channel migration, and sediment deposition. For example, germination of cottonwoods and some willow species generally occurs on locally scoured beds following channel adjustments such as channel narrowing caused by reduced flows (Figure 2-20B) or a flood pulse (Figure 2-20D). Whether the seedlings are established and reproduce depends on future disturbance events; plants may perish under drought conditions or they could be scoured away during subsequent floods or winter ice flows. Only plant species capable of tolerating these disturbances are likely to survive over the long term. For example, the vertical accretion of flood-deposited sediment may result in burial of the root crowns of trees (Scott et al., 1996). This is a selective process that allows survival of tree populations that are capable of producing new root systems when their stems become buried. For actively meandering rivers (Figure 2-20C), large-scale sloughing occurs when flows locally undercut forest vegetation positioned on the outside of meander curves. At the same time, point bar formation and vertical accretion on the inside of meander curves provide substrates for seedling, and ultimately forest establishment (Friedman and Auble, 2000). Where lateral channel migration is active, the rate of movement can be calculated from increases in tree age along transects perpendicular to the inside of the meander curve (Everitt, 1968).

The microtopographic variation created by diverse fluvial processes supports a species richness that would not otherwise occur (Gregory et al., 1991). In more humid areas where precipitation is higher and more evenly distributed throughout the year, flood intensity is generally lower and flood duration longer than in arid climates. In such situations, vegetation tends to be more effective in

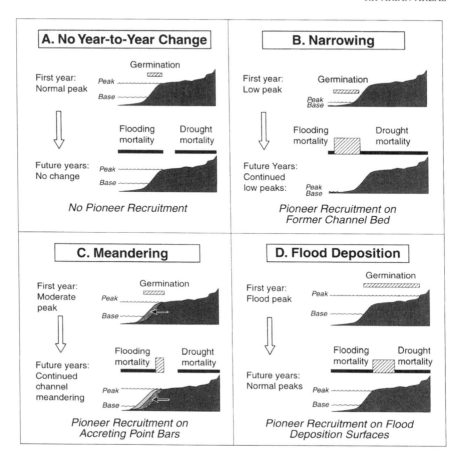

FIGURE 2-20 Hydrogeomorphic control of recruitment of woody pioneer species. Seed germination, early seedling mortality, and tree recruitment are shown in relation to annual high and low flow lines along four bottomland cross sections (A–D). In each of the four situations, the cross-hatched area in the upper part indicates the zone of seedling establishment, and the cross-hatched area in the lower part indicates the zone of long-term survival. (A) In the absence of interannual flow variability and channel movement, there is little or no tree recruitment. (B) On a narrowing channel, there is recruitment on the former channel bed. (C) Recruitment on point bars is typical of a meandering river. (D) Tree recruitment at high elevations is associated with infrequent floods and no channel movement. SOURCE: Reprinted, with permission, from Friedman and Auble (2000). © 2000 by Cambridge University Press.

stabilizing channel migrations, thus leading to a diminished zone of disturbance (Figure 2-21A). In contrast, rivers in more arid regions experience flood discharges that can be orders of magnitude greater than their base flows (which may be zero in some cases), resulting in significantly greater flow variation and physical disturbance. In arid regions disturbance-tolerant trees may be found along the entire cross section of riparian areas (Figure 2-21B).

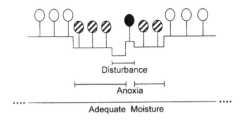

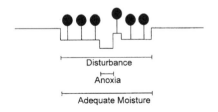

FIGURE 2-21 Influences of environmental stresses on the distribution of riparian trees. (A) Cross section of typical bottomlands in the low-gradient streams of the Coastal Plain of the southeastern United States. (B) Cross section typical of the bottomlands in the western Great Plains. Bars under cross sections show the extent of physical disturbance, anoxia, and adequate moisture for survival of trees. Although the three tree strategies are presented as distinct, many species combine strategies. The term "shade tolerant" refers to trees that are able to reproduce in the shade of other trees in the absence of physical disturbance. SOURCE: Reprinted, with permission, from Friedman and Auble (2000). © 2000 by Cambridge University Press.

Physical disturbance from trees, ice floes, boulders, and other solid objects in floodwaters can be effective in limiting vegetation to those species of very short stature, especially immediately adjacent to channels. Ice floes have a profound effect in boreal and arctic Alaska where ice floating downstream removes vegetation (Bliss and Cantlon, 1957; MacDonald and Lewis, 1973). During hurricane-generated flows that pass directly over the Luquillo Experimental Forest in Puerto Rico, slope failures and torrential stream discharge transport whole trees and meter-diameter boulders to riparian channels, creating disturbance that makes riparian forests less species-rich than surrounding upland forests (Ahmad et al., 1993; Scatena and Lugo, 1995). Nonetheless, it is clear that floodwater by itself can effectively eliminate vegetation, as evidenced by areas of sparsely colonized floodplains in the arid Southwest (Zimmerman, 1969). An extreme example of the long-term influence of fluvial processes is the elimination of forest areas by channel migration and the subsequent isolation of these forests in the tectonically active regions of upper Amazonia (Kalliola et al., 1991). These processes have been proposed as a speciation mechanism for forest plants on evolutionary time scales (Råsånen et al., 1987).

Fire is a physical disturbance predominantly in uplands, but it can become significant in arid riparian areas when drought conditions develop. In fact, fire is second only to flood events as a disturbance agent affecting riparian vegetation in northern Montana (Lee, 1983). In riparian areas along the southern California coast, where fire was historically only of minor importance, the introduction of a highly fire-tolerant grass (*Arundo donax*) in floodplains has led to the eventual elimination of the more fire-intolerant valley oaks that originally dominated (Boose and Holt, 1999). Such replacement tends to reduce the complexity of riparian forests with predictable consequences for nesting birds, shade-adapted plants, and other tree-dependent organisms.

Change in relative sea level is yet another, albeit more chronic, natural disturbance that causes changes in coastal riparian areas. Where sea level is rising in relation to the land surface, both inundation and salinity intrusion cause coastal riparian areas to change from forest to salt marsh vegetation (Brinson et al., 1995; Williams et al., 1999). As sea level rises, riparian areas migrate landward and replace upland forest while the estuarine shoreline erodes. Depending on the relative rates of forest replacement and shoreline erosion, the width of riparian areas can increase, decrease, or remain at steady-state. On islands or on coastlines where impediments to landward migration interfere, eroding shorelines caused by rising sea level will eventually eliminate riparian areas (Figure 2-22). The Mississippi alluvial valley represents a large-scale and long-time span case of downcutting and alluvial filling forced by multiple fluctuations in sea level during the Pleistocene (Fisk and McFarlan, 1955). Although the Mississippi example is largely irrelevant to most current ecological and socioeconomic concerns, it emphasizes how great a role sea level changes and tectonic activities played in eventually establishing today's distribution of vegetation in estuarine

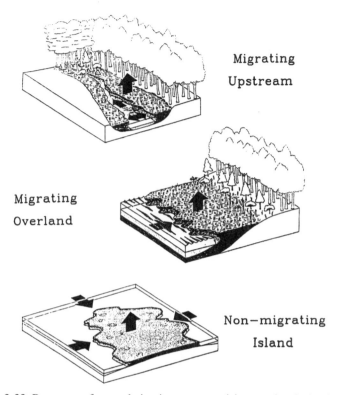

FIGURE 2-22 Response of coastal riparian areas to rising sea level. As the shoreline moves landward, riparian areas may migrate upstream or overland, or be eliminated as shorelines of islands erode. Vertical arrows represent the accumulation of sediment in pace with rising sea level, which must occur or riparian areas will be eliminated; horizontal arrows indicate shoreline erosion. SOURCE: Reprinted, with permission, from Brinson (1991). © 1991 by Dr. Douglas A. Wilcox, Editor-in-Chief, Wetlands.

riparian areas. Sea level changes and tectonic activity continue to shape the vegetation of high-latitude regions where continental margins are rising relative to sea level (Jordan, 2001). In these cases, coastal vegetation becomes increasingly isolated from tidal and salinity effects—the opposite of the more common landward "migration" of riparian areas.

Moisture Availability and Anoxia

Although physical disturbances are important in higher-order streams with strong flows and in headwater streams with steep gradients, soil moisture and depth to the water table also influence the composition of plant communities. Riparian areas in humid climates at the lowest elevations commonly experience such frequent flooding that adequate moisture for plant growth is generally found across the entire riparian area (Figure 2-21A). In such cases, soil anoxia brought on by persistent flooding or saturation is a greater factor in controlling riparian species composition than is physical disturbance. Anoxia is particularly prevalent in the fine sediment grains typical of low-gradient downstream reaches, which tend to retain moisture much more efficiently than coarse sediment. Some of the most flood-tolerant species in the United States (e.g., *Taxodium distichum*, *Nyssa aquatica*) can tolerate constant soil saturation, although they require a drawdown period for seed germination and seedling establishment. Where trees cannot establish because of continuous flooding, marshes and submerged plant communities become established. Beaver may facilitate this process by both felling trees and creating continuous inundation. Thus, while structural differences in vegetation between riparian areas and uplands may not be readily apparent in humid climates, species composition is very different.

Riparian areas in arid regions, on the other hand, are in close proximity to the only significant sources of water available for tree growth. Thus, anoxia is extremely limited and is of negligible importance in influencing species composition compared to overall moisture availability and disturbance patterns (Figure 2-21B). The narrow band of available water along streams in arid regions results in a forested riparian area of pioneer species, such as willow and cottonwood, that stands in stark contrast to the surrounding more sparsely vegetated uplands.

The different gradients of plant species richness brought about by moisture availability are illustrated in Figure 2-23. In this case, the number of species found within the riparian area of arid Sierra Nevada streams is markedly higher than in adjacent uplands, while species richness throughout the more humid Cascades riparian areas and uplands is relatively uniform (although species composition is not identical).

Regional Riparian Vegetation

The characteristics of riparian vegetation vary substantially across the United States and correspond with geographic variation in climate, hydrologic regime, and associated geomorphology. For reasons discussed earlier, there are major differences between the riparian vegetation found in humid versus arid regions. In addition, within most of the geographic regions described, riparian vegetation types at low elevations differ from those in high-elevation mountainous regions. The vegetation of riparian areas also varies according to the kinds of terrestrial and aquatic ecosystems they connect. Salinity and tidal inundation influence riparian vegetation adjacent to estuarine and marine waterbodies. Shorelines of lakes often undergo interannual fluctuations in water level, forcing riparian plants to constantly shift their positions. In riverine settings, sedimentation and erosion,

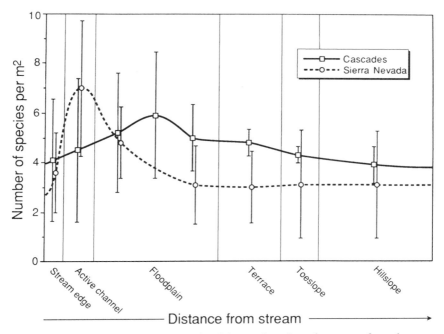

FIGURE 2-23 Gradients of plant species richness along lateral transects from the stream channel to upper hillslopes along three streams on the west slope of the Cascade Mountains of Oregon (more humid) and three streams in the Sierra Nevada of California (more arid). SOURCE: Reprinted, with permission, from Gregory et al. (1991). © 1991 by American Institute of Biological Sciences.

moisture availability, and the frequency, duration, and intensity of floods are selective factors that help shape the spatial patterns and species composition of riparian plant communities. The dynamic nature of many stream channels leads to cyclic changes in vegetation as the result of floodplain erosion and point bar deposition.

Riparian areas often support woody vegetation, and forests are a common community structure. There are notable exceptions, however, particularly along arid-zone, high-altitude, and high-latitude streams, where graminoids and shrubs are prevalent. Further, many forested riparian areas have herbaceous wetlands embedded within them and as a component of the forest floor. It is this complexity of vegetation types that contributes to the enormous richness of riparian areas in comparison to uplands, particularly in arid regions (Gregory et al., 1991; Brinson and Verhoeven, 1999). Thus, in the Great Plains and the arid and semiarid West, the contrast in species composition and physiognomic structure between riparian areas and uplands is large (Figure 2-23). In north-central Oklahoma, riparian forests have a much higher species diversity than adjacent upland forests, while the opposite is true for corresponding studies in the eastern deciduous forest region (Collins et al., 1981). Outside the United States, riparian areas have also been shown to harbor more diverse plant communities than uplands on a regional scale. For example, at least 260 species (or 13 percent) of the total Swedish flora occur along the Vindel River, which holds the country's record of 131 species per 200 m length (Nilsson, 1992). The diversity in riparian vegetation stems not so much from species richness within individual plant communities but rather from the existence of broad plant community types as well as age and structural diversity within plant communities. This large-scale heterogeneity is a result of the extreme conditions of disturbance, wetness and dryness, and other fluvial processes characteristic of riparian areas. For example, in the Saskatchewan River delta in Canada, none of the plant communities is particularly species rich (Dirschl and Coupland, 1972). Yet the delta vegetation was classified into 11 broad plant community types according to physiognomy, with six of these consisting of woody vegetation (white spruce-hardwoods forest, black spruce-tamarack forest, tall willow-alder shrub, medium willow shrub, low willow shrub, and bog birch shrub). The remaining five community types were dominated by herbaceous wetland vegetation. Even in arid regions where riparian vegetation represents the only forest structure in the landscape, tree species diversity is low, but age class and structural diversity are typically high.

Because of these sources of biodiversity, regional differences among riparian areas are best illustrated by comparing the plant community and species compositions rather than determining that one region is more species-rich than another. The following section focuses on the dominant woody component of riparian areas to illustrate regional differences, organized around the broad geographical areas depicted in Figure 2-24. Because of the lack of equivalent and collated information on other riparian plant types, only trees are covered comprehensively. It should be noted that most riparian areas exhibit lateral zonation of species, something that cannot be addressed at large regional scales and as such is not part of this summary.

Boreal and Arctic Alaska

Boreal (interior) Alaska extends north from the maritime climate of the southeastern part of the state to the Brooks Range, which separates the interior from the Arctic, or North Slope. Although precipitation in the interior and the North Slope is relatively low, a combination of a short growing season, low evapotranspiration, and low topographic gradients results in excess moisture and saturated soil. Black spruce dominates the uplands of interior Alaska (Post, 1996), while riparian vegetation fluctuates in response to the dynamics of ice scour, active channel meandering, and sediment deposition. Where ice floes have removed vegetation, willow and poplar form low-growing communities (though

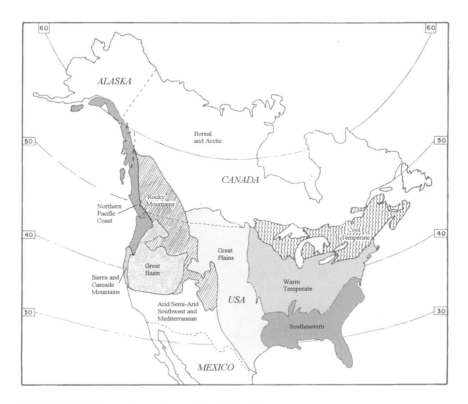

FIGURE 2-24 Vegetation regions of the United States.

willow is short-lived) (Walker et al., 1986). Alder also colonizes bare soil, but is eventually replaced by a continuous deciduous canopy of balsam poplar in the absence of further disturbances. Next, productive stands of white spruce develop on the rich alluvial soils, but eventually organic matter and a moss ground cover accumulate. This insulating layer subsequently modifies the soil microclimate so that permafrost develops and reduces rates of nutrient cycling, at which point, given enough time, a black spruce forest replaces white spruce, superficially resembling the upland forests (Van Cleve et al., 1991).

Riparian vegetation in arctic Alaska is generally shorter than that of the boreal zone. A range of community types are possible: a pioneer herbaceous community; a feltleaf willow stage (up to 6 m in height); a zone of deteriorating feltleaf willow with increasing prominence of greenleaf willows, mosses, and herbs; and an alder–willow stage, representing the oldest and most elevated condition (Bliss and Cantlon, 1957). Unvegetated sand and gravel bars are extensive because of disturbance by ice floes, channel meandering, and rapid sediment deposition. Species richness of vegetation along arctic river corridors increases in a downstream direction, principally due to increased habitat heterogeneity (Gould and Walker, 1999).

Pacific Northwest and Coastal Mountains

The Pacific Northwest and coastal mountains support a diverse woody plant flora with a high percentage of endemic species, particularly in the California Floristic Province. The species dominating riparian forests in California are related to air temperature, groundwater depth and aeration, and frequency and intensity of disturbance (Holstein, 1984). At lower elevations and in the south, Fremont cottonwood is often the dominant riparian tree. At higher elevations or in the north, dominance shifts to black cottonwood. Where disturbance is more severe, however, narrowleaf willow or Goodding's willow is common. White alder is abundant in areas of shallow, well-aerated groundwater in central California, particularly along montane streams. Further north, red alder forms dense riparian stands, but it can also grow in the uplands in the wetter climate. Where groundwater is deeper but still aerated, the dominant tree is often California sycamore. Where groundwater is deeper and not well aerated, California white oak—a deciduous oak endemic to the Central Valley and adjacent foothills—once grew in large stands.

Several other tree species grow in riparian areas of California and give the region a distinctive floristic character (Holstein, 1984). Boxelder grows primarily as a riparian species, while bigleaf maple, which is common in riparian areas, also grows in uplands. There are two species of riparian ash—velvet ash in the south and Oregon ash in the north. Another riparian species endemic to California is Hind's walnut. In several areas, California laurel and coastal redwoods are abundant riparian species.

The riparian vegetation of western Oregon and Washington is closely related to that of northern California and southeast Alaska; one of the most characteristic trees is black cottonwood. Black cottonwood forests are especially prominent along large rivers, where several species of willow form a thick forest understory (Franklin and Dyrness, 1988). Oregon ash is also abundant in the poorly drained soils of riparian areas in the interior valleys of Oregon; these areas also include populations of bigleaf maple, Oregon white oak, red alder, and localized populations of California laurel. Buttonbush, which grows in riparian areas of the humid eastern United States, is represented by relictual riparian populations along the Cosumnes River in the southern Sacramento Valley. Water birch is a common riparian tree in a few restricted areas of the Klamath Mountains and the southeastern Sierra Nevadas.

Great Basin

The Great Basin is a vast area of the western United States characterized by predominantly basin topography, relatively low precipitation, and low streamflow. Warm summers and cold winters result in an extensive cover of sagebrush and associated shrubs, sometimes referred to as cold desert. In the lower elevations of the Great Basin, riparian forests are the only forests, and their structure varies substantially with latitude (Stine et al., 1984; Minshall, 1989). In the northwestern regions, riparian forests included black cottonwood, willow, hawthorn, water birch, chokecherry, and gray alder. In the southern Great Basin, Fremont cottonwood grows in association with several willow species and mesquite. Eastern Oregon and central Washington also have little-known riparian communities of willow and hawthorn or chokecherry (Lytjen, 1998). The riparian areas of the southern Great Basin have been increasingly invaded by saltcedar.

Arid and Semiarid Southwest

In the Mojave and Sonoran deserts and in Southern California, precipitation limits upland vegetation to scattered shrubs and occasional grasslands in the lowlands and evergreen woodlands and forests in the mountains. The structure, species composition, diversity, and productivity of riparian areas contrast sharply with adjacent uplands. The riparian vegetation in much of Arizona is dominated by a few species of trees (Lowe, 1961, 1964). "Wet" riparian areas in Arizona and across most of the arid Southwest support primarily Fremont cottonwood, Goodding's willow, Arizona sycamore, velvet ash, and Arizona walnut. Netleaf hackberry is found to be a significant riparian tree species in the Catalina Mountains (Whittaker and Niering, 1965). Similar sets of species dominate the lower-elevation riparian areas of New Mexico and Southern California.

Stromberg et al. (1996) described a strong association between the distributions of these species along the San Pedro River of southern Arizona and depth to groundwater. One result of their analysis is a classification of species by the degree to which they are dependent on wetland environments. Goodding's willow was classified as an obligate wetland species, Arizona walnut and Fremont cottonwood as facultative wetland species, velvet ash as a facultative species, and netleaf hackberry, Texas mulberry, and mesquite as facultative upland species.[1] These species represent a sequence of drought tolerance, with Goodding's willow being the least tolerant.

In a recent classification of New Mexico wetlands, 99 different forested and shrub–scrub wetland communities are described (Muldavin et al., 2000). It is

---

[1] Obligate, facultative wetland, and facultative species are associated with wetlands in 99 percent, 67 percent to 99 percent, and 33 percent to 67 percent of their occurrences, respectively.

interesting that many of the same genera found in New Mexico are also found in the humid eastern United States, i.e., alder, willow, maple, dogwood, cottonwood, sycamore, and walnut. Species common to the two regions are boxelder and plains, or eastern, cottonwood.

Riparian areas of the arid and semiarid West are increasingly dominated by exotic tree species, especially saltcedar, Russian olive, Siberian elm, tree of heaven, and white mulberry (Crawford et al., 1993; Muldavin et al., 2000). The dry riparian areas along ephemeral washes in Arizona and Mexico are dominated by a very different group of woody plant species. In Arizona, these include mesquite, desert ironwood, blue paloverde, and desert willow (Johnson et al., 1989).

Rocky Mountains

The Rocky Mountains extend from Canada south to Mexico and are characterized by snowmelt-dominated stream flows. The riparian vegetation of the Rocky Mountains, as elsewhere, varies with elevation and latitude. The low valleys once had well-developed forests of narrowleaf cottonwood and strapleaf willow. The riparian association in the montane zone of the Colorado Rockies includes narrowleaf cottonwood, blue spruce, gray alder, water birch, and redosier dogwood (Baker, 1989). This association also occurs in eastern Idaho, western Wyoming, and southern Utah. In southern Colorado and northern New Mexico, the dominant riparian association consists of white fir, blue spruce, narrowleaf cottonwood, Rocky Mountain maple, redosier dogwood, and gray alder. Narrowleaf cottonwood continues as an important riparian tree into the subalpine region of the central and southern Rocky Mountains. In the northern Rockies of Montana, black cottonwood is the dominant riparian tree. Other riparian trees or shrubs in the Rocky Mountains of regional importance include serviceberry, chokecherry, hackberry, and hawthorn. In the northern Rocky Mountains willows replace trees as the dominant riparian woody species on sites with gravelly silt loams, less than 42 percent coarse fragments, and a groundwater table within 0.6 m of the surface (Law et al., 2000).

Great Plains

The Great Plains, extending from the midsection of the country to the Front Range of the Rocky Mountains, are characterized by a strong moisture gradient from east to west and a steep change in temperature from south to north. Although thunderstorms dominate runoff in most of the region, the amount and timing are highly variable in comparison with winter storms. The riparian, or gallery, forests of the Great Plains show a great structural contrast with the adjacent uplands, which in their pre-agricultural state were dominated by short, mixed, or tall grass prairies. The western Great Plains show reduced diversity in woody plants. For instance, on the Canadian River of central Oklahoma, the riparian forest was found to be composed of two native species—plains cottonwood and sandbar willow—and one introduced species—saltcedar (Ware and Penfound, 1949). The low diversity within the riparian forest was attributed to frequent flash flooding, shifting sands, high rates of evaporation, and intense heat. Tree diversity within Great Plains riparian areas appears to increase northward. Along Plum Creek, Colorado, the riparian vegetation is dominated by plains cottonwood and four willow species (Hupp and Osterkamp, 1996). These sites also support narrowleaf cottonwood and gray alder.

The eastern Great Plains show much greater riparian tree diversity, much like riparian communities farther east. For instance, the riparian forests of eastern Oklahoma support a rich tree flora, including black birch, sweet gum, water tupelo, American sycamore, bald cypress, hackberry, honey locust, elm, hickory, and boxelder (Bruner, 1931). In the northern Great Plains, the diversity of the riparian tree community is also greater than in the western plains. The riparian forests of the Missouri River in central North Dakota are composed predominantly of plains cottonwood, peachleaf willow, chokecherry, green ash, boxelder, American elm, netleaf hackberry, and bur oak (Johnson et al., 1976).

Cool Temperate East

Riparian forests from the western Great Lakes to New England are somewhat less diverse than those farther to the south (Heinselman, 1970; Wells and Thompson, 1974; Morris, 1977; Brinson, 1990). The trees include northern white cedar, black ash, tamarack, balsam fir, and white spruce. Speckled alder grows in forests where sufficient light penetrates that canopy. In riparian areas with poorly drained silty soils of high organic matter content, the dominant trees are slippery maple, eastern and black cottonwood, black cherry, and several species of willow. Well-drained silts generally support ash and American sycamore. On the sandy silt soils of frequently flooded point bars exposed to high flood energy, there is an association of slippery maple, sugar maple, black locust, and American elm. Bitternut hickory grows along with slippery elm and sugar maple on point bars of coarse sandy soils that are flooded with lower frequency. On riparian terraces exposed to the least flooding are forests of red maple, pines, black cherry, and oak.

An extensive listing of wetland tree species by the Minnesota Department of Natural Resources adds several tree species to the riparian flora characteristic of the cool temperate East (Henderson et al., 1998). The wettest areas of Minnesota support buttonbush and narrowleaf willow. In riparian areas that are flooded for shorter periods, tree communities include maple, alder, birch, dogwood, common winterberry, ash, tamarack, black cottonwood, eastern cottonwood, swamp white oak, several willow species, and northern white cedar.

Warm Temperate East

The warm temperate East includes the regions from southern New York and Delaware to Iowa and northern Missouri. In the westernmost part of this region, riparian forests of the Southeast (described below) extend up the Mississippi, Ohio, and Wabash Rivers where they are replaced gradually by those of the warm temperate East (Buell and Wistendahl, 1955; Wistendahl, 1958; Lindsey et al., 1961; Wolfe and Pittillo, 1977; Hughes and Cass, 1997). In the lower portions of floodplains, the historical forest was dominated by cottonwood, black willow, silver maple, ash, and American elm. The structure of these communities, however, was significantly altered with the introduction of Dutch elm disease. Less frequently flooded associations include hackberry, silver maple, sugar maple, redbud, American beech, and American elm. Other important riparian trees in this region include red maple, tulip tree, white ash, American hornbeam, black walnut, spice bush (a shrub), oak, and American sycamore. Along streams with steeper flow gradients and higher water velocities, the woody vegetation is dominated by hazel alder, silky dogwood, American witch hazel, possum haw, and black willow. On the narrow floodplains along streams in the western Appalachian Mountains, riparian forests are dominated by river birch or a more diverse association of bitternut hickory, red maple, tulip tree, black cherry, and black locust (Wolfe and Pittillo, 1977).

Southeast

The southeastern riparian region corresponds broadly to the distribution of bald cypress, though within the region, bald cypress is largely limited to the wettest and most deeply flooded sites (Mattoon, 1915). In these sites, bald cypress often occurs with water tupelo. Moving toward uplands from the wettest sites, bald cypress and water tupelo are replaced by a diverse forest that includes hickory, ash, willow, birch, cottonwood, sweet gum, hackberry, red maple, and oak (Brinson, 1990). Pines and oaks often dominate the forests in the least flooded sections of riparian areas.

Hupp and Osterkamp (1985) analyzed the distribution of woody species across the riparian landscape along Passage Creek, Virginia, dividing the riparian area of this high-gradient stream into depositional bars, channel shelf, floodplain bank, and floodplain. Species found mainly or largely restricted to channel shelves included hazel alder, silky dogwood, American sycamore, common winterberry, slippery elm, and southern arrowwood. Species restricted to the floodplains of the study area included black walnut, bitternut hickory, and American elm. A third group of species, found broadly across riparian areas, included boxelder, American hornbeam, green ash, eastern cottonwood, and American bladdernut. Only one tree species—black willow—occurred on depositional bars. In studying lower-gradient rivers in Virginia, Hupp and Osterkamp (1996) found that the composition of riparian woody plant communities was similar to those found along the higher-gradient Passage Creek. However, the riparian areas of lower-gradient rivers included additional species—sweetgum, water hickory, silver maple, American beech, tulip tree, bald cypress, and water tupelo. Flood frequency and intensity were hypothesized to be the most important factors influencing the distribution of woody plant species within riparian landscapes.

Summary

The regional variation in riparian tree communities across North America is summarized in Table 2-1, which corresponds with the broad geographic areas in Figure 2-24. The only tree species listed are native trees with a wetland indicator status of obligate species, facultative wetland species, or facultative species for a particular region. One of the patterns discernable from the table is the higher regional diversity—from 26 to 33 genera—in riparian trees from the Great Plains eastward. West of the Great Plains, the number of riparian tree genera within regions ranges from 9 to 22. These results, however, are skewed toward areas where species reach tree stature, and thus do not thoroughly take into account riparian areas in arid climates and high altitudes that are dominated by shrubs and grasses. The purpose of making the comparison is not to determine which geographic areas have the highest species richness, but to show the distribution patterns of dominant genera. Despite the wide regional variation in riparian tree genera represented, a core of genera—*alders*, *cottonwoods*, and *willows*—grows in riparian areas across the continent. A second group of very widely distributed genera includes *ash*, *birch*, *hackberry*, *hawthorn*, and *maple*. These primary and secondary core genera are also associated with riparian areas across Europe and Asia. The remaining 37 genera in Table 2-1 are of regional significance and often dominate riparian tree biomass within their characteristic regions.

Table 2-2 attempts to identify the most important tree genera within each region in terms of their relative biomass or frequency of occurrence within riparian forests. The result is a list of five or six genera for eight of the regions and ten genera for the southeastern United States. The southeastern flora are more difficult to compress because riparian vegetation ranges from bald cypress and tupelo in the wettest portions of floodplains to less flood-tolerant taxa in the upper reaches of the same drainage system. In this respect, the Southeast is more diverse than other regions included in the analysis.

Influences of Riparian Vegetation

Riparian vegetation has profound effects on the microclimate of streams and floodplain forests. We have previously discussed how vegetation affects water cycling through evapotranspiration—an inevitable consequence of the growth of

TABLE 2-1 Distribution of Native Riparian Tree Genera Across the United States[a]

Primary core genera	X
Secondary core genera	X
Regionally important genera	X

Riparian Tree Genera	Boreal and Arctic	Pacific Northwest and Coastal Mountains	Great Basin	Arid and Semiarid Southwest	Rocky Mountains	Great Plains	Cool Temperate East	Warm Temperate East	Southeast
Alder, *Alnus*	X	X	X	X	X	X	X	X	X
Ash, *Fraxinus*		X	X	X	X	X	X	X	X
Bald cypress, *Taxodium*						X		X	X
Beech, *Fagus*								X	X
Birch, *Betula*	X	X	X		X	X	X	X	X
Bladdernut, *Staphylea*						X	X	X	X
Buckeye, *Aesculus*						X		X	X
Buttonbush, *Cephalanthus*		X		X		X	X	X	
Calif laurel, *Umbellularia*		X							
Cedar/arborvitae, *Thuja*	X	X			X		X	X	
Cherry, *Prunus*		X		X	X	X	X	X	
Coastal redwood, *Sequoia*		X							
Cottonwood, *Populus*[b]	X	X	X	X	X	X	X	X	X
Desert ironwood, *Olneya*				X					
Desert palm, *Washingtonia*				X					
Desert willow, *Chilopsis*				X					
Dogwood, *Cornus*	X	X	X	X	X	X	X	X	X
Elderberry, *Sambucus*		X	X	X	X	X	X	X	
Elm, *Ulmus*						X	X	X	X
Fir, *Abies*		X			X		X		
Hackberry, *Celtis*		X	X	X	X	X	X	X	X
Hawthorn, *Crataegus*		X	X	X	X	X	X	X	X
Hemlock, *Tsuga*	X								

continues

Riparian Tree Genera	Boreal and Arctic	Pacific Northwest and Coastal Mountains	Great Basin	Arid and Semiarid Southwest	Rocky Mountains	Great Plains	Cool Temperate East	Warm Temperate East	Southeast
Hickory, *Carya*						X	X	X	X
Honey locust, *Gleditsia*			X		X	X	X	X	X
Hornbeam, *Carpinus*						X	X	X	X
Maple, *Acer*		X	X	X	X	X	X	X	X
Mesquite, *Prosopis*		X	X	X		X			
Mulberry, *Morus*			X	X		X		X	X
Oak, *Quercus*		X				X	X	X	
Paloverde, *Parkinsonia*				X					
Planertree, *Planera*						X			X
Redbud, *Cercis*				X				X	
Serviceberry, *Amelanchier*		X	X	X	X	X	X	X	X
Spicebush, *Lindera*						X	X	X	X
Spruce, *Picea*	X	X			X		X		
Sweetgum, *Liquidambar*						X		X	X
Sycamore, *Platanus*		X		X		X	X	X	X
Tamarack, *Larix*	X						X		
Tulip tree, *Liriodendron*									X
Tupelo, *Nyssa*						X		X	X
Viburnum						X	X	X	X
Walnut, *Juglans*		X		X			X		
Willow, *Salix*	X	X	X	X	X	X	X	X	X
Winterberry, *Ilex*						X	X	X	X
Witchhazel, *Hamamelis*						X		X	
Total Genera	9	22	13	21	16	31	26	33	29

[a] Data abstracted from Brinson (1990) and the USDA NRCS Plants Database (2001), which includes a FWS Wetland Indicator Status for plant species. Although the analysis was based on species, only genera are listed since a species-level analysis would be too detailed for the present report.

[b] All *Populus* are considered in this row, not just cottonwood.

TABLE 2-2 Dominant Tree Genera in Riparian Areas Across the United States[a]

Riparian Tree Genera	Boreal and Arctic	Pacific Northwest and Coastal Mountains	Great Basin	Arid and Semiarid Southwest	Rocky Mountains	Great Plains	Cool Temperate East	Warm Temperate East	Southeast
Alder, *Alnus*	X	X	X	X	X				
Ash, *Fraxinus*		X				X	X	X	X
Bald cypress, *Taxodium*									X
Birch, *Betula*	X		X		X				
Cottonwood, *Populus*[b]	X	X	X	X	X	X	X	X	X
Dogwood, *Cornus*	X				X				
Elm, *Ulmus*						X	X	X	
Hawthorn, *Crataegus*			X						
Hackberry, *Celtis*						X			
Hickory, *Carya*									X
Maple, *Acer*		X			X	X	X	X	X
Mesquite, *Prosopis*				X					
Oak, *Quercus*									X
Cedar/arborvitae, *Thuja*	X						X		
Sweetgum, *Liquidambar*									X
Sycamore, *Platanus*		X		X				X	
Tulip tree, *Liriodendron*									X
Tupelo, *Nyssa*									X
Walnut, *Juglans*				X					X
Willow, *Salix*	X	X	X	X	X	X	X	X	X

[a] Data abstracted from the review of Brinson (1990).
[b] All *Populus* are considered in this row, not just cottonwood.

riparian vegetation. As discussed below, there are numerous important ecological and geomorphic benefits of riparian vegetation, such as shading and temperature amelioration, provision of large wood, and stabilization of alluvial sediments that constitute the floodplain.

Thermal Regulation

Temperature in environmental systems can have important and profound effects upon a wide range of physical, chemical, and biological processes. For example, in aquatic systems the solubility of oxygen gas is inversely dependent upon water temperature; hence, higher water temperatures often lead to decreased levels of dissolved oxygen and increased stress for those organisms requiring adequate amounts of instream dissolved oxygen. The rates at which many chemical reactions proceed are also strongly influenced by temperature. As a result, basic processes such as transpiration, respiration, outgoing long-wave radiation, and others are strongly dependent upon environmental temperature regimes.

Vegetation has an important role in influencing the local thermal regime of streams and their adjacent riparian areas. It has been well established that riparian plant canopies—particularly those of forest vegetation—are effective interceptors of incoming solar radiation and can thus greatly reduce the amount of solar energy available to a stream or river (Brown, 1969). The shading effects of riparian plant canopies can prevent or retard the rate of stream warming during clear-sky conditions in summer months when streamflows are often low and high instream temperatures are of concern to aquatic biota (especially cold-water fish) (Gregory et al., 1991; McCullough, 1999; Naiman et al., 2000). For example, the only measured environmental variable that clearly distinguished between trout and non-trout streams in Ontario, Canada, was weekly maximum water temperature, which was inversely related to the percentage of upstream banks covered by forest (Barton et al., 1985). In Bear Creek, Oregon, the partial recovery of riparian vegetation following a change in grazing practices led to the trapping of sediments, increased storage of moisture in riparian soils, and conversion to perennial flow (see Box 5-1). Rainbow trout, a cold-water species that had been eliminated possibly by high temperatures and degraded habitat, returned to the stream and is present year round. On hot summer days in Glacier National Park, Montana, Hauer et al. (2000) observed temperatures as high as 25 °C in exposed channels of an alpine stream segment, whereas the stream did not exceed 10 °C further downstream where the stream was completely canopied by riparian forest. A large number of studies from various portions of the world have shown an increase in stream temperature as a result of decreased riparian vegetation (Figure 2-25).

Riparian vegetation can also influence local wind patterns, conductive and convective heat transfer, outgoing long-wave radiation, and other energy transfer processes in aquatic systems and the land occupied by riparian plants. The result

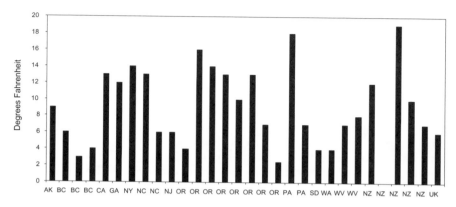

FIGURE 2-25 Changes in stream temperature in response to differences in riparian shade in empirical field studies throughout the world. Temperature change is expressed as the change in maximum temperature before and after vegetation removal or between reaches with and without vegetation. Locations are U.S. States, British Columbia (BC), New Zealand (NZ), and United Kingdom (UK).

is a local microclimate along streams and adjacent riparian systems that is more thermally moderate than adjacent upland environments. Although the effects of riparian plants upon stream temperatures has been widely studied (e.g., see review by Beschta et al., 1987b), their influences on other microclimate variables (atmospheric humidity, air temperatures, soil temperatures) have not.

In addition to the direct thermal moderation associated with riparian plant communities, there is an indirect, yet important, role that these plants can have on instream temperatures. The root strength associated with riparian plants often allows channels to remain relatively narrow (i.e., a small width-to-depth ratio). If such vegetation is removed and the root strength of the streambank is lost, channels over time may experience widening (i.e., a larger width-to-depth ratio) and become shallower. Given a specific amount of energy absorbed per unit surface area, a wide, shallow channel will experience a greater temperature increase than a narrow, deep channel. Furthermore, channel morphology can play an important role in the connectivity of instream flows with the hyporheic zone and groundwater. For example, alterations to riparian plant communities that contribute to channel incision may sufficiently modify hyporheic and groundwater connectivity such that thermal regimes and low-flow periods are significantly altered. Because streamside vegetation can influence both channel morphology and energy transfers in a variety of ways, alterations to riparian plant communities often have major influences on local microclimates and the thermal regimes of aquatic systems. These effects occur not only in local reaches, but may also contribute cumulatively to downstream warming of a stream or river (Bach, 1995; Boyd et al., 1998).

In some areas, the hyporheic zone is also important in temperature regulation in streams. Mosley et al. (1983) observed no response of stream temperature to removal of riparian vegetation in a New Zealand stream (see Figure 2-25). The authors concluded that subsurface exchange of water was the major factor determining water temperature in this alluvial system. Schloz (2001) found that reaches of the Wenatchee River in Washington exhibited little response to riparian shading and that hyporheic exchange had a much greater influence on stream temperatures. Such studies and findings are scarce, but they indicate the potential for subsurface exchange, either from hyporheic flow or groundwater, to influence stream temperatures and moderate the influence of riparian shade on stream temperature. Emerging studies of hyporheic processes will provide the empirical basis for modeling the potential influences of alluvial and groundwater exchange for stream temperature.

Because of the presence of water (surface and/or subsurface) and potentially high transpiration rates, riparian areas typically have microclimates (e.g., thermal and relative humidity regimes) that are different than adjacent terrestrial environments. Nonetheless, investigation of the microclimate characteristics of such landscapes prior to the 1990s was limited. In the Pacific Northwest, forested riparian areas during summertime conditions have been found to have lower air temperatures (and less variance in air temperature) and higher daytime humidity (and less variance in humidity) than nearby terrestrial or upland areas (Danehy and Kirpes, 2000). This issue is particularly important to the use of riparian buffer strips alongside forestry operations for maintaining many of the functions of stream and riparian systems. That is, the effectiveness of such buffers with regard to protecting riparian microclimates has not been demonstrated (Brosofske et al., 1997; Dong et al., 1998). In general, little is known about how changes in riparian microclimates that might occur because of adjacent timber harvests may or may not affect riparian plant and animal communities. Perhaps even less is known regarding the microclimates of riparian areas in range and agricultural settings and the extent to which they have been altered by historical management practices. Additional microclimate research is clearly needed for a wide range of conditions.

Large Wood

Forested riparian areas contribute wood to streams, lakes, and wetlands. Several processes can recruit trees from riparian forests into adjacent bodies of water, including individual tree mortality, blowdown, bank erosion, and, in steep mountainous terrain, landslides. Large-wood accumulation in streams and rivers has historically been an important feature of forested riparian systems for all stream sizes and in all geographical areas of North America (Sedell and Luchessa, 1982; Maser et al., 1988).

Large wood plays a critical role in maintaining and restoring physical habitat, biodiversity, and ecosystem processes in stream and river ecosystems at both local and landscape scales. Large wood creates roughness elements in flowing water, which shape pools and riffles, create variable velocities, and increase the residence time of water (Sedell and Beschta, 1991). Large wood helps streams and rivers slow the downstream routing of sediment and organic matter by providing increased hydraulic resistance to flow and encouraging the local deposition of these materials. Large wood also increases retention of dissolved nutrients and particulate material, and it supports microbial processes and nutrient transformation (Bisson et al., 1987; Maser et al., 1988). Wildlife benefit from the presence of large wood because it can serve as habitat, including refuge during floods and cover, and it provides an abundant but low-nutritional-quality food supply (i.e., it is a potential carbon source to some streams).

The amount of wood found in aquatic and riparian systems is a function of the rate of input, downstream transport, and rate of breakdown from decay and abrasion. As a result, standing stocks of wood differ from small streams to large rivers (Harmon et al., 1986; Maser and Sedell, 1994; Ralph et al., 1994) (Figure 2-26). Amounts of wood per unit area of streambed are greatest in small headwater streams where lateral input rates are high, flows are inadequate to transport

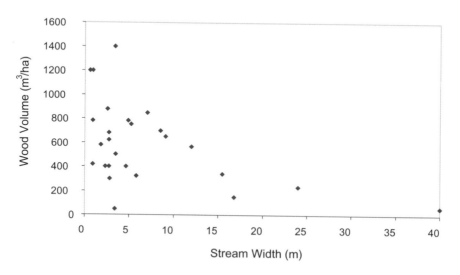

FIGURE 2-26 Standing stocks of wood in streams of different size in the McKenzie River Basin of Oregon. SOURCE: Reprinted, with permission, from Harmon et al. (1986). © 1986 by Harcourt International.

large pieces, and narrow channels resist movement of wood. In these small streams, wood plays a critical role in bed formation, storage of sediment and organic matter, habitat provision, and flood refuge. As streams increase in discharge and width downstream, flows can transport larger pieces, and wider channels provide less resistance to movement. The greater width of these streams also means that lateral inputs are distributed across a larger area, leading to lower standing stocks per unit area. In these larger systems, wood functions laterally along the river margins, and as part of major dams that accumulate during transport events. In floodplain rivers, high flows can move even the largest logs, and the extremely wide channels provide little resistance to movement. Wood outside the channel dissipates the power of overbank flows, traps sediment during flooding, serves as wildlife habitat and refuge, provides nursery logs for certain tree species that typically regenerate on downed wood, and supports microbial processes and transformation of critical nutrients (Maser et al., 1988).

Historically, large wood has often been cleared from streams, rivers, and riparian areas in order to protect bridges and roads from the effects of wood being transported downstream during high flows and to make rivers more navigable. As discussed in Chapter 5, the reintroduction of large wood to streams has become attractive and popular, particularly in the Pacific Northwest, as part of efforts to restore anadromous fish runs. These activities have seldom been evaluated in a systematic manner, however, and their level of success remains largely unknown.

Bank Stabilization

As early as 1885, it was understood that vegetation in riparian areas plays a critical role in stabilizing streambanks and thus supporting aquatic habitats (Van Cleef, 1885). Nearly a century later, Hickin (1984) identified five important mechanisms whereby vegetation influenced rivers in British Columbia: flow resistance, bank strengthening, bar sedimentation, formation of logjams, and concave bank (point bar) deposition. He identified a need "for studies to isolate the influence and quantity and, particularly, quality of vegetation on channel morphology and on lateral migration rates"—a need that continues today. Although the interconnection between soil, hydrology, and vegetation has widely been recognized in the jurisdictional delineation and understanding of wetland functions, an appreciation of how these components are intrinsically linked in riparian areas, particularly with regard to bank stability, has developed much more slowly (Sedell and Beschta, 1991).

For most streams and rivers, native plant communities are generally effective at protecting banks from the erosive effects of moving water. Aboveground plant parts (stems of trees, shrubs, graminoids, and forbs) tend to dampen turbulence and slow velocities at the water–streambank interface (Klingeman and Bradley, 1976). Root systems and soil organic matter help bind soil particles and alluvial sediments (Rutherford et al., 1999). Thus, streambanks can remain relatively stable even though they may experience considerable shear stress during periods of high flow. Although the shear stress imparted from flowing water is highly variable across a given channel cross section, such forces tend to be relatively high along the outside banks of meandering channels (Bathurst, 1979).

Different riparian plant communities may influence bank stabilization in different ways. For example, Trimble (1997) found that forested banks of some streams are more prone to periodic bank erosion compared with banks vegetated by herbaceous species, particularly grasses and sedges. Vegetated streambanks of any kind are typically more stable than unvegetated ones (Hicken, 1984).

Not only does vegetation provide stability to banks directly via its root system, but its ability to dissipate stream energy along a reach is also an extremely important function. As discussed previously, riparian vegetation plays a major role in creating relatively high roughness values for streambanks, thus retarding the potential for a bank to erode or the channel to migrate laterally. The establishment, growth, and succession of individual plants and groups of plants represent a mechanism by which riparian plant communities can continually respond to natural changes in streamflow and sediment loads and to local adjustments in channel morphology (Kalliola and Puhakka, 1988; Kalliola et al., 1991).

In addition to the immediate effects of resisting fluvial erosion of streambanks, vegetation plays a similarly important role with regard to floodplain development. Floodplain vegetation (including canopies, branches, stems, roots, and litter) not only protects the soil from direct rainfall impact and reduces the velocity of overbank flows (thereby preventing scour), but it also promotes deposition of fine sediments. In years without floods, riparian vegetation becomes established on exposed areas of floodplains, providing stability to these areas and promoting the vertical accretion of sediment during subsequent floods (Friedman et al., 1996a,b). Graminoids and other low-growing plants are particularly effective at protecting soils from surface erosion at high flows and are important for causing deposition of silts, although the amount of roughness associated with graminoids depends on species, stage of growth, and degree of disturbance (Wilson, 1967). Thus, riparian vegetation has had a fundamental role in the long-term maintenance of streambanks and floodplain landforms.

Other Effects of Vegetation

Other effects of vegetation on water cycling include the effects of root growth and decay in creating soil macroaggregates and consequently macropores, which increases soil permeability. Trees also modify the delivery of precipitation to the ground by intercepting, storing, and evaporating a portion of the incoming precipitation. In other instances, precipitation that is temporarily detained in the forest canopy but subsequently routed to the ground may be spatially redistributed by tree leaves, branches, and trunks. Root wads along the banks of streams reduce the rate of erosion, but also create channel complexity and instream habitat through overhangs and shading.

RIPARIAN AREAS AS HABITAT

The role of riparian areas in maintaining biodiversity is well known; their relative contributions greatly exceed the proportion of the landscape they occupy (Naiman et al., 1993, 2000; Crow et al., 2000). Scientific documentation of the importance of these areas for plants and animals comes from studies across the continent. In the Pacific Coast ecoregion, 60 percent of amphibian species, 16 percent of reptiles, 34 percent of birds, and 12 percent of mammals can be classified as riparian obligates (Kelsey and West, 1998). In the arid Southwest, 70 percent of threatened and endangered vertebrate species are listed as riparian obligates (Johnson, 1989); 60 percent of all vertebrate species are so defined (Ohmart and Anderson, 1982). Even in the relatively mesic environment of Wisconsin, 80 percent of plant and animal species on the state's endangered list live all or part of their lives in the riparian area of lakes (Korth and Cunningham, 1999).

Habitat, which refers to the place an organism lives, comprises both abiotic and biotic factors (Odum, 1971) including the basic provisions of food and shelter. Clearly, many animals require more than one habitat type throughout their life cycle. In popular accounts and wildlife management literature, habitat commonly refers to places providing reproductive requirements. In the case of migratory animal species, habitat also is used to refer to requirements of the animal in the non-breeding season, including stopover and wintering sites. Some animals use riparian areas as part of their home ranges or territories, moving through them on a short-term basis and over periods of hours and days. Indeed, the short-term movements of small mammals and birds within riparian areas have been shown to aid in the cycling of nutrients between the aquatic environment and adjacent uplands (Dobrowolski et al., 1993).

Riparian areas provide food resources for animals throughout the food web. Riparian vegetation along streams and lakes is critical as a primary food source to invertebrates from all the guilds (filter feeders, shredders, scrapers, and predators). It also provides a landing substrate for adult insects (such as midges, stoneflies, and mayflies) emerging from a waterbody (Benke and Wallace, 1990). As discussed below, the importance of this flux of insects in riparian areas to migratory and resident birds has received increased attention from ornithologists. The structural diversity of plant species in riparian areas creates a wide variety of feeding niches for herbivores and carnivores alike.

Various types of dispersal occur in riparian areas, including immigration, emigration, and migration. When applied to animals, migration connotes a cyclical movement, such as the spring and autumn migrations of North American songbirds. Applied to plants, migration refers to contraction or expansion of a population through time (Sauer, 1988). Hunter (1996) speaks of four basic types of movement: (1) daily movements of animals among patches of preferred habitat within their home range, (2) annual migrations, (3) dispersal movements of young animals and plant propagules, and (4) shifts in range in response to climate changes.

Although the following examples are limited to vertebrate animals and plants, it is well known that riparian areas also provide valuable habitat for invertebrates. The thermal regulation of streams and the supply of large wood afforded by riparian vegetation lead to characteristic invertebrate species both within streams and in their associated riparian areas (Hawkins et al., 1982; Anderson and Wallace, 1984; Benke and Wallace, 1990; Ward et al., 1999).

Fish

Much of the concern about riparian areas, particularly in the Pacific Northwest, had its genesis in maintaining viable fish populations (Hall and Lantz, 1969; Budd et al., 1987; Hall et al., 1987; Hartman et al., 1987). The major fish habitat elements that are influenced by riparian areas are water temperature, food supply, large wood, channel structure, and sediment (Wilzbach, 1985; Budd et al., 1987; Gillilan and Brown, 1997; Bisson et al., 2000). Temperature changes caused by the presence or absence of riparian vegetation have been shown to account for variability in trout populations (Barton et al., 1985; Wesche et al., 1987). Subsurface exchange between groundwater underneath riparian areas and streams also plays a role in creating critical cold-water fish habitats (Stanford and Ward, 1993).

At least three decades of studies have emphasized the connection between aquatic nutrient input and processing and riparian areas. Studies have shown that in forested regions over 99 percent of energy and organic carbon in food webs of small headwater streams originate in adjacent forest ecosystems (Fisher and Likens, 1973). These terrestrial sources of food influence the growth and abundance of fish communities (Chapman, 1965, 1966; Mundie, 1969). Riparian vegetation also influences light availability, which affects the efficiency of prey capture by stream fish (Wilzbach and Hall, 1985). As forested streams increase in size from headwaters to large rivers, instream primary production contributes more to the food base of stream ecosystems, although floodplains and their forests still deliver large portions of the energy supply to large lowland rivers (Vannote et al., 1980; Minshall, 1988; Junk et al., 1989). Riparian forests with multiple canopy layers, species, and age classes offer a wider array of food resources and physical habitats than do simple uniform plant communities (Gregory et al., 1991; Swanson et al., 1992; Osborne and Kovacic, 1993; Moyle and Yoshiama, 1994). In particular, the diversity and density of aquatic macroinvertebrates—a primary fish food resource—are higher in streams with wider riparian areas (Newbold et al., 1980). Similar observations have been made in lakeshore fish communities (Bryan and Scarnecchia, 1992; Christiansen et al., 1996).

Riparian areas and floodplains are major habitat elements for fish assemblages in streams and rivers (Schlosser, 1991) and have a pronounced influence on fish movement, particularly on migration of anadromous fishes (Gillilan and Brown, 1997). River networks and their floodplains create distinctive landforms that lead to biological diversity among fish species. Tributary junctions and complex channels are some of the most diverse areas within the river network because of their array of depths, velocities, and channel edges (Gorman and Karr, 1978; Bayley and Li, 1992); they also support different fish species than those found in the main channel and adjoining tributary system. Floodplains, side channels, and alcoves can serve as lower-velocity refuges during floods and provide access to abundant food resources for fish communities during flooding (Schlosser, 1982; Angermeier, 1987; Bayley and Li, 1992).

Given the many roles that riparian areas play in supporting fish communities, it is not surprising that the abundance and distributions of fish populations have been shown to be affected by land-use practices and riparian forest conditions (Hall and Lantz, 1969; Murphy et al., 1981; Bisson et al., 1992; Frissell, 1993; Gregory and Bisson, 1996; Naiman et al., 2000). Habitat degradation and impaired riparian conditions have been associated with 90 percent of the observed extinctions and declines in anadromous salmonids in the Pacific Northwest (Nehlsen et al., 1991). Thus, it is clear that healthy and functional riparian areas are essential for the abundance and diversity of fish (NRC, 1996).

Herptiles

Most amphibian species and many reptiles are intimately associated with riparian areas and their waterbodies. A recent summary of herptile (reptile and amphibian) species in the eastern United States lists 121 species found in riparian areas of springs, streams, rivers, lakes, ponds, bogs, and ephemeral pools (Pauley et al., 2000). Because of this close association, herptiles have been used, in some cases, as indicators of riparian condition (Lowe, 1989; Wake, 1991).

Amphibians. Few terrestrial vertebrates are as closely tied to the presence of water as amphibians, which require water bodies for completion of their life cycles. Those amphibians living entirely within small streams, such as salamander species of the genera *Desmognathus* and *Plethodon* in the Southeast and the genera *Dicamptodon* and *Rhyacotriton* in the Pacific Northwest, depend (as do fishes) on riparian functions that perpetuate healthy aquatic ecosystems. Frogs, toads, and salamanders are closely associated with riparian areas year-round throughout the continent, and intact riparian areas between upland habitat and aquatic breeding habitat are crucial to their viability. Adults must effectively reach breeding areas in their seasonal migration, and emerging young need to disperse from the breeding waters into surrounding uplands. Analysis of data from five eastern states found that a buffer extending at least 164 meters from the water's edge was required to perpetuate salamander populations (Semlitsch, 1998). For managed riparian areas to be effective in conserving pond-breeding species, they need to include a longitudinal component as well as sufficient width (Dodd and Cade, 1998). Microhabitat attributes of riparian areas are also important for meeting habitat requirements of herptiles. Work in eastern Texas revealed a positive relationship between a closed canopy and leaf litter ground cover and amphibian and reptile abundance in narrow streamside zones (Rudolph and Dickson, 1990). Degradation of riparian areas has led to the decline of some herptile species. For example, the tarahumara frog, which is restricted to southern Arizona streams with an intact community of sycamores and willows, is currently in danger of extinction (Ohmart and Anderson, 1978).

Reptiles. Most turtle species require functioning riparian areas to complete their life cycles. The wood turtle of the northeastern United States uses a variety of riparian habitats lying within about 360 meters of its home river for foraging, basking, and nesting (Vogt, 1981; Ewert, 1985). Other species of turtles, such as the snapping turtle, are almost exclusively aquatic but need the riparian area for nesting (Harding, 1997). Sea turtles of various families and species are completely dependent on coastal riparian dunes and beaches for nesting.

Snake species provide a less clear-cut example of obligatory dependence on riparian areas, although they are frequently seen hunting these biologically rich ecotones or seeking the cooler microclimate associated with them. In the Mid-

west, the queen snake is a riparian obligate that finds its major prey within riparian areas of streams, ponds, lakes, marshes, ditches, and canals (Harding, 1997). In the West, the western hognose snake is a denizen of river floodplains and even mountain canyons where there are alluvial deposits suitable for burrowing (Stebbins, 1966). Several species of lizards are strongly associated with the vegetative cover and organic material of southwestern riparian areas (Jones and Glinski, 1985; Warren and Schwalbe, 1985).

Birds

The importance of riparian areas as breeding habitat for birds is well known to birdwatchers and professional ornithologists alike. Because riparian areas are inherently diverse in plant species and varied in vertical and horizontal structure, they provide a variety of niches for birds, as documented by numerous studies (MacArthur, 1964; James, 1971; Karr and Roth, 1971; Whitmore, 1975; Rice et al., 1983, 1984). Riparian areas also provide a ready source of insects for breeding and migrating birds. Indeed, because of their close association with vegetation, and the relative ease of counting them, birds are often studied in assessments of riparian community functioning.

Perhaps nowhere is the relative importance of riparian areas vs. uplands to bird species as dramatic as in the arid southwestern United States. Avian density,

species richness, biodiversity, number of rare species, number of breeding pairs of birds, and biomass are extremely high in riparian areas compared to uplands (Szaro, 1991; Ohmart, 1996). These patterns are observed, sometimes to a lesser extent, across the country. In Iowa, wooded riparian areas provide habitat for 32 bird species as contrasted with only eight species in herbaceous areas (Stauffer and Best, 1980). Intact riparian areas of 75–175 m included 90 percent to 95 percent of Vermont's bird species (Spackman and Hughes, 1995). Even in mesic forests of Canada, boreal riparian conifer stands have higher avian diversity and abundance than do the adjacent coniferous uplands (Larue et al., 1995). In some landscapes lacking contrast between riparian and upland vegetation, such as in parts of the Pacific Northwest, avian diversity in riparian areas has not been found to be significantly greater than in uplands, although some differences in community composition have been documented (McGarigal and McComb, 1992; Murray and Stauffer, 1995).

Migrating birds use riparian areas as navigational aids and for stopover sites (Faaborg, 1988; Gill, 1990; Helmers, 1992). Indeed, their suitability as stopover sites in terms of food availability, safety from predators, and vulnerability to environmental stresses is receiving increasing attention by researchers concerned about population declines (Moore and Simons, 1992; Ewert and Hamas, 1996). Midcontinent banding studies along the North Platte River have demonstrated the role of large-scale riparian areas in the migration of songbirds (Brown et al., 1996; Scharf and Kren, 1997). Even very tiny riparian areas can be crucial to birds, as was illustrated in Wisconsin by the observation of 20 migratory avian species around vernal pools in a landscape that was otherwise virtually devoid of birds (Premo et al., 1992, and similar studies such as Long and Long, 1992).

Riparian areas have value not only for migratory birds, but also as habitat for juvenile birds in the post-fledging, premigration period (Faaborg et al., 1996) and for the dispersal of juveniles at the end of the breeding season (Machtans et al., 1996). As illustrated in Box 2-2, riparian areas in the sagebrush steppe have been found to be critically important to sage grouse species as brood-rearing habitat. Work along the lower Colorado River valley has highlighted the importance of riparian vegetative structure to wintering avian species, an often-overlooked habitat component (Anderson and Ohmart, 1977). Based on our knowledge of bird migration, riparian areas are critical to the conservation of avian diversity at the scale of regions, continents, and even hemispheres.

Mammals

Mammalian species with semi-aquatic habits, such as water shrew, starnosed mole, beaver, river otter, and mink, clearly incorporate riparian areas as part of their habitats, finding in such areas critical food and shelter resources (e.g., see DeGraaf and Yamasaki, 2000). Examples of mammalian presence in riparian areas abound in the literature. The greatest activity for several species in Michigan, including white-tailed deer, bobcat, red squirrel, and snowshoe hare, occurred within 400 feet of a third-order stream (Rogers et al., 1992). Similarly, in Vermont "most movement" of white-tailed deer, coyote, raccoon, red fox, gray fox, snowshoe hare, and voles occurs within a few meters of the high-water mark (Spackman and Hughes, 1995). In the Southwest, over half the small mammals trapped in a larger desert study were from a desert wash (Szaro, 1991). In such arid areas, species like the cotton rat and harvest mouse are perhaps not so much attracted to the water itself as to the denser vegetation found where water is more abundant (Ohmart and Anderson, 1978).

Mammals also use riparian areas at all scales for movement, either as part of their home ranges on a short-term movement basis, or during dispersal. Studies of culvert use by small and medium-sized mammals underlie the importance of even small drainages as crossing points along highways (Clevenger and Waltho, 2000). Conservation biologists have long argued for protecting riparian areas as a mechanism for conserving far-ranging large mammals in landscapes with fragmented habitats, as they provide dispersal avenues and means for gene flow (Beier, 1993; Noss et al., 1996). Indeed, Florida continues to work on designing a linked reserve system for species such as Florida panther and Florida black bear (Hoctor et al., 2000).

Although many mammals depend on riparian areas to survive and thrive, perhaps no North American mammal is as influential in riparian areas as beaver, described in Box 2-3.

BOX 2-2
Importance of Riparian Areas as Sage Grouse Habitat

At the time of European settlement, sage grouse were abundant through most of the west. Since settlement, distribution of greater sage grouse (*Centrocercus urophasianus*) has decreased by about 50 percent, with abundance declining by 45 percent to 80 percent (Braun, 1998). In July 2000, Gunnison's sage grouse (*C. minimus*), currently represented by only a few remnant populations in Colorado and southeastern Utah, was recognized as a distinct species (Young et al., 2000) and promptly became a candidate for federal listing as threatened. The U.S. Fish and Wildlife Service determined that listing as threatened is warranted but is precluded by other, higher administrative priorities. Similarly, a distinct population of greater sage grouse in the state of Washington recently was identified as a candidate for threatened status. There is some debate regarding the need to federally list the greater sage grouse throughout its range (Tweit, 2000).

Concerns about precipitous population declines have resulted in considerable research on habitat requirements of both sage grouse species at all phases of their life cycles. Both greater and Gunnison's sage grouse are critically dependent on riparian areas and wet sites within the sagebrush steppe for brood-rearing habitat. Within hours of hatch, females take their young broods from the sagebrush-dominated stands of their nesting sites to more mesic habitat, such as riparian areas. During their first two to four weeks, chicks forage for insects in riparian areas (Klebenow and Gray, 1968; Johnson and Boyce, 1990). Broods often use hay meadows in major drainages and frequently use the interface of two habitat types, such as sagebrush/wet meadow or willow–alder/wet meadow (Young, 1994). In addition to wet meadows, females with broods also use smaller mesic riparian areas around natural seeps and springs. Forbs that occur commonly in these areas constitute a significant portion of the diet of chicks between the ages of 2 and at least 12 weeks (Klebenow and Gray, 1968; Peterson, 1970; Johnson and Boyce, 1990). The importance of riparian areas as brood-rearing habitat appears to be further accentuated in drought years, when these moist areas may provide the only source of succulent forbs in the landscape (J. A. Crawford, Oregon State University, personal communication, 2001). Sufficient escape cover in the form of live, taller sagebrush that grows in riparian areas is important in reducing the dep-

redation of sage grouse that use these sites (C. E. Braun, retired from Colorado Division of Wildlife, now Grouse, Inc., personal communication, 2001). Successful management for sage grouse will undoubtedly require a landscape-scale approach that includes perpetuation of ecologically functioning riparian areas.

BOX 2-3
Beaver—The Quintessential Riparian Animal

Few would argue with the characterization of beaver (*Castor canadensis*), with its profound influence on riparian lands and aquatic ecosystems, as the quintessential riparian mammal. Its historic impact on North American economies and exploration eclipses that of most other wildlife. In addition, numerous scientific studies illustrate the critical role beaver play in riparian areas of North America at a wide range of temporal and spatial scales. By all available estimates, this influence was even greater in the past. Historical data from trapping harvests put the continental population of beaver before European settlement as high as 60–400 x 10^6 individuals (Seton, 1929), with a geographic range extending from the deserts of northern Mexico to the arctic tundra (Burt and Grossenheider, 1976; Jenkins and Busher, 1979). The explorer David Thompson, after crossing North America in 1784, stated that the continent "may be said to have been in the possession of two distinct races of beings, man and the beaver—with man occupying the highlands and the beaver in solid possession of the lowlands" (Rezendes and Roy, 1996). The present-day population has probably been reduced by a factor of ten-fold or more (Naiman et al., 1986), even after rebounding from extirpation in many portions of the species' range (Bergerud and Miller, 1977).

Beavers modify riparian areas and aquatic ecosystems by feeding and dam-building activities. These modifications create habitat for a variety of plant and animal species (Cook, 1940; Kirby, 1975; Hair et al., 1978; Malanson, 1993). For example, newly flooded forest ponds allow the formation of heron rookeries. The impounded nutrients result in a particularly rich aquatic system that lasts well beyond the life of the dam. In Oregon, the growth rate and size of juvenile salmonids in beaver ponds has been found to be superior to those of their same age cohorts in stream channel locations (Leidholt-Bruner et al., 1992).

The influence of beaver on landscape patch dynamics is profound and well documented (Slough and Sadleir, 1977; Naiman et al., 1986; Remillard et al., 1987), causing more than one researcher to assign beaver the role of keystone species. The impounding of water (primarily on first- through fourth-order streams) has multiple effects, including modifications of channel geomorphology and hydrology, changes in riparian wetland types and vegetation, alterations in the aquatic invertebrate community, modifications in nutrient cycling, and changes in water chemistry (McDowell and Naiman, 1986; Naiman et al., 1986; Ford and Naiman, 1988). The positive effects of beavers on headwater streams (e.g., retention of runoff, groundwater recharge, and sediment trapping) have been particularly appreciated in the western states where beaver have been reestablished by natural resources managers as an integral component of watershed restoration efforts (Federal Interagency Working Group, 1998). Studies in the Southwest suggest that beaver may be an especially important factor in resisting erosional perturbations in first-order streams stemming from overgrazing at the turn of the twentieth century and that they should thus be used as a restoration management tool (Parker et al., 1987).

Beavers affect the successional changes and species composition in the riparian vegetative community through differential feeding on woody plant species

continues

(Barnes and Dibble, 1988; Johnston and Naiman, 1990; Nolet et al., 1994). This influence is limited to the area in proximity to their aquatic homes since beavers generally do not forage for food much beyond 50–100 meters from their aquatic safety zone (Bradt, 1938; Jenkins, 1980; Belovsky, 1984). Their foraging distance from water has been shown to increase greatly, however, when large predators such as wolves are removed from the system, shrinking again when large predators return (R. Naiman, personal communication, University of Washington, 2000). In the Great Lakes region, early research found that a typical interval between abandonment and reoccupation of beaver ponds ranged from a few years to about 20 years (Lawrence, 1952). Other more recent work recognizes ontogenies of beaver ponds that can range from a year to many centuries (Naiman et al., 1988).

As beaver populations have rebounded in some portions of the country, conflicts with humans have often ensued. In spite of the demonstrated vital ecological roles played by beaver, the species' ability to modify the environment frequently runs afoul of human goals. Most obviously and dramatically, beaver cut down vegetation that is desirable to people for its aesthetic or economic value. In addition, beaver activities flood lands that are economically valuable to people, such as forested or arable land. Road culverts provide perfect settings for beaver dams, resulting in flooded roads and added costs in maintenance and repair.

Perhaps no conflict, however, matches the fervor of that between fish managers and beaver. In the Great Lakes states, beaver are often targeted for their deleterious impacts on populations of native trout, allegedly arising from effects of their dams on fish passage, streamflow, and water temperature. Yet few quantitative data are available to support this premise. A field study in Wisconsin (Avery, 1992), quantifying the effects of the removal of 546 beaver dams via "before and after" studies on a 9.8-mile branch of a major trout stream (Pemebonwon River) and on 22.7 miles of 14 smaller tributaries, had inconclusive findings that were statistically untested. For example, the effects of beaver dam removal on aquatic invertebrates were equivocal, with no detectable change as measured by the Hilsenhoff Biotic Index. Removal of dams appeared to have a cooling effect that was more noticeable on the tributaries, although no statistical testing was done. Finally, the study provided no direct evidence that dams acted as a physical barrier to trout movement in the river, although there appeared to be a trend (untested) toward increased trout numbers in the tributaries after dam removal. A more statistical study on the effects of beaver dams on streams in the Pestigo River watershed of northeastern Wisconsin, which examined effects of dam removal on downstream water temperature, found no significant simple reductions in temperature (McRae and Edwards, 1994). Instead, it found that large beaver impoundments may actually have a beneficial effect of dampening daily fluctuations in water temperatures downstream. The researchers concluded that local differences in the degree of vegetative and streambank shading, groundwater inflow, and stream volume make it difficult to generalize about the effects of beaver dams. They advocated that any beaver dam removal be selective and take such factors into account. Human land uses throughout a watershed also have a measurable influence on stream health (as measured by the Index of Biotic Integrity) (Wang et al., 1997), suggesting the need for larger-scale and more inclusive approaches to fish management.

Overarching all beaver–human conflicts is the fact that most human activities and goals occur in a spatial-temporal framework that is out of sync with the patch dynamics of a beaver-influenced watershed. Beaver co-evolved with salmonids, making improbable an essentially negative relationship between these organisms. However, long-term population cycles of beaver and their predators, as well as ongoing beaver-induced successional changes in riparian vegetation and stream geomorphology, are not easily incorporated into short-term resource management plans. Even our longest-term baseline data are quite recent (within approximately the last 50 years) and come exclusively from a time after presettlement beaver populations were decimated and the larger landscapes were already greatly altered by agriculture and extensive logging. There certainly are some streams or reaches where beaver activity has been concentrated for an extended period of time by land management practices and thus has become a problem with regard to fish populations. And there are undoubtedly marginal salmonid streams where beaver activities, combined with factors such as riparian vegetative modification or siltation, may impair salmonid habitat. On some streams, fish passage may be temporarily prevented by beaver dams, and fish populations in a particular reach may decline in the short term. But in the long term, beavers eventually will deplete food resources in the area, their dams will degrade and be breached, and young salmonids will access the high-productivity areas (with respect to invertebrates, plants, and nutrients) upstream of dams as part of an ongoing cycle. As was pointed out over a half century ago by fisheries biologist Cook (1940), negative effects on salmonids are influenced by many specific circumstances (such as over-fishing, sedimentation, and channelization) associated with a particular stream, although beavers are often assigned the blame. A new management paradigm is needed that seeks to understand the beaver's influence in terms of spatial-temporal patch dynamics and that recognizes the key and valuable ecological role played by this obligate riparian mammal (Naiman and Rogers, 1997).

Plants

In addition to being characterized by unique assemblages of plant species (as discussed earlier), riparian areas frequently harbor rare plant species. Waterbodies and their riparian areas have microclimates and disturbance regimes that allow species to persist at the edges of their ranges. Furthermore, movement and regeneration of plant species are facilitated by floodplain disturbances, further enhancing the potential for perpetuation of plant diversity in riparian areas (Spackman and Hughes, 1995).

Riparian areas provide refugia for populations that may be relicts of former climatic periods. Prairie plant communities recorded along rivers in Miami County, Ohio, since the time of western settlement are illustrative of this phenomenon (Huston, 1978). Work in Europe has shown how landscape elements such as ditch edges are critical to the persistence of endangered plant species even in a highly modified landscape (Ruthsatz and Haber, 1981). Non-vascular plants such as bryophytes are disproportionately found in moist sites such as riparian areas (Andrus, 1990). In an extreme example of rarity associated with riparian areas, the Virginia round-leaf birch (*Betula uber*) is endemic to two creeks in Smythe County in southwestern Virginia (Ogle and Mazzeo, 1976). Rare and relic plant species found only along the Great Lakes shoreline, in addition to the many animal species using riparian areas in this region, are described in Box 2-4.

ENVIRONMENTAL SERVICES OF RIPARIAN AREAS

The biosphere is often called the human "life support" system because of its paramount importance to maintaining the atmosphere, oceans, and land resources that support human societies. Riparian areas are a subset of this life-support system. On a global scale, they are a major component of the conduits that transport water from the continents to the ocean (Schlesinger and Melack, 1981). They are highly valued by society because they provide sites for human settlement near ports, proximity to water supplies, convenient sites for waste disposal, and opportunities for water-based recreation.

Riparian areas associated with ephemeral channels and small intermittent streams contribute to society in more subtle ways, and they are often disregarded—in part because of the ease with which they can be converted to alternative uses. However, riparian areas of small streams warrant attention not only for their local contributions, but also for their great collective length (see Figure 2-15).

Because the fundamental ecological processes that riparian areas perform occur whether or not humans are present to take advantage of them, they can be discussed somewhat independently of human values that change over time and differ among cultures. Functions fall into three major categories: (1) hydrology and sediment dynamics, (2) biogeochemistry and nutrient cycling, and (3) habitat

BOX 2-4
Biodiversity of Riparian Areas in the Great Lakes Region

The extensive Great Lakes riparian shoreline is a dominant feature on a continental scale. The state of Michigan alone has a coastal shoreline encompassing 3,222 miles—second only to Alaska. Islands contribute a striking proportion of coastal riparian area, as exemplified by the 948 miles of the Isle Royale archipelago shoreline, which is noted for its unusual assemblage of disjunct plant species (Voss, 1972). Small islands, in fact, may be nearly 100 percent riparian in composition, according to this report's definition. In addition, the Great Lakes region is well known for its glaciated landscape punctuated with numerous lakes and streams of all sizes. In effect, large portions of this region's landscape can be thought of as being almost entirely riparian. It is no wonder, then, that riparian areas play a central role in this region's biodiversity.

Great Lakes coastal riparian areas provide habitat for a variety of endemic plant species. Houghton's goldenrod (*Solidago houghtonii*), first collected by Douglass Houghton in 1839, occurs nowhere in the world except along the northern shores of Lakes Michigan and Huron (Voss, 1996). Dwarf lake iris (*Iris lacustris*), another northern Great Lakes endemic, is particularly abundant on the rubble of old glacial beach ridges (Voss, 1972). Lake Huron tansy (*Tanacetum huronense*), first described by Thomas Nuttall in 1810 at the Straits of Mackinac, inhabits Great Lakes riparian dunes and upper beaches, tracking the fluctuations in lake levels (Voss, 1996). In the Apostle Islands National Lakeshore (Wisconsin), botanists have recorded 809 plant species. A European rush (*Juncus squarrosus*), known previously only from Greenland, a Eurasian sedge new to the Great Lakes (*Carex ovalis*), and an eastern North American sedge (*C. tincta*) are listed among its disjuncts and rarities (Judziewicz and Koch, 1993). In the upper peninsula of Michigan, major river systems such as the Menominee and Escanaba provide refugia for prairie plant species believed to be relicts from the widespread savannas of the warm, dry hypsithermal period (D. Henson, Independent Consulting Biologist, Tamarack Studios, personal communication, 2000) occurring roughly 6,000 years ago (Pielou, 1991; Davis et al., 2000). In general, relict and disjunct populations and endemic species represent pockets of genetic variability (Utter and Hurst, 1990) that may become even more important during periods of future climate change. Pollen analysis from coastal riparian areas of the Great Lakes (national parks and national lakeshores) has demonstrated a climate-modifying capacity of large water bodies that apparently has translated into refugia for plant species during past climate changes and would likely do the same for future changes predicted as a result of global warming (Davis et al., 2000).

Great Lakes riparian areas are likewise important to the region's animals. Work along the Lake Huron shoreline has emphasized the importance of Great Lakes shorelines as stopover points for early-spring migrant birds; birds are substantially more abundant and diverse at coastal sites than they are inland. The main attraction appears to be the rich source of food in the form of emerging aquatic insects such as midges and mayflies that light on the riparian vegetation (Dallman and Smith, 1995; Ewert and Hamas, 1996; Seefelt, 1997; Hyde, 1998; Smith et al., 1998). Riparian areas of rivers and tributaries flowing into the Great Lakes have

continues

> enhanced diversity and habitat values resulting from their connectivity to these larger waterbodies. For example, small intermittent and first-order streams provide critical links between major river systems and numerous lakes and rivers on the landscape. Species as diverse as gray wolf (*Canis lupus*), coyote (*Canis latrans*), black bear (*Ursus americanus*), river otter (*Lutra canadensis*), red fox (*Vulpes vulpes*), and muskrat (*Ondatra zibethicus*) have been seen using seasonal drainages that connect with a third-order stream that is a major tributary of the Menominee River, the largest catchment in Michigan's upper peninsula (E. Rogers and D. J. Tiller, White Water Associates, Inc., personal communication, 2000). Riparian areas of third-order streams flowing into Lake Huron likewise are important to foraging neotropical migrants as stopover sites (Wilson, 2000).

and food web maintenance. Nine common examples are presented in Table 2-3, and many more could be added or derived from those listed. Present knowledge about these functions derives from an emerging but relatively large body of research conducted in the United States and elsewhere. In many cases, knowledge is sufficiently well developed such that indicators can be used as shortcuts to judge whether the functions are occurring at appropriate levels (see Chapter 5 for an in-depth discussion of assessment methods).

Riparian functions have both on-site and off-site effects, some of which may be expressed as goods and services available to society (Table 2-3). (This approach has been used to interpret the societal values of the functions of other ecosystem types [Christensen et al., 1996]). For example, functions related to hydrology and sediment dynamics include storage of surface water and sediment, which reduces damage from floodwaters downstream from the riparian area. Similarly, the function of cycling and accumulating chemical constituents has been measured in a number of studies on nitrogen and phosphorus cycling. These studies have shown that nutrients are intercepted, to varying degrees, as runoff passes through managed and natural riparian zones. The societal benefit is the buffering effect of pollutant removal, a service that has been a major motivation for protecting and managing riparian areas. Functions related to habitat and food web maintenance are the basis for many valued fisheries, not to mention that they contribute to human activities such as bird watching and wildlife enjoyment.

The hydrologic, nutrient cycling, and habitat/food web functions of riparian areas correspond to goods and services such as support of biodiversity, flood peak reduction, and removal of pollutants from runoff. Except for support of biodiversity, some of the environmental services of riparian areas can be provided by technologies, such as reservoirs for flood peak reduction and wastewater treatment plants for pollutant removal. However, these substitutions are directed at single functions rather than the multiple functions that riparian areas carry out simultaneously and with little direct costs to society.

CONCLUSIONS

Much of our understanding of riparian ecology (as reflected in this chapter) is based upon studies of streams and small shallow rivers because they are easier to study and are prevalent features of all landscapes. Nonetheless, large alluvial rivers have expansive riparian areas that are important as habitat for an enormous array of biota including endangered species, as natural sites for nutrient retention and pollution detoxification, and for other functions and values. Because large river floodplain–riparian areas have been substantially altered by human activities (see Chapter 3), their restoration will require a firm understanding of riparian structure and function at much larger watershed scales.

In a similar vein, our current understanding of the relationships between riparian vegetation and local environmental conditions is quite good. Even though the fundamental principles are universal, transferring specific information on riparian structure and function from one region to another can be problematic because of differences in vegetative composition and stature, discharge regimes, climate, geology and soils, and river network structure. Comparisons at the scale of climatic and geographic zones are seldom attempted even though such information is critical in making nationwide management decisions about riparian resources. Thus, in addition to the specific research recommendations noted throughout this chapter, it will be important to extend our knowledge of riparian functioning into large river systems and to those physical factors that strongly influence riparian systems (e.g., climate, geomorphology, and runoff) and the responses of biological components.

Riparian areas perform important hydrologic, geomorphic, and biological functions. These areas encompass complex above- and below-ground habitats created by the convergence of biophysical processes in the transition zone between aquatic and terrestrial ecosystems. Riparian areas encompass interactions in all three spatial dimensions of drainage corridors: laterally between stream channel and adjacent terrestrial zone, vertically between the surface and subsurface, and longitudinally between upstream and downstream reaches.

Riparian areas cannot be thought of in isolation from stream channels. The characteristic geomorphology, plant communities, and associated aquatic and wildlife species of riparian systems are intrinsically linked to the role of water as both an agent of disturbance and a critical requirement of biota. Although it is widely recognized that the aquatic portion of a riverine or lake system can have a pronounced effect on associated riparian areas, these riparian areas can have important influences on their aquatic systems.

Natural riparian systems have adapted to specific hydrologic disturbance regimes (i.e., flow frequency, magnitude, duration, timing and rate of

TABLE 2-3 Functions of Riparian Areas and Their Relationship to Environmental Services[a]

Examples of Functions	Indicators that Functions Exist
Hydrology and Sediment Dynamics	
Stores surface water over the short term	Floodplain connected to stream channel
Maintains a high water table	Presence of flood-tolerant and drought-intolerant plant species
Accumulates and transports sediments	Riffle-pool sequences, point bars, and other features
Biogeochemistry and Nutrient Cycling	
Produces organic carbon	A balanced biotic community
Contributes to overall biodiversity	High species richness of plants and animals
Cycles and accumulates chemical constituents	Good chemical and biotic indicators
Sequesters carbon in soil	Organic-rich soils
Habitat and Food Web Maintenance	
Maintains streamside vegetation	Presence of shade-producing forest canopy
Supports characteristic terrestrial vertebrate populations	Appropriate species having access to riparian area
Supports characteristic aquatic vertebrate populations	Migrations and population maintenance of fish

[a]Effects of functions sometimes are expressed off-site. Indicators are often used to evaluate whether or not a function exists, and are commonly used as shortcuts for evaluating the condition of riparian areas. The functions listed are examples only and are not comprehensive. Modified from NRC (1995).

On-site or off-site Effects of Functions	Goods and Services Valued by Society
Attenuates downstream flood peaks	Reduces damage from floodwaters (Daily, 1997)
Maintains vegetation structure in arid climates	Contributes to regional biodiversity through habitat (e.g., forest canopy) provision (Szaro, 1991; Ohmart, 1996; James et al., 2001)
Contributes to fluvial geomorphology	Creates predictable yet dynamic channel and floodplain dynamics (Beschta et al., 1987a; Klingeman et al., 1999)
Provides energy to maintain aquatic and terrestrial food webs	Supports populations of organisms (Gregory et al., 1991; Meyer and Wallace, 2001)
Provides reservoirs for genetic diversity	Contributes to biocomplexity (Szaro, 1991; Naiman and Rogers, 1997; Pollock et al., 1998)
Intercepts nutrients and toxicants from runoff	Removes pollutants from runoff (Bhowmilk et al., 1980; Peterjohn and Correll, 1984)
Contributes to nutrient retention and to sequestration of carbon dioxide from the atmosphere	Potentially ameliorates global warming (Van Cleve et al., 1991)
Provides shade to stream during warm season	Creates habitat for cold-water fish (Beschta et al., 1987b; McCullough, 1999)
Allows daily movements to annual migrations	Supplies objects for bird watching, wildlife enjoyment, and game hunting (Green and Tunstall, 1992; Flather and Cordell, 1995)
Allows migratory fish to complete life cycles	Provides fish for food and recreation (Nehlsen et al, 1991; Naiman et al., 2000)

change) and corresponding sediment regimes (e.g., transport frequency and magnitude, particle sizes). Particularly in arid regions, the types of riparian vegetation present are largely determined by their tolerance of and adaptation to hydrologic disturbances. Managing riparian areas without regard to how they are influenced by the dynamic patterns of adjacent waterbodies ignores a fundamental aspect of how these systems function.

In addition to disturbance regimes, soil moisture plays a significant role in shaping the vegetative structure of riparian areas. In humid regions of the country, vegetation is abundant in both uplands and riparian areas, with species tolerant to high soil moisture and anoxia being more prevalent in riparian areas. In arid regions, plants are often concentrated along the band of adequate soil moisture provided by riparian areas.

A core of tree genera—*alders, cottonwoods,* and *willows*—grows in riparian areas across the continent. Gallery stands of these trees are predominant features of many riparian areas throughout the country. In much of the American West, the integrity of riparian areas in low-gradient meadow systems requires intact communities of shrubs (e.g., willows) and hydrophytic graminoids (sedges, rushes, and grasses).

Riparian areas, in proportion to their area within a watershed, perform more biologically productive functions than do uplands. Riparian areas provide a wide range of functions such as microclimate modification and shade, bank stabilization and modification of sedimentation processes, contributions of organic litter and large wood to aquatic systems, nutrient retention and cycling, wildlife habitat, and general food-web support for a wide range of aquatic and terrestrial organisms. Thus, even though they occupy only a small proportion of the total land base in most watersheds, they are uniquely positioned between the aquatic and terrestrial ecosystems to provide a wide range of functions critical for many aquatic and terrestrial species, for maintenance of water quality, for aesthetics, for the production of goods and services, and for a wide range of social and cultural values.

Riparian areas are effective in filtering and transforming materials (such as dissolved and particulate nonpoint source pollutants) from hillslope runoff. Pollutant removal in riparian areas is most effective along first- and second-order streams because a greater percentage of their flow derives from hillslope rather than upstream sources. However, where flows are concentrated into topographic depressions prior to entering a riparian area, the effectiveness of the riparian area for pollutant removal will be greatly reduced.

Because riparian areas are located at the convergence of terrestrial and aquatic ecosystems, they are regional hot spots of biodiversity and often exhibit high rates of biological productivity in marked contrast to the larger landscape. This is particularly dramatic in arid regions, as evidenced by the high number of plant and animal species that find crucial habitats along watercourses and washes. Riparian areas provide connectivity at all spatial and temporal scales, helping maintain landscape biodiversity by countering the negative ecological effects of habitat fragmentation.

REFERENCES

Abbe, T. B., and D. R. Montgomery. 1996. Large woody debris jams, channel hydraulics and habitat formation in large rivers. Regulated Rivers: Research & Management 12(2–3):201–221.

Addy, K. L., A. J. Gold, P. M. Groffman, and P. A. Jacinthe. 1999. Ground water nitrate removal in subsoil of forested and mowed riparian buffer zones. J. Environmental Quality 28(3):962–970.

Ahmad, R., F. N. Scatena, and A. Gupta. 1993. Morphology and sedimentation in Caribbean montane streams: examples from Jamaica and Puerto Rico. Sedimentary Geology 85:157–169.

Amoros, C., J. Rostan, G. Pautou, and J. Bravard. 1987. The reversible process concept applied to the environmental management of large river systems. Environ. Manage. 11(5):607–617.

Anderson, B. W., and R. D. Ohmart. 1977. Vegetation structure and bird use in the lower Colorado River Valley. Paper presented at the symposium on importance, preservation, and management of riparian habitat, Tucson, AZ.

Anderson, N. H., and J. B. Wallace. 1984. Habitat, life history, and behavioral adaptations of aquatic insects. Pp. 38–58 In: An introduction to the aquatic insects of North America. R. W. Merritt and K. W. Cummins (eds.). Dubuque, IA: Kendall/Hunt Publishing Co.

Andrus, R. E. 1990. Why rare and endangered byophytes? In: Ecosystem management: rare species and significant habitats. Mitchell, R. S., Sheviak, C. J. and D. J. Leopold (eds.). Proceedings of the 15th Annual Natural Areas Conference. New York State Museum Bulletin No. 471. Albany, NY.

Angermeier, P. L. 1987. Spatiotemporal variation in habitat selection by fishes in small Illinois streams. Pp. 52–60 In: Community and evolutionary ecology of North American stream fishes. W. J. Matthews and D. C. Heins (eds.). Norman, OK: University of Oklahoma Press.

Avery, E. L. 1992. Effects of removing beaver dams upon a northern Wisconsin brook trout stream. Study No. 406. Madison, WI: Wisconsin Department of Natural Resources, Bureau of Research, Fish Research Section.

Bach, L. 1995. River basin assessment: Upper/middle Grande Ronde River and Catherine Creek. Portland, OR: Oregon Department of Environmental Quality.

Baker, W. L. 1989. Classification of the riparian vegetation of the montane and sub-alpine zones in western Colorado. Great Basin Naturalist 49:214–228.

Bard, A. J., and R. L. Wilby. 1999. Eco-hydrology: plants and water in terrestrial environments. Routledge, London.

Barnes, W. J., and E. Dibble. 1988. The effects of beaver in riverbank forest succession. Can. J. Bot. 66:40–44.

Barton, D. R., W. D. Taylor, and R. M. Biette. 1985. Dimensions of riparian buffer strips required to maintain trout habitat in southern Ontario streams. North American Journal of Fisheries Management 5:364–378.

Bathurst, J. C. 1979. Distribution of boundary shear stress in rivers. Pp. 95–116 In: Adjustment of fluvial systems. D. D. Rhodes and G. P. Williams (eds.). Dubuque, IA: Kendall/Hunt.

Bayley, P. B. 1991. The flood pulse advantage and the restoration of river–floodplain systems. Regulated Rivers: Research and Management 6:75–86.

Bayley, P. B., and H. W. Li. 1992. Riverine fishes. Pp. 252–281 In: The rivers handbook: hydrological and ecological principles, Vol. 1. P. Calow and G. E. Petts (eds.). Oxford: Blackwell Scientific Publications.

Beier, P. 1993. Determining minimum habitat areas and habitat corridors for cougars. Conservation Biology 7:94–108.

Belovsky, G. E. 1984. Summer diet optimization by beaver. The American Midland Naturalist 111:209–222.

Benke, A. C., and J. B. Wallace. 1990. Wood dynamics in coastal plain blackwater streams. Can. J. Fish. Aquat. Sci. 47:92–99.

Bergerud, A. T., and D. R. Miller. 1977. Population dynamics of Newfoundland beaver. Can. J. Zool. 55:1480–1492.

Beschta, R. L., T. Blinn, G. E. Grant, G. G. Ice, and F. G. Swanson (editors). 1987a. Erosion and sedimentation in the Pacific Rim. International Association of Hydrological Sciences, Publication No. 165. 510 pp.

Beschta, R. L., R. E. Bilby, G. W. Brown, L. B. Holtby, and T. D. Hofstra. 1987b. Stream temperature and aquatic habitat: fisheries and forestry interactions. Pp. 191–232 In: Streamside management: forestry and fisheries comparisons. E. O. Salo and T. W. Cundy (eds.). Seattle, WA: University of Washington, Institute of Forest Resources, Contribution No. 57. 471 pp.

Bhowmilk, N. G., Bonini, A. P., Bogner, W. C., and Byrne, R. P. 1980. Hydraulics of flow and sediment transport to the Kankakee River in Illinois. Illinois State Water Survey Rep. of Investigation 98, Champaign, Ill., 170 p.

Bisson, P. A., T. P. Quinn, G. H. Reeves, and S. V. Gregory. 1992. Best management practices, cumulative effects, long-term trends in fish abundance in Pacific Northwest river systems. Pp. 189–232 In: Watershed management: balancing sustainability and environmental change. R. J. Naiman (ed.). New York: Springer-Verlag. 542 p.

Bisson, P.A., R. E. Bilby, M. D. Bryant, C. A. Dolloff, G. B. Grette, R. A. House, M. L. Murphy, K. V. Koski, and J. R. Sedell. 1987. Large woody debris in forested streams in the Pacific Northwest: Past, present, and future. Pp. 143–190 In: Streamside management: forestry and fishery interactions. E. O. Salo and T. W. Cundy (eds.). Seattle, WA: University of Washington, Institute of Forest Resources, Contribution No. 57.

Bisson, P. A., S. V. Gregory, and T. Nickelson. 2002. Long-term variation in anadromous and resident salmonid populations. In: Ecological and Environmental Research in the Alsea Watershed. J. Stednick (ed.). Springer-Verlag.

Bliss, L. C., and J. E. Cantlon. 1957. Succession on river alluvium in northern Alaska. American Midland Naturalist 58:452–469.

Bohlke, J. K., and J. M. Denver. 1995. Combined use of ground-water dating, chemical and isotopic analyses to resolve the history and fate of ground-water nitrate contamination in two agricultural watersheds, Atlantic coastal plain, Maryland. Water Resources Research 31: 2319–2339.

Boose, A. B., and J. S. Holt. 1999. Environmental effects on asexual reproduction in Arundo donax. Weed Research 39:117–127.

Boyd, M., B. Kasper, and A. Hamel. 1998 (draft). Tillamook basin temperature total maximum daily load (TMDL). Oregon Department of Environmental Quality, Portland. 42 pp. + appendices.

Boyer, E. W., G. M. Hornberger, K. E. Bencala, and D. M. McKnight. 1997. Response characteristics of DOC flushing in an alpine catchment. Hydrological Processes 11:1635–1647.

Bradt, G. W. 1938. A study of beaver colonies in Michigan. J. Mammalogy 19:139–162.

Braun, C. E. 1998. Sage grouse declines in western North America: What are the problems? Proceedings of the Western Association of Fish and Wildlife Agencies 78:139–156.

Brinson, M. M. 1990. Riverine forests. In: Forested wetlands. A. E. Lugo, M. Brinson, and S. Brown (eds.). New York: Elsevier.

Brinson, M. M. 1991. Landscape properties of pocosins and associated wetlands. Wetlands 11:441–465.
Brinson, M. M. 1993. Changes in the functioning of wetlands along environmental gradients. Wetlands 13:65–7.
Brinson, M. M. and J. Verhoeven. 1999. Riparian forests. In: Maintaining biodiversity in forested ecosystems. M. L. Hunter (ed.). Cambridge, England: Cambridge University Press.
Brinson, M. M., R. R. Christian, and L. K. Blum. 1995. Multiple states in the sea-level induced transition from terrestrial forest to estuary. Estuaries 18:648–659.
Brock, T. D., and M. T. Madigan. 1991. Biology of Microorganisms. Englewood Cliffs, NJ: Prentice Hall. 873 pp.
Brosofske, K. D., J.Chen, R. J. Naiman, and J. F. Franklin. 1997. Effects of harvesting on microclimaic gradients from small streams to uplands in western Washington. Ecological applications 7(4):1188–1200.
Brown, C. R., M. B. Brown, P. A. Johnsgard, J. Kren, W. C. Scharf. 1996. Birds of the Cedar Point Biological Station area, Keith and Garden counties, Nebraska: Season occurrence and breeding data. Transactions of the Nebraska Academy of Sciences 23:91–108.
Brown, G. W. 1969. Predicting temperatures of small streams. Water Resources Research 5:68–75.
Bruner, W. E. 1931. The vegetation of Oklahoma. Ecological Monographs 1:99–188.
Bryan, M. D. and D. L. Scarnecchia. 1992. Species richness, composition, and abundance of fish larvae and juveniles inhabiting natural and developed shorelines of a glacial Iowa lake. Environ. Biol. of Fishes 35:329–341.
Budd, W. W., P. L. Cohen, and P. R. Saunders. 1987. Stream corridor management in the Pacific Northwest: I. Determination of stream corridor widths. Environmental Management 11:587–597.
Buell, M. F., and W. A. Wistendahl. 1955. Flood plain forests of the Raritan River. Bulletin of the Torrey Botanical Club 82:463–472.
Burnett, A. W., and S. A. Schumm. 1983. Alluvial-river response to neotectonic deformation in Louisiana and Mississippi. Science 222:49–50.
Burt, W. H., and Grossenheider. 1976. A field guide to the mammals. Boston, MA: Houghton Mifflin Company.
Carpenter, S. R., E. van Donk, and R. G. Wetzel. 1998. Effects of the nutrient loading gradient in shallow lakes. Pp. 393–396 In: The role of submersed macrophytes in structuring the biological community and biogeochemical dynamics in shallow lakes. Dordrecht, Netherlands: Kluwer Publishers.
Chapin, D. M., R. L. Beschta, and J. W. Shen. 2000. Flood frequencies required to sustain riparian plant communities in the Upper Klamath Basin, Oregon. Pp. 17–22 In: International conference on riparian ecology and management in multi-land use watersheds. P. J. Wiggington and R. L. Beschta (eds.). Middleburg, VA: American Water Resources Association.
Chapman, D. W. 1965. Net production of juvenile coho salmon in three Oregon streams. Trans. Amer. Fish. Soc. 94:40–52.
Chapman, D. W. 1966. Food and space as regulators of salmonid populations in streams. Amer. Nat. 100:345–357.
Cheng, J. D. 1989. Streamflow changes after clear-cut logging of a pine beetle-infested watershed in southern British Columbia, Canada. Water Resources Research 25:449–456.
Choi, J., and J. Harvey. 2000. Characterizing multiple timescales of stream and storage zone interaction that affect solute fate and transport in streams. Water Resources Research 36(6):1511–1518.
Christensen, N. L., A. M. Bartuska, J. H. Brown, S. Carpenter, C. D'Antonio, R. Francis, J. F. Franklin, J. A. MacMahon, R. F. Noss, D. J. Parsons, C. H. Peterson, M. G. Turner, and R. G. Woodmansee. 1996. The report of the Ecological Society of America committee on the scientific basis for ecosystem management. Ecological Applications 6:665–691.

Clevenger, A. P., and N. Waltho. 2000. Factors influencing the effectiveness of wildlife underpasses in Banff National Park, Alberta, Canada. Conservation Biology 14:47–55.
Collins, S. L., P. G. Risser, and E. L. Rice. 1981. Ordination and classification of mature bottomland forests in North Central Oklahoma. Bulletin of the Torrey Botanical Club 108:152–165.
Cook, D. B. 1940. Beaver-trout relations. J. Mammalogy 21:397–401.
Correll, D. L. 1997. Buffer zones and water quality protection. Pp. 7–20 In: Buffer zones: their processes and potential in water protection. N. E. Haycock,T. P. Burt, K. W. T. Goulding and G. Pinay (eds.). Harpenden, Hertfordshire, UK: Quest Environmental.
Cowan. W. L. 1956. Estimating hydraulic roughness coefficients. Agricultural Engineering 37:473–475.
Crawford C. S., A. C. Culley, R. Leutheuser, M. S. Sifuentes, L. H. White, and J. P. Wilber. 1993. Middle Rio Grande Ecosystem: Bosque Biological Management Plan. Albuquerque, NM: U.S. Fish and Wildlife Service, District 2. 291 pp.
Crow, T. R., M. E. Baker, and B. V. Barnes. 2000. Diversity in riparian landscapes. Pp. 34–66 In: Riparian management in forests of the continental eastern United States. E. S. Verry, J. W. Hornbeck, and C. A. Dolloff (eds.). New York: Lewis Publishers.
Dahm, C. N., N. B. Grimm, P. Marmonier, H. M. Valett, and P. Vervier. 1998. Nutrient dynamics at the interface between surface waters and groundwaters. Freshwater Biology 40:427–451.
Daily, G. C. (ed.). 1997. Nature's services: Societal dependence on natural ecosystems. Washington, DC: Island Press.
Dallman, M., and R. J. Smith. 1995. Avian predation on Chironomids along the nearshore waters of Lake Huron. Michigan Birds and Natural History 2:201–204.
Daly, C., R. P. Neilson, and D. L. Phillips. 1994. A statistical-topographic model for mapping climatological precipitation over mountainous terrain. J. Appl. Meteor. 33:140–158.
Danehy, R. J., and B. J. Kirpes. 2000. Relative humidity gradients across riparian areas in eastern Oregon and Washington forests. Northwest Science 74(3):224–233.
Daniel, J. F. 1976. Estimating groundwater evapotranspiration from streamflow records. Water Resources Research 12(3):360–364.
Davis, M., C. Douglas, R. Calcote, K. L. Cole, M. G. Winkler, and R. Flakne. 2000. Holocene climate in the western Great Lakes national parks and lakeshores: implications for future climate change. Conservation Biology 14:968–983.
Dawson, T. E., and J. R. Ehleringer. 1993. Streamside trees that do not use streamwater. Nature 350:335–337.
DeGraaf, R. M., and M. Yamasaki. 2000. Bird and mammal habitat in riparian areas. Pp. 139–156 In: Riparian management in forests of the continental eastern United States. E. S. Verry, J. W. Hornbeck, and C.A. Dolloff (eds.). New York: Lewis Publishers.
Dillaha, T. A., R. B. Reneau, S. Mostaghimi, and V. O. Shanholtz. 1989. Vegetative filter strips for nonpoint source pollution control. Transactions of the ASAE 32(2):491–496.
Dingman, S. L. 1984. Fluvial hydrology. New York: W. H. Freeman.
Dirschl, H. J., and R. T. Coupland. 1972. Vegetation patterns and site relationships in the Saskatchewan River delta. Canadian Journal of Botany 50:647–675.
Dobrowolski, K. A., A. Kozakiewicz, and B. Leznicka. 1993. The role of small mammals and birds in transport of matter through the shore zone of lakes. Hydrobiologia 251:81–93.
Dodd, C. K., Jr., and B. S. Cade. 1998. Movement patterns and the conservation of amphibians breeding in small, temporary wetlands. Conservation Biology 12:331–339.
Dong, J., J. Chen, K. D. Brosofske, and R. J. Naiman. 1998. Modeling air temperature gradients across managed small streams in western Washington. Journal of Environmental Management 53:309–321.
Dunne, T., and R. D. Black. 1970. Partial area contributions to storm runoff in a small New England watershed. Water Resources Research 6: 1296–1311.

Dunne, T. 1978. Field studies of hillslope flow processes. Pp. 227–293 In: Hillslope hydrology. M. J. Kirkley (ed.). Chicester: John Wiley.

Everitt, B. L. 1968. Use of cottonwood in an investigation of the recent history of a flood plain. American Journal of Science 266: 417–439.

Ewert, M. A. 1985. Second Year Report: Assessment of the current distribution and abundance of the Wood Turtle (*Clemmys insculpta*) in Minnesota and along the St. Croix National Scenic Riverway in Wisconsin. Unpublished report for the Minnesota Nongame Wildlife Program and Minnesota Field Office of the Nature Conservancy, 37 p.

Ewert, D. N., and M. J. Hamas. 1996. Ecology of migratory landbirds during migration in the Midwest. Pp. 200–208 In: Management of Midwestern landscapes for the conservation of neotropical migratory birds. Gen. Tech. Rep. NC–187. F. R. Thompson III (ed.). St. Paul, MN: USDA Forest Service North Central Experiment Station.

Faaborg, J. 1988. Ornithology, an ecological approach. Englewood Cliffs, NJ: Prentice Hall. 470 pp.

Faaborg, J., A. D. Anders, M. E. Baltz, and W. K. Gram. 1996. Non-breeding season considerations for the conservation of migratory birds in the Midwest: Post breeding and wintering species. Pp. 189–199 In: Management of Midwestern landscapes for the conservation of neotropical migratory birds. Gen. Tech. Rep. NC–187. F. R. Thompson III (ed.). St. Paul, MN: USDA Forest Service North Central Experiment Station.

Federal Interagency Working Group. 1998. Stream corridor restoration. principles, processes, and practices. Washington, DC: National Technical Information Service, U.S. Department of Commerce.

Findlay, S. 1995. Importance of surface–subsurface exchange in stream ecosystems: the hyporheic zone. Limnology and Oceanography 40: 159–164.

Fisher, S. G., N. B. Grimm, E. Marti, R. M. Holmes, and J. B. Jones. 1998. Material spiraling in stream corridors: a telescoping ecosystem model. Ecosystems 1:19–34.

Fisher, S. G., and G. E. Likens. 1973. Energy flow in Bear Brook, New Hampshire: an integrative approach to stream ecosystem metabolism. Ecological Monographs 43(4): 421–439.

Fisk, H .N., and E. McFarland, Jr. 1955. Later Quaternary deltaic deposits of the Mississippi River. Geological Society of America Special Paper 62:279–302.

Flanagan, L. B., J. R. Ehleringer and T. E. Dawson. 1992. Water sources of plants growing in woodland, desert and riparian communities: Evidence from stable isotope analysis. In: Proceedings of the symposium on ecology and management of riparian shrub communities. Gen. Tech. Rep. INT-289. Clary, W. P., E. D. Durant McArthur, D. Bedunah and C. L. Wambolt (eds.). Ogden, UT: USDA Forest Service Intermountain Research Station.

Flather, C. H., and H. K. Cordell. 1995. Outdoor recreation: historical and anticipated trends. Pp. 3–16 In: Wildlife and recreationists: historical and anticipated trends. R. L. Knight and K. J. Gutzwiller (eds.). Washington, DC: Island Press.

Ford, T. E., and R. J. Naiman. 1988. Alteration of carbon cycling by beaver: methane evasion rates from boreal forest streams and rivers. Can. J. Zool. 66:529–533.

Franklin, J. F., and C. T. Dyrness. 1988. Natural vegetation of Oregon and Washington. Corvallis, OR: Oregon State University Press.

Friedman, J. M., and G. T. Auble. 2000. Floods, flood control, and bottomland vegetation. Pp. 219–237 In: Inland flood hazards: human, riparian, and aquatic communities. E. Wohl (ed.), Cambridge, England: Cambridge University Press.

Friedman, J. M., W. R. Osterkamp, and W. M. Lewis, Jr. 1996a. Channel narrowing and vegetation development following a Great Plains flood. Ecology 77(7):2167–2181.

Friedman, J. M., W. R. Osterkamp, and W. M. Lewis, Jr. 1996b. The role of vegetation and bed-level fluctuations in the process of channel narrowing. Geomorphology 14:341–351.

Frissell, C. A. 1993. A new strategy for watershed restoration and recovery of Pacific Salmon in the Pacific Northwest. Eugene, OR: Pacific Rivers Council, Inc.

Garten, C. T., Huston, M. A., Thomes, C. A. 1994. Topographic variation of soil nitrogen dynamics at Walker Branch Watershed, Tennessee. Forest Science 40(3):497–512.

Gatewood, J. S., J. W. Robinson, B. R. Colby, J. D. Hem, and L. C. Halpenny. 1950. Use of water by bottomland vegetation in the Lower Safford Valley, Arizona. U.S. Geological Survey Water-Supply Paper 1103.

Geraghty, J. J., D. W. Miller, F. van der Leeden, and F. L. Troise. 1973. Water Atlas of the United States. Third edition. Port Washington, NY: Water Information Center.

Gibert, J., J. A. Stanford, M.-J. Dole-Oliver, and J. V. Ward. 1994. Basic attributes of groundwater ecosystems and prospects for research. Pp. 7–40 In: Groundwater ecology. J. Gibert, D. L. Danielopol, and J. A. Stanford (eds.). San Diego, CA: Academic Press, Inc.

Gill, F. B. 1990. Ornithology. New York: W. H. Freeman and Co. 660 pp.

Gilliam, J. W., J. E. Parsons, and R. L. Mikkelsen. 1997. Nitrogen dynamics and buffer zones. Pp. 54–61 In: Buffer zones: their processes and potential in water protection. N. E. Haycock, T. P. Burt, K. W. T. Goulding and G. Pinay (eds.). Harpenden, Hertfordshire, UK: Quest Environmental.

Gillilan, D., and T. Brown. 1997. Instream Flow Protection: Seeking a Balance in Western Water Uses. Washington, DC: Island Press.

Goodrich, D. C. et al. 2000. Seasonal estimates of riparian evapotranspiration using remote and in situ measurements. Agricultural and Forest Meteorology 105(1–3):281–310.

Gorman, O. T., and J. R. Karr. 1978. Habitat structure and stream fish communities. Ecology 59:507–515.

Gould, W. A., and M. D. Walker. 1999. Plant communities and landscape diversity along a Canadian arctic river. Journal of Vegetation Science 10:537–548.

Green, C. H., and S. M Tunstall. 1992. The amenity and environmental value of river corridors in Britain. In: River conservation and management. New York: John Wiley & Sons.

Gregory, S. V., and P. A. Bisson. 1996. Degradation and loss of anadromous salmonid habitat in the Pacific Northwest. Pp. 277–314. In: Pacific salmon and their ecosystems: status and future options. D. Stouder and R. J. Naiman (eds.). New York: Springer-Verlag.

Gregory, K. J., and D. E. Walling. 1973. Drainage basin form and process. New York: John Wiley and Sons. 456 pp.

Gregory, S. V., F. J. Swanson, W. A. McKee, and K. W. Cummins. 1991. An ecosystem perspective of riparian zones. BioScience 41:540–551.

Groffman, P. M., A. J. Gold, and R. C. Simmons. 1992. Nitrate dynamics in riparian forests: microbial studies. J. Environ. Qual. 21:666–671.

Hair, J. D., G. T. Hepp, L. M. Luckett, K. P. Reese, and D. K. Woodward. 1978. Beaver pond ecosystems and their relationships to multi-use natural resource management. Pp. 80–92 In: Strategies for protection and management of floodplain wetlands and other riparian ecosystems. Proceedings of the Symposium, Dec. 11–13, 1978, Callaway Gardens, GA. Washington, DC: USDA and USFS.

Hall, F. R. 1968. Base-flow recessions—a review. Water Resources Research 4(5):973–983.

Hall, J. D., and R. L. Lantz. 1969. Effects of logging on the habitat of coho salmon and cutthroat trout in coastal streams. Pp. 355–375 In: Salmon and trout in streams. T. G. Northcote (ed.). H. R. MacMillan Lectures in Fisheries. Vancouver, British Columbia: University of British Columbia.

Hall, J. D., G. W. Brown, and R. L. Lantz. 1987. The Alsea watershed study: a retrospective. Pp. 399–416 In: Streamside management: forestry-fisheries interactions. E. O. Salo and T. W. Cundy (eds.). Seattle, WA: University of Washington Institute of Forest Resources.

Harding, J. H. 1997. Amphibians and reptiles of the Great Lakes region. Ann Arbor, MI: The University of Michigan Press. 378 pp.

Harmon, M. E., J. F. Franklin, F. J. Swanson, P. Sollins, S. V. Gregory, J. D. Lattin, N. H. Anderson, S. P. Cline, N. G. Aumen, J. R. Sedell, G. W. Lienkaemper, K. Cromack, Jr., and K. W. Cummins. 1986. Ecology of coarse woody debris in temperate ecosystems. Advances in Ecological Research 15:133–302.

Hartman, G. F., J. C. Scrivener, L. B. Holtby, and L. Powell. 1987. Some effects of different streamside treatments on physical conditions and fish population processes in Carnation Creek, a coastal rain forest stream in British Columbia. Pp. 330–372 In: Streamside management: forestry-fisheries interactions. E. O. Salo and T. W. Cundy (eds.). Seattle, WA: University of Washington Institute of Forest Resources.

Harvey, J. W., et al. 1991. Preliminary investigation of the effect of hillslope hydrology on the mechanics of solute exchange between streams and subsurface gravel zones. Pp. 413–418 In: U.S. Geological Survey Toxic Substances Hydrology Program—Proceedings of the technical meeting, Monterey, California, March 11–15, 1991. G. E. Mallard and D. A. Aronson (eds.). Water Resources Investigations Report 91-4034.

Harvey, J. W., and K. E. Bencala. 1993. The effect of streambed topography on surface-subsurface water interactions in mountain catchments. Water Resources Research 29:89–98.

Hauer, F. R., J. A. Stanford, J. J. Giersch, and W. H. Lowe. 2000. Distribution and abundance patterns of macroinvertebrates in a mountain stream: an analysis along multiple environmental gradients. Verh. Internat. Verein. Limnol. 27:1–4.

Hawkins, C. P., M. L. Murphy, and N. H. Anderson. 1982. Effect of canopy, substrate composition, and gradient on the structure of macroinvertebrate communities in Cascade Range streams of Oregon. Ecology 63:1840–1856.

Hayes, J. C., and J. E. Hairston. 1983. Modeling the long-term effectiveness of vegetative filters as on site sediment controls. ASAE Paper No. 83-2081. St. Joseph, MI: Am. Soc. Agric. Engrs. 27 pp.

Heinselman, M. L. 1970. Landscape evolution, peatland types, and the environment in the Lake Agassiz Peatlands Natural Area, Minnesota. Ecological Monographs 40:235–261.

Helmers, D. L. 1992. Shorebird management manual. Manomet, MA: Western Hemisphere Shorebird Reserve Network. 58 pp.

Henderson, C. L., C. C. Dindorf, F. J. Rozumalski. 1998. Landscaping for Wildlife and Water Quality. Minneapolis-St. Paul, MN: Minnesota Department of Natural Resources.

Hewlett, J. K., and A. R. Hibbert. 1967. Factors affecting the response of small watersheds to precipitation in humid areas. Pp. 275–290 In: International symposium on forest hydrology. W. E. Sopper and H. W. Lull (eds.). Oxford, England: Pergamon Press. 813 pp.

Hickin, E. J. 1984. Vegetation and river channel dynamics. Canadian Geographer 28:111–126.

Hicks, B. J., R. L. Beschta, and D. Harr. 1991. Long-term changes in streamflow following logging—western Oregon and associated fisheries implications. Water Resources Bulletin 27(2):217–226.

Hinkle, S. R., J. H. Duff, F. J. Triska, A. Laenen, E. B. Gates, K. E. Bencala, D. A. Wentz, and S. R. Silva. 2001. Linking hyporheic flow and nitrogen cycling near the Willamette River—a large river in Oregon, USA. J. Hydrol. 244:157–180.

Hoctor, T. S., M. H. Carr, and P. D. Zwick. 2000. Identifying a linked reserve system using a regional landscape approach: the Florida ecological network. Conservation Biology 14:984–1000.

Holstein, G. 1984. California riparian forests: deciduous islands in an evergreen sea. Pp. 2–22 In: California riparian systems. R. E. Warner and K. M. Hendrix (eds.). Berkeley, CA: University of California Press.

Hughes, J. W., and W. B. Cass. 1997. Pattern and process of a floodplain forest, Vermont, USA: predicted responses of vegetation to perturbation. Journal of Applied Ecology 34:594–612.

Hunt, R. J., D. P. Krabbenhoft, and M. P. Anderson. 1996. Groundwater inflow measurements in wetland systems. Water Resources Research 32(3):495–507.

Hunter, M. L., Jr. 1996. Fundamentals of Conservation Biology. Cambridge, MA: Blackwell Science. 482 pp.
Hupp, C. R. 2000. Hydrology, geomorphology and vegetation of Coastal Plain rivers in the southeastern USA. Hydrological Processes 14:2991–3010.
Hupp, C. R., and W. R. Osterkamp. 1985. Bottomland vegetation distribution along Passage Creek, Virginia, in relation to fluvial landforms. Ecology 66:670–681.
Hupp, C. R., and W. R. Osterkamp. 1996. Riparian vegetation and fluvial geomorphic processes. Geomorphology 14:277–295.
Huston, S. L. 1978. Prairie remnants along the Stillwater River in Miami County, Ohio. In: The Prairie Peninsula—in the "Shadow" of Transeau. Proceedings of the Sixth North American Prairie Conference, Ohio State University, Columbus, Ohio. Ohio Biological Survey Biological Notes No. 15.
Hyde, D. 1998. Stopover ecology of migratory birds along the central Lake Michigan shoreline. M.S. thesis, Central Michigan University, Mt. Pleasant, MI.
Inamdar, S. P., R. R. Lowrance, L. S. Altier, R. G. Williams and R. K. Hubbard. 1999. Riparian ecosystem management model (REMM): II. Testing of the water quality and nutrient cycling component for a Coastal Plain riparian system. Trans. Am. Soc. Agr. Eng. 42:1691–1707.
James, A., K. J. Gaston, and A. Balmford. 2001. Can we afford to conserve biodiversity? BioScience 51:43–52.
James, F. C. 1971. Ordination of habitat relationships among breeding birds. Wils. Bull. 83:215–236.
Jenkins, S. H. 1980. A size-distance relation in food selection by beavers. Ecology 61:740–746.
Jenkins, S. H., and P. E. Busher. 1979. Castor canadensis. Mammalian Species 120:1–9.
Johnson, A. S. 1989. The thin green line: riparian corridors and endangered species in Arizona and New Mexico. Pp. 35–46 In: Preserving communities and corridors. G. Mackintosh (ed.). Washington, DC: Defenders of Wildlife.
Johnson, G. D., and M. S. Boyce. 1990. Feeding trials with insects in the diet of sage grouse chicks. J. Wildl. Manage. 54:89–91.
Johnson, R. R., P. S. Bennett, and L. Haight. 1989. Southwestern woody riparian vegetation and succession: an evolutionary approach. Pp. 135–139 In: Proceedings of the California Riparian Systems Conference. D. L. Abell (ed.). USDA Forest Service General Technical Report PSW—110.
Johnson, W. C., R. L. Burgess, and W. R. Keammerer. 1976. Forest overstory vegetation and environment on the Missouri River floodplain in North Dakota. Ecological Monographs 46:59–84.
Johnston, C. A., and R. J. Naiman. 1990. Browse selection by beaver: effects on riparian forest composition. Can. J. For. Res. 20:1036–1043.
Jones, K. B., and P. C. Glinski. 1985. Microhabitats of lizards in a southwestern riparian community. Pp. 342–346 In: Riparian ecosystems and their management: reconciling conflicting uses. USDA Forest Service Gen. Tech. Bull. RM-120.
Jones, J. B., and P. J. Mulholland. 2000. Streams and ground waters. San Diego, CA: Academic Press.
Jordan, J. W. 2001. Late Quaternary sea level change in Southern Beringia: postglacial emergence of the Western Alaska Peninsula. Quaternary Science Reviews 20:509–523.
Judziewicz, E. J., and R. G. Koch. 1993. Flora and vegetation of the Apostle Islands National Lakeshore and Madeline Island, Ashland and Bayfield Counties, Wisconsin. The Michigan Botanist 32 (2):43–193.
Junk, W. J., Bayley, P. B., and Sparks, R. E. 1989. The flood pulse concept in river floodplain systems. Pp. 110–127 In: Proceedings of the International Large River Symposium. D. P. Dodge (ed.). Can. Spec. Publ. Fish. Aquat. Sci. 106.
Kalliola, R., and M. Puhakka. 1988. River dynamics and vegetation mosaicism: a case study of the River Kamajohka, northernmost Finland. Journal of Biogeography 15:703–719.

Kalliola, R., J. Salo, M. Puhakka, and M. Rajasilta. 1991. New site formation and colonizing vegetation in primary succession on the western Amazon floodplains. Journal of Ecology 79:877–901.

Karr, J. R., and R. R. Roth. 1971. Vegetation structure and avian diversity in several new world areas. Am. Nat. 105:423–435.

Kauffman, J. B., and W. C. Krueger. 1984. Livestock impacts on riparian ecosystems and streamside management implications. Journal of Range Management 37:430–437.

Kelsey, K. A., and S. D. West. 1998. Riparian wildlife. Pp. 235–258 In: River ecology and management: lessons from the Pacific Coastal ecoregion. R. J. Naiman and R. E. Bilby (eds.). New York: Springer-Verlag.

Keppeler, E. T., and R. R. Ziemer. 1990. Logging effects on streamflow-water yield and summer low flows at Caspar Creek in northwestern California. Water Resources Research 26:1669–1679.

Kirby, R. E. 1975. Wildlife utilization of beaver flowages on the Chippewa National Forest, North Central Minnesota. The Loon Winter:180–181.

Klebenow, D. A., and G. M. Gray. 1968. Food habits of juvenile sage grouse. J. Range Manage. 21:80–83.

Klingeman, P. C., R. L. Beschta, P. D. Komar, and J. B. Bradley. 1999. In: Gravel-bed rivers in the environment. Highland Ranch, CO: Water Resources Publications. 832 pp.

Klingeman, P. C., and J. B. Bradley. 1976. Willamette River basin streambank stabilization by natural means. Corvallis, OR: Oregon State University Water Resources Research Institute.

Korth, R., and P. Cunningham. 1999. Margin of error? Human influence on Wisconsin shores. Wisconsin Lakes Partnership (Wisconsin Association of Lakes, Wisconsin Department of Natural Resources, and University of Wisconsin-Extension). UWEX Lakes, University of Wisconsin, Stevens Point, WI.

Krabbenhoft et al., 1990. Estimating groundwater exchange with lakes—1. the stable isotope mass balance method. Water Resources Research 26(10):2445–2453.

Kratz, T. K., K. E. Webster, C. J. Bowser, J. J. Magnuson, and B. J. Benson. 1997. The influence of landscape position on lakes in northern Wisconsin. Freshwater Biology 37:290–17.

Labaugh, J. W., Winter, T. C., Swanson, G. A., Rosenberry, D. O., Nelson, R. D., and Euliss, N. H. 1996. Changes in atmospheric circulation patterns affect midcontinent wetlands sensitive to climate. Limnology and Oceanography 41(5):864–870.

Larkin, R. G., and J. M. Sharp. 1992. On the relationship between river-basin geomorphology, aquifer hydraulics, and ground-water flow direction in alluvial aquifers. Geological Society of America Bulletin 104:1608–1620.

Larue, P. L. Belanger, and J. Huot. 1995. Riparian edge effects on boreal balsam fir bird communities. Can. J. For. Res. 25:555–566.

Law, D. J., C. B. Marlow, J. C. Mosley, S. Custer, P. Hook and B. Leinard. 2000. Water table dynamics and soil texture of three riparian plant communities. Northwest Science 74(3):233–241.

Lawrence, W. H. 1952. Evidence of the age of beaver ponds. J. Wildlife Management 16:69–78.

Lee, D., T. A. Dillaha, and J. H. Sherrard. 1989. Modeling phosphorus transport in grass buffer strips. J. Environ. Eng. 115:409–427.

Lee, L. C. 1983. The floodplain and wetland vegetation of two Pacific Northwest river ecosystems. Ph.D. Dissertation, University of Washington, Seattle, WA.

Leidholt-Bruner, K., D. E. Hibbs, and W. C. McComb. 1992. Beaver dam locations and their effects on distribution and abundance of coho salmon fry in two coastal Oregon streams. Northwest Science 66:218–223.

Leopold, L. B., M. G. Wolman, and J. P. Miller. 1964. Fluvial Processes in Geomorphology. San Francisco, CA: W. H. Freeman and Co.

Li, R., and H. W. Shen. 1973. Effect of tall vegetation on flow and sediment. American Society of Civil Engineers Journal of the Hydraulics Division 99:793–814.

Lindsey, A. A., R. O. Petty, D. K Sterlin, and W. Van Asdall. 1961. Vegetation and environment along the Wabash and Tippecanoe Rivers. Ecological Monographs 31:105–156.

Long, C. A., and C. F. Long. 1992. Some effects of land use on avian diversity in a Wisconsin's oak–pine savanna and riparian forest. Passenger Pigeon 54:125–136.

Lowe, C. H. 1961. Biotic communities in the sub-Mogollon region of the inland Southwest. Journal of the Arizona Academy of Science 2:40–49.

Lowe, C. H. 1964. The Vertebrates of Arizona. Tucson, AZ: The University of Arizona Press. 270 pp.

Lowe, C. H. 1989. The riparianness of a desert herptofauna. USDA Forest Service General Technical Report PSW-110. Washington, DC: USDA Forest Service.

Lowrance et al., 1995. Water quality functions of riparian forest buffer systems in the Chesapeake Bay watershed. Report No. EPA-R-95-004. Annapolis, MD: Chesapeake Bay Program. 67 pp.

Lugo, A. E., S. L .Brown, R. Dodson, T. S. Smith, and H. H. Shugart. 1999. The Holdridge life zones on the conterminous United States in relation to ecosystem mapping. Journal of Biogeography 26:1025–1038.

Lytjen, D. 1998. Ecology of woody riparian vegetation in tributaries of the upper Grande Ronde River basin, Oregon. M.S. thesis, Oregon State University, Corvallis. 76 pp.

MacArthur, R. H. 1964. Environmental factors affecting bird species diversity. Am. Nat. 98:387–397.

MacDonald, B. C., and C. P. Lewis. 1973. Geomorphic and sedimentologic processes of rivers and coast, Yukon Coastal Plain. Geological Survey of Canada. Cat. No. R72-12173. Ottawa, Canada: Information Canada.

Machtans, C., M. Villard, and S. J. Hannon. 1996. Use of riparian buffer strips as movement corridors by forest birds. Conservation Biology 1366–1379.

MacNish, R. D., C. L. Unkrich, E. Smythe, D. C. Goodrich, and T. Maddock, III. 2000. Comparison of riparian evapotranspiration estimates based on a water balance approach and sap flow measurements. Agricultural and Forest Meteorology 105(1–3):271–280.

Malanson, G. P. 1993. Riparian Landscapes. Cambridge Studies in Ecology. Cambridge, UK: Cambridge University Press.

Maser, C., and J. R. Sedell. 1994. From the forest to the sea: the ecology of wood in streams. Delray Beach, FL: St. Lucie Press. 200 p.

Maser, C., R. F. Tarrant, J. M. Trappe, and J. F. Franklin (eds.). 1988. From the forest to the sea: a story of fallen trees. General Technical Report PNW-GTR-229. Portland, OR: USDA Forest Service. 153 pp.

Mattoon, W. R. 1915. The Southern Cypress. Bulletin 272. Washington DC: USDA. 74 pp.

McCullough, D. A. 1999. A review and synthesis of effects of alterations to the water temperature regime of freshwater life stages of salmonids, with special reference to chinook salmon. Water Resource Assessment EPA 910-R-99-010. Portland, OR: Columbia River Inter-Tribal Fish Commission. 291 pp.

McDowell, D. M., and R. J. Naiman. 1986. Structure and function of a benthic invertebrate stream community as influenced by beaver (*Castor canadensis*). Oecologia 68:481–489.

McGarigal, K., and W. C. McComb. 1992. Streamside versus upslope breeding bird communities in the central Oregon coast range. Journal of Wildlife Management 56:10–23.

McRae, G., and C. J. Edwards. 1994. Thermal characteristics of Wisconsin headwater streams occupied by beaver: implications for brook trout habitat. Transactions of the American Fisheries Society 123:641–656.

Meade, R. H., T. R. Yuzyk, and T. J. Day. 1990. Movement and storage of sediment in rivers of the United States and Canada. Pp. 255–280 In: Surface water hydrology. M. G. Wolman and H. C. Riggs (eds.). Boulder, CO; Geological Society of America.

Meyer, J. L., and J. B. Wallace. 2001. Lost linkages and lotic ecology: rediscovering small streams. Pp. 295–317 In: Ecology: achievement and challenge. M. C. Press, N. J. Huntly, and S. Levin (eds.). Oxford: Blackwell Scientific Publications.

Midwest Plan Service. 1985. Livestock Waste Facilities Handbook. Ames, IA: Iowa State University.

Minshall, G. W. 1988. Stream ecosystem theory: a global perspective. Journal of the North American Benthological Society 7:263–288.

Minshall, G. W. 1989. The ecology of stream and riparian habitats of the Great Basin region: a community profile. U.S. Fish and Wildlife Service Biological Report 85(7.24).

Moench, A. F., Sauer, V. B., and M. E. Jennings. 1974. Modification of routed streamflow by channel loss and base flow. Water Resources Research 10(5):963–968.

Montgomery, D. R., T. B. Abbe, J. M. Buffington, N. P. Peterson, K. M. Schmidt, and J. D. Stock. 1996. Distribution of bedrock and alluvial channels in forested mountain drainage basins. Nature 381:587–589.

Moore, F. R., and T. R. Simons. 1992. Habitat suitability and stopover ecology of neotropical landbird migrants. Pp. 345–355 In: Ecology and conservation of neotropical migrant landbirds. J. M. Hagan, III, and D. W. Johnston (eds.). Washington, DC: Smithsonian Institution Press.

Moran, M. S., and P. Heilman (eds.). 2000. Special Issue: Semi-Arid Land-Surface-Atmosphere (SALSA) Program. Agricultural and Forest Meteorology 105(1–3). 323 pp.

Morris, L. A. 1977. Evaluation, classification and management of the floodplain forest of south central New York. M.S. Thesis, SUNY College Environmental Science and Forestry, Syracuse, N.Y.

Mosley, M. P. 1983. Variability of water temperatures in the braided Ashley and Rakaia rivers. New Zealand Journal of Marine and Freshwater Research 17:331–342.

Moyle, P. B., and R. M. Yoshiama. 1994. Protection of aquatic biodiversity in California: a five-tiered approach. Fisheries 19:6–19.

Muldavin, E., P. Durkin, M. Bradley, M. Stuever, and P. Melhop. 2000. Handbook of wetland vegetation communities of New Mexico. Albuquerque, NM: New Mexico Heritage Program.

Mundie, J. H. 1969. Ecological implications of diet of juvenile coho in streams. Pp. 135–152 In: Symposium on salmon and trout in streams. T. G. Northcote (ed.). H. R. MacMillan Lectures in Fisheries, Univ. of British Columbia.

Murphy, M. L., C. P. Hawkins, and N. H. Anderson. 1981. Effects of canopy modification and accumulated sediment on stream communities. Transactions of the American Fisheries Society 110:469–478.

Murray, N. L., and Stauffer, D. F. 1995. Nongame bird use of habitat in central Appalachian riparian forests. J. Wildl. Manage. 59:78–88.

Naiman, R. J., and K. H. Rogers. 1997. Large animals and system-level characteristics in river corridors. BioScience 47:521–529.

Naiman, R. J., C. A. Johnston, J. C. Kelley. 1988. Alteration of North American streams by beaver. BioScience 38:753–761.

Naiman, R. J., D. G. Lonzarich, T. J. Beechie, and S. C. Ralph. 1992. General principles of classification and the assessment of conservation potential in rivers. Pp. 93–123 In: River conservation and management. P. Boon, P. Calow, and G. Petts (eds.). Chichester, UK: Wiley and Sons.

Naiman, R. J., H. Décamps, and M. Pollock. 1993. The role of riparian corridors in maintaining regional biodiversity. Ecological Applications 3:209–212.

Naiman, R. J., J. M. Melillo, and J. E. Hobbie. 1986. Ecosystem alteration of boreal forest streams by beaver (*Castor canadensis*). Ecology 67:1254–1269.

Naiman, R. J., R. E. Bilby, and P. A. Bisson. 2000. Riparian ecology and management in the Pacific coastal rain forest. BioScience 50:996–1011.

National Research Council (NRC). 1995. Wetlands: characteristics and boundaries. Washington, DC: National Academy Press.

NRC. 1996. Upstream: salmon and society in the Pacific Northwest. Washington, DC: National Academy Press.
Nehlsen, W., J. E. Williams, and J. A. Lichatowich. 1991. Pacific salmon at the crossroads: stocks at risk from California, Oregon, Idaho, and Washington. Fisheries 16:4–21.
Newbold, J. D., D. C. Erman, and K. B. Roby. 1980. Effects of logging on macroinvertebrates in streams with and without buffer strips. Can. J. Fish. Aquat. Sci. 37:1077–1085.
Newbold, J. D., J. W. Elwood, R. V. O'Neill, and W. Van Winkle. 1982. Nutrient spiraling in streams: implications for nutrient limitation and invertebrate activity. American Naturalist 120:628–652.
Nilsson, C. 1992. Conservation management of riparian communities. Pp. 352–372 In: Ecological principles of nature conservation. L. Hansson (ed.). London, England: Elsevier Applied Science.
Nilsson, C., R. Jansson, and U. Zinko. 1997. Long-term responses of river-margin vegetation to water-level regulation. Science 276:798–800.
Nolet, B. A., A. Hoekstra, and M. M. Ottenheim. 1994. Selective foraging on woody species by the beaver *Castor fiber*, and its impact on a riparian willow forest. Biological Conservation 70:117–128.
Noss, R. F., H. B. Quigley, M. G. Hornocker, T. Merrill, and P. C. Paquet. 1996. Conservation biology and carnivore conservation in the Rocky Mountains. Conservation Biology 10:949–963.
Odum, E. P. 1971. Fundamentals of Ecology. Philadelphia, PA: W. B. Saunders Company.
Ogle, D. W., and P. M. Mazzeo. 1976. *Betula uber*, the Virginia round-leaf birch, rediscovered in southwest Virginia. Castanea 41:248–256.
Ohmart, R. D. 1996. Historical and present impacts of livestock grazing on fish and wildlife resources in western riparian habitats. Pp. 245–279 In: Rangeland wildlife. P. R. Krausman (ed.). Denver, CO: Society for Range Management. 440 pp.
Ohmart, R. D., and B. W. Anderson. 1978. Wildlife use values of wetlands in the arid southwestern United States. Pp. 278–295 In: Wetland functions and values: the state of our understanding. Proceedings of the National Symposium on Wetlands. P. E. Greeson, J. R. Clark, and J. E. Clark (eds.). Minneapolis, MN: American Water Resources Association.
Ohmart, R. D., and B. W. Anderson. 1982. North American desert riparian ecosystems. Pp. 433–479 In: Reference handbook on the deserts of North America. G. L. Bender (ed.). Westport, CT: Greenwood Press. 594 pp.
Osborne, L. L., and D. A. Kovacic. 1993. Riparian vegetated buffer strips in water-quality restoration and stream management. Freshwater Biology 29:243–258.
Parker, M., F. J. Wood, Jr., B. H. Smith, and R. G. Elder. 1987. Erosional downcutting in lower order riparian ecosystems: have historical changes been caused by removal of beaver? Pp. 35–38 In: Riparian ecosystems and their management: reconciling conflicting uses. USDA Forest Service Gen. Tech. Bull. RM-120.
Parkin, T. B. 1987. Soil microsites as a source of denitrification variability. Soil Science Society of America Journal 51(5):1194–1199.
Patrick, R. 1995. Rivers of the United States. Volume II: Chemical and Physical Characteristics. New York: John Wiley and Sons, Inc.
Pauley, T. K. J. C. Mitchell, R. R. Buech, and J. J. Moriarty. 2000. Ecology and management of riparian habitats for amphibians and reptiles. Pp. 169–206 In: Riparian management in forests of the continental eastern united States. E. S. Verry, J. W. Hornbeck, and C. A. Dolloff (eds.). New York: Lewis Publishers.
Peterjohn, W. T., and D. L. Correll. 1984. Nutrient dynamics in an agricultural watershed: observations on the role of a riparian forest. Ecology 65:1466–1475.
Peterson, J. G. 1970. The food habits and summer distribution of juvenile sage grouse in central Montana. J. Wildl. Manage. 34:147–154.

Petryk, S., and G. Bosmajian III. 1975. Analysis of flow through vegetation. American Society of Civil Engineers Journal of the Hydraulics Division 101:871–884.

Phillips, J. D., Denver, J. M., Shedlock, R. J., and P. A. Hamilton. 1993. Effect of forested wetlands on nitrate concentrations in groundwater and surface water of the Delmarva Peninsula. Wetlands 13:75–83.

Piegay, H., and A. M. Gurnell. 1997. Large woody debris and river geomorphological pattern: examples from S.E. France and S. England. Geomorphology 19(1–2):99–116.

Pielou, E. C. 1991. After the ice age. Chicago: University of Chicago Press. 366 pp.

Pinay, G., A. Fabre, P. Vervier, and F. Gazelle. 1992. Control of C, N and P distribution in soils of riparian forests. Landscape Ecology 6:121–132.

Pinder, G. F., and S. P. Sauer. 1971. Numerical simulation of flood wave modification due to bank storage effects. Water Resources Research 7(1):63–70.

Poff, N. L., J. D. Allan, M. B. Bain, J. R. Karr, K. L. Prestegaard, B. D. Richter, R. E. Sparks, and J. C. Stromberg. 1997. The natural flow regime: a paradigm for river conservation and restoration. BioScience 47(11):769–784.

Pollock, M. M., R. J. Naiman, and T. A. Hanley. 1998. Plant species richness in riparian wetlands—a test of biodiversity theory. Ecology 79:94–105.

Post, R. A. 1996. Functional profile of black spruce wetlands in Alaska. EPA 910/R-96-006. Seattle, WA: U.S. Environmental Protection Agency.

Premo, D., Premo, B, Rogers, E. I., Tiller, D. J. 1992. The woodland vernal pond: an oasis of diversity. In: Total ecosystem management strategies Vol. 1(3), Amasa, MI: White Water Associates, Inc.

Ralph, S. C., G. C. Poole, L. L. Conquest, and R. J. Naiman. 1994. Stream channel morphology and woody debris in logged and unlogged basins of western Washington. Canadian Journal of Fisheries and Aquatic Sciences 51:37–51.

Råsånen, M. E., J. S. Salo, and R. J. Kalliola. 1987. Fluvial perturbance in the western Amazon basin: regulation by long-term sub-Andean tectonics. Science 238:1398–1401.

Reifsnyder, W. E., and H. W. Lull. 1965. Radian energy in relation to forests. USDA Forest Service, Technical Bull. No. 1344. 111 pp.

Remillard, M. M., G. K. Gruendling, and D. J. Bogucki. 1987. Disturbance by beaver (*Castor canadensis*) and increased landscape heterogeneity. Pp. 103–122 In: Landscape heterogeneity and disturbance. Ecological Studies, Vol. 64. M. G. Turner (ed.). New York: Springer-Verlag.

Rezendes, P., and P. Roy. 1996. Wetlands, the web of life. A Sierra Club Book. Burlington, VT: Verve Editions.

Rice, J., B. W. Anderson, and R. D. Ohmart. 1984. Comparison of the importance of different habitat attributes to avian community organization. J. Wildl. Manage. 48:895–911.

Rice, J., R. D. Ohmart, and B. W. Anderson. 1983. Habitat selection attributes of an avian community: a discriminant analysis investigation. Ecological Monographs 5:263–290.

Rogers, E. I., D. Tiller, and D. Premo. 1992. Mammal tracking study along the West Fence River in Iron County, Michigan. TEMS Research Brief. Total Ecosystem Management Program Annual Report. (In house report.) Amasa, MI: White Water Associates, Inc.

Rothacher, J. 1970. Increases in water yield following clear-cut logging in the Pacific Northwest. Water Resources Research 6(2): 653–658.

Rudolph, D. C., and J. G. Dickson. 1990. Streamside zone width and amphibian and reptile abundance. The Southwestern Naturalist 35:472–476.

Rutherford, I., B. Abernethy, and I. Prosser. 1999. Pp. 61–78 In: Riparian land management technical guidelines, volume one: principles of sound management. S. Lovett and P. Price (eds.). Canberra, Australia: Land and Water Resources Research Development Corporation.

Ruthsatz, B., and W. Haber. 1981. The significance of small-scale landscape elements in rural areas as refuges for endangered plant species. Proc. Int. Congr. Neth. Soc. Landscape Ecol. Veldhoven, Pudoc, Wageningen.

Rutledge, A. T., and T. O. Mesko. 1996. Estimated hydrologic characteristics of shallow aquifer systems in the Valley and Ridge, the Blue Ridge, and the Piedmont Physiographic Provinces based on analysis of streamflow recession and base flow. U.S. Geological Survey Professional Paper 1422-B. 54 pp.

Satterlund, D. R., and P. W. Adams. 1992. Wildland watershed management. New York: John Wiley & Sons, Inc. 436 pp.

Sauer, J. D. 1988. Plant migration: the dynamics of geographic patterning in seed plant species. Berkeley, CA: University of California Press. 298 pp.

Scatena, F. N., and A. E. Lugo. 1995. Geomorphology, disturbance, and the soil and vegetation of two subtropical wet steepland watersheds of Puerto Rico. Geomorphology 13:199–213.

Schade, J. D., S. G. Fisher, N. B. Grimm, and J. A. Seddon. 2001. The influence of a riparian shrub on nitrogen cycling in a Sonoran desert stream. Ecology 82:3363–3376.

Scharf, W. C., and J. Kren. 1997. Summer diet of Orchard Orioles in southwestern Nebraska. Southwestern Naturalist 42:127–131.

Schlesinger, W. H., and J. M. Melack. 1981. Transport of organic carbon in the world's rivers. Tellus 33:172–187.

Schlosser, I. J. 1982. Fish community structure and function along two habitat gradients in a headwater stream. Ecological Monographs 52(4):395–414.

Schlosser, I. J. 1991. Stream fish ecology: a landscape perspective. BioScience 41:704–712.

Schloz, J. 2001. The variability in stream temperatures in the Wenatchee National Forest and their relationship to physical, geological, and land management factors. M.S. Thesis. University of Washington, Seattle, WA.

Schultz, R. C., J. P. Colletti, T. M. Isenhart, W. W. Simpkins, C. W. Mize, and M. L. Thompson. 1995. Design and placement of multi-species riparian buffer strip system. Agroforestry Systems 29:201–226.

Schultz, R. C., J. P. Colletti, T. M. Isenhart, C. O. Marquez, W. W. Simpkins, and C. J. Ball. 2000. Riparian forest buffer practices. Pp. 189–281 In: North American agroforestry: an integrated science and practice. Madison, WI: American Society of Agronomy.

Schumm, S. A. 1960. The shape of alluvial channels in relation to sediment type. U.S. Geological Survey Professional Paper 352B, 17–30.

Scott, M. L., J. M. Friedman, and G. T. Auble. 1996. Fluvial process and the establishment of bottomland trees. Geomorphology 14:327–339.

Sebestyen, S. D., and R. L. Schneider. 2001. Dynamic temporal patterns of nearshore seepage flux in a headwater Adirondack lake. Journal of Hydrology 247(3–4):137–150.

Sedell, J. R., and K. J. Luchessa. 1982. Using the historical record as an aid to salmonid habitat enhancement. Pp. 210–223 In: Acquisition and utilization of aquatic habitat inventory information. N. B. Armantrout (ed.). Portland, OR: Western Division, American Fisheries Society.

Sedell, J. R., and R. L. Beschta. 1991. Bringing back the "bio" in bioengineering. American Fisheries Society Symposium 10:160–175.

Seefelt, N. 1997. Foraging behaviors and attack rates of American redstarts and black–throated green warblers. M.S. Thesis, Central Michigan University, Mt. Pleasant, MI.

Semlitsch, R. D. 1998. Biological delineation of terrestrial buffer zones for pond-breeding salamanders. Conservation Biology 12:1113–1119.

Seton, E. T. 1929. Lives of game animals. Vol. 4, Part 2. Rodents, etc. Garden City, New York: Doubleday.

Slough, B. G., and R. M. F. S. Sadleir. 1977. A land capability classification system for beaver (*Castor canadensis Kuhl*). Can. J. Zool. 55:1324–1335.

Smith, R., M. Hamas, M. Dallman, and D. Ewert. 1998. Spatial variation in the foraging of the black-throated green warbler along the shoreline of northern Lake Huron. The Condor 100:474–484.

Spackman, S. C., and J. W. Hughes. 1995. Assessment of minimum stream corridor width for biological conservation: species richness and distribution along mid-order streams in Vermont, USA. Biological Conservation 71:325–332.

Squillace, P. J. 1996. Observed and simulated movement of bank-storage water. Ground Water 34(1):121–134.

Stanford, J. A., and F. R. Hauer. 1992. Mitigating the impacts of stream and lake regulation in the Flathead River catchment, Montana, USA: An ecosystem perspective. Aquatic Conservation: Marine & Freshwater Ecosystems 2(1):35–63.

Stanford, J. A., and J. V. Ward. 1993. An ecosystem perspective of alluvial rivers: connectivity and the hyporheic corridor. Journal of the North American Benthological Society 12:48–60.

Stanford, J. A. 1998. Rivers in the landscape: introduction to the special issue on riparian and groundwater ecology. Freshwater Biology 40(3):402–406.

Stauffer, D. F., and L. B. Best. 1980. Habitat selection by birds of riparian communities: evaluating effects of habitat alterations. J. Wildl. Manage. 44:1–15.

Stebbins, R. C. 1966. A field guide to western reptiles and amphibians. Boston, MA: Houghton Mifflin Company. 279 pp.

Stine, S., D. Gaines, and P. Vorster. 1984. Destruction of riparian systems due to water development in the Mono Lake watershed. Pp. 528–533 In: California riparian systems. R. E. Warner and K. M. Hendrix (eds.). Berkeley, CA: University of California Press.

Strahler, A. N. 1952. Hypsometric (area-altitude) analysis of erosional topography. Bulletin of the Geological Society of America 63:1117–1142.

Stromberg, J. C., R. Tiller, and B. Richter. 1996. Effects of groundwater decline on riparian vegetation of semiarid regions: the San Pedro, Arizona. Ecological Applications 6:113–131.

Swanson, F. J., S. H. Johnson, S. V. Gregory, and S. A. Acker. 1998. Flood disturbance in a forested mountain landscape. BioScience 48(9):681–689.

Swanson, F. J., S. M. Wondzell, and G. E. Grant. 1992. Landforms, disturbance, and ecotones. Pp. 304–323 In: Landscape boundaries: consequences for biotic diversity and ecological flows. A. J. Hansen and F. di Castri (eds.). New York: Springer Verlag.

Szaro, R. C. 1991. Wildlife communities of southwestern riparian ecosystems. Pp. 174–200 In: Wildlife habitats in managed landscapes. J. E. Rodiek and E. G. Bolen (eds.). Washington, DC: Island Press.

Tockner, K., F. Mallard, and J. V. Ward. 2000. An extension of the flood pulse concept. Hydrological Processes 14:2861–2883.

Trimble, S. W. 1997. Stream channel erosion and change resulting from riparian forests. Geology 25(5):467–469.

Triska, F. J., V. C. Kennedy, R. J. Avanzino, G. W. Zellweger, and K. E. Bencala. 1989. Retention and transport of nutrients in a third-order stream in northwestern California: hyporheic processes. Ecology 70:1893–1905.

Troendle, C. A. 1983. The potential of water yield augmentation from forest management in the Rocky Mountain Region. Water Resources Bulletin 19:359–373.

Tweit, S. J. 2000. The next spotted owl? Audubon Nov.–Dec.

U.S. Department of Agriculture (USDA). 2000. Conservation buffers to reduce pesticide losses. Washington, DC: USDA Natural Resources Conservation Service. 21 pp.

USDA NRCS. 2001. Plants Database. Baton Rouge, LA: National Plant Data Center.

Utter, J. M., and A. W. Hurst. 1990. The significance and management of relict populations of *Chamaelirium luteum* (L.) Gray. In: Ecosystem management: rare species and significant habitats. R. S. Mitchell, C. J. Sheviak, and D. J. Leopold (eds.). Proceedings of the 15th Annual Natural Areas Conference. New York State Museum Bulletin No. 471. Albany, NY.

Van Cleef, J. S. 1885. How to restore our trout streams. Transactions of the American Fisheries Society 14:50–55.

Van Cleve, K., F. S. Chapin III, C. T. Dyrness, and L. A. Viereck. 1991. Element cycling in taiga forests: state-factor control. BioScience 41:78–88.

Vannote, R. L., G. W. Minshall, K. W. Cummins, J. R. Sedell, and C. E. Cushing. 1980. The river continuum concept. Canadian Journal of Fisheries and Aquatic Sciences 37:130–137.

Vervier, P. 1990. Hydrochemical characterization of the water dynamics of a karst system. Journal of Hydrology 121:103–117.

Vervier, P., M. Dobson, and G. Pinay. 1993. Role of interaction zone between surface and ground waters in DOC transport and processing: considerations for river restoration. Freshwater Biology 29:275–284.

Vervier, P., J. Gibert, P. Marmonier, and M.-L. Dole-Olivier. 1992. A perspective on the permeability of the surface freshwater–groundwater ecotone. Journal of the North American Benthological Society 11:93–102.

Vogt, R. C. 1981. Natural history of amphibians and reptiles in Wisconsin. Milwaukee, WI: The Milwaukee Public Museum. 205 pp.

Voss, E. G. 1972. Michigan flora, Part I, gymnosperms and monocots. Cranbrook Institute of Science Bulletin 55 and University of Michigan Herbarium.

Voss, E. G. 1996. Michigan Flora, Part III, Dicots (Pyrolaceae-Compositae). Cranbrook Institute of Science Bulletin 61 and University of Michigan Herbarium.

Wake, D. B. 1991. Declining amphibian populations. Science 253.

Walker, L. R., J. C. Zasada, and F. S. Chapin, III. 1986. The role of life history processes in primary succession on an Alaskan floodplain. Ecology 67:1243–1253.

Wang, L., Lyons, J., Kanehl, P, Gatti, R. 1997. Influences of watershed land use on habitat quality and biotic integrity in Wisconsin streams. Fisheries 22:6–12.

Ward, J. R., K. Tockner, and F. Schiemer. 1999. Biodiversity of floodplain river ecosystems: ecotones and connectivity. Regulated Rivers: Research and Management 15:125–139.

Ward, J. V. 1998. A running water perspective of ecotones, boundaries and connectivity. Verh. Internat. Verein. Limnol. 26:1165–1168.

Ward, J. V., K. Tockner, P. J. Edwards, J. Kollmann, A. M. Gurnell, G. E. Petts, G. Bretschko, and B. Rossaro. 2000. Potential role of island dynamics in river ecosystems. Verh. Internat. Verein. Limnol. 27.

Ware, G. H., and W. T. Penfound. 1949. The vegetation of the lower levels of the floodplain of the south Canadian River in Central Oklahoma. Ecology 30:478–484.

Warren, P. L., and C. R. Schwalbe. 1985. Herpetofauna in riparian habitats along the Colorado River in Grand Canyon. Pp. 347–354 In: Riparian ecosystems and their management: reconciling conflicting uses. USDA Forest Service Gen. Tech. Bull. RM-120.

Wells, J. R., and P. W. Thompson. 1974. Vegetation and flora of Keweenaw County, Michigan. The Michigan Botanist 13:107–151.

Wesche, T. A., C. M. Goertler, and C. B. Frye. 1987. Contribution of riparian vegetation to trout cover in small streams. North American Journal of Fisheries Management 7:151–153.

Wetzel, R. G. 1999. Plants and water in and adjacent to lakes. In: Eco-hydrology: plants and water in terrestrial and aquatic environments. A. J. Baird and R. L. Wilby (eds.). London: Routledge.

Wetzel, R. G., and M. Søndergaard. 1998. Role of submersed macrophytes for the microbial community and dynamics of dissolved organic carbon in aquatic ecosystems. Pp. 133–148 In: Role of submersed macrophytes in structuring the biological community and biogeochemical dynamics in shallow lakes. Dordrecht, Netherlands: Kluwer Publishers.

Wetzel, R. G. 2001. Limnology: lake and river ecosystems. San Diego, CA: Academic Press.

Wharton, C. H., Kitchens, W. M., and Sipe, T. W. 1982. The ecology of bottomland hardwood swamps of the southeast: a community profile. FWS/OBS-81/37. U.S. Fish and Wildlife Service. 133 pp.

Whitmore, R. C. 1975. Habitat ordination of passerine birds of the Virgin River Valley, southwestern Utah. Wils. Bull. 87:65–74.

Whittaker, R. H., and W. A. Niering. 1965. Vegetation of the Santa Catalina Mountains, Arizona: a gradient analysis of the south slope. Ecology 46:429–452.

Williams, K. K. C. Ewel, R. P. Stumpf, F. E. Putz, and T. W. Workman. 1999. Sea-level rise and coastal forest retreat on the west coast of Florida, USA. Ecology 80:(6)2045–2063.

Williamson, K. J., D. A. Bella, R. L. Beschta, G. Grant, P. C. Klingeman, H. W. Li, and P. O. Nelson. 1995. Gravel disturbance impacts on salmon habitat and stream health, Volume 1: Summary report. Corvallis, OR: Oregon Water Resources Research Institute, Oregon State University. 52 pp.

Wilson, L. G. 1967. Sediment removal from flood water by grass filtration. Transactions of the American Society of Civil Engineers 10:35–37.

Wilson, T. 2000. Vernal use of riparian habitats by neotropical migrants. M.S. Thesis, Central Michigan University, Mt. Pleasant, MI.

Wilzbach, M. A. 1985. Relative roles of food abundance and cover in determining the habitat distribution of stream-dwelling cutthroat trout (*Salmo clarki*). Can. J. Fish. Aquat. Sci. 42:1668–1672.

Wilzbach, M. A., and J. D. Hall. 1985. Prey availability and foraging behavior of cutthroat trout in an open and forested section of stream. Verh. Int. Ver. Limnol. 22:2516–2522.

Winter, T. C., J. W. Harvey, O. L. Franke, and W. M. Alley. 1998. Ground water and surface water: a single resource. USGS Circular 1139. Denver, CO: USGS.

Winter, T. C. 2001. The concept of hydrologic landscapes. Journal of the American Water Resources Association 37(2):335–349.

Wistendahl, W. A. 1958. The flood plain of the Raritan River, New Jersey. Ecological Monographs 28:129–153.

Wolfe, C. B., Jr., and J. D. Pittillo. 1977. Some ecological factors influencing the distribution of *Betula nigra* L. in western North Carolina. Castanea 42:18–32.

Wolock, D. M. 2001. USGS. Personal Communication. Reston, VA.

Wroblickly et al. 1998. Seasonal variation in surface-subsurface water exchange and lateral hyporheic area of two stream-aquifer systems. Water Resources Research 34(3):317–328.

Young, J. R. 1994. The influence of sexual selection on phenotypic and genetic divergence of Sage Grouse. Ph.D. diss. Purdue Univ., West Lafayette, Indiana.

Young, J. R., Braun, C. E., Oyler-McCance, S. J., Hupp, J. W., and T. W. Quinn. 2000. A new species of sage-grouse from southwestern Colorado. Wils. Bull. 112:445–453.

Zimmerman, R. C. 1969. Plant ecology of an arid basin: Tres Alamos-Redington area, southeastern Arizona. U.S. Geological Survey Professional Paper 485-D. 47 pp.

3

Human Alterations of Riparian Areas

Because humans worldwide now use more than half (~54 percent) of the geographically and temporally accessible river runoff (Postel et al., 1996), it is not surprising that we have had a significant impact on the structure and functioning of riparian areas. Human effects range from changes in the hydrology of rivers and riparian areas and alteration of geomorphic structure to the removal of riparian vegetation. Drastic declines in the acreage and condition of riparian lands in the United States since European settlement are testimony to these effects.

Manipulation of the hydrologic regimes that influence the physical and biological character of riparian systems has often occurred via the construction of dams, interbasin diversion, and irrigation. As discussed below, these activities disconnect rivers from their floodplains. A second major impact is related to the initial harvest of riparian areas, followed by subsequent conversion to other plant species via forestry, agriculture, livestock grazing, residential development, and urbanization. The removal of streamside vegetation not only removes the binding effects of roots upon the soil, but also causes a reduction in the hydraulic roughness of the bank and an increase in flow velocities near the bank (Sedell and Beschta, 1991). Such situations invariably lead to accelerated channel erosion during subsequent periods of high flow. Although degradation of native riparian plant communities by forestry, agriculture, and grazing can often be reversed, other practices such as drainage modifications and structural developments in urban areas generally lead to irreversible changes in riparian areas over long time periods.

The impacts to riparian areas are manifested in the quality of adjacent waterbodies throughout the United States. Only about two percent of the nation's streams and rivers are classified as having high water quality (Benke, 1990). A 1998 summary of polluted waters for all 50 states indicates there are more than 300,000 miles of rivers and streams and more than 5 million acres of lakes that do not meet state water-quality standards (EPA, 2000).

HYDROLOGIC AND GEOMORPHIC ALTERATIONS

Throughout history, societies have sought to regulate water resources. Today, over three-fourths of the 139 largest river ecosystems in the northern third of the earth are strongly or moderately fragmented by dams, interbasin diversions, and irrigation (Dynesius and Nilsson, 1994). In the contiguous 48 states, all large rivers greater than 1,000 km in length, except the Yellowstone River of Montana, have been severely altered for hydropower and/or navigation, and only 42 free-flowing river segments greater than 200 km in length remain (Benke, 1990). Disconnection of river systems from their historical floodplains is a severe problem worldwide about which there is limited but growing understanding (Naiman and Décamps, 1990).

Changes in natural hydrologic disturbance regimes and patterns of sediment transport include alteration of the timing of downstream flow, attenuation of peak flows, and other effects. Such alterations can result from dam construction, from transbasin diversions, or by water removal from rivers for irrigation or other consumptive uses, often in combination. For example, along the mainstem Columbia River in the Pacific Northwest, snowmelt peak flows have been suppressed by upriver storage facilities and the management of the river system for both power generation and flood control (NRC, 1996). Similarly, the Willamette River in Oregon has a reduced frequency of overbank flows, disconnected side channels, and greatly reduced potential for maintaining riparian and floodplain forests because of extensive bank stabilization and dam construction (Figure 3-1). Box 3-1 gives an example of the effects of various hydrologic manipulations on riparian plant communities and ecosystem processes in the arid Southwest.

The following sections discuss the specific effects of dams, bank-stabilizing structures, levees, and groundwater withdrawal on riparian structure and functioning. The extent to which downstream riparian areas are affected by these changes depends upon the degree of flow and sediment alteration plus the capability of the riparian plant communities to respond to these changing environmental conditions.

Dams

The vast majority of dam building and associated water resources development in the contiguous United States occurred during the middle portion of the

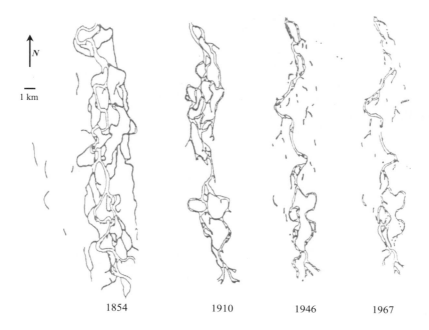

FIGURE 3-1 Channelization of the Willamette River since the 1800s has reduced channel complexity, riparian trees, and off-channel habitat. SOURCE: Reprinted, with permission, from Sedell and Froggatt (1984). © 1984 by Science Publishers.

twentieth century—an extremely short time period compared to the many thousands of years over which riparian plant communities have adapted to shifting climatic regimes, runoff patterns, and adjustments in channel morphology. There are currently 75,000 dams on the streams and rivers of the United States (Meyer, 1996; Graf, 1999), and large dams[1] worldwide are being completed at an estimated rate of 160 to 320 per year (World Commission on Dams, 2000). Dams have been constructed for hydropower generation, irrigation, flood control, domestic and industrial water use, recreational use, improved navigation, or some combination of these uses. Although detailed methods for the design of dams (e.g., Bureau of Reclamation, 1977) have been available for many years, such methods have provided little or no context for understanding the potential impacts such structures might have on other portions of a river and its riparian system.

---

[1]A large dam is 15 meters or more high (from the foundation). A dam 5–15 meters deep with a reservoir volume over 3 million cubic meters is also classified as a large dam. Using this definition, there are more than 45,000 large dams worldwide. (World Commission on Dams website: www.dams.org)

BOX 3-1
Effects of Multiple Hydrologic Changes

The effects of hydrologic manipulation on riparian area functioning have been particularly well documented along the middle Rio Grande (Shaw and Finch, 1996; Molles et al., 1998). Historically, the middle Rio Grande was a flood-dominated ecosystem. Spring snowmelt from the mountains of southern Colorado and northern New Mexico produced peak discharges between mid-May and mid-June, based on analysis of more than 100 years of flow records prior to impoundment (Slack et al., 1993). As in other floodplain systems, overbank flooding was an integral component controlling the structure of the riparian forest.

Given the relatively frequent flooding of the middle Rio Grande floodplain systems, the riparian area was a complex mosaic of vegetation types, including cottonwood (*Populous deltoides* ssp. *wislizenii*), Goodding willow (*Salex gooddingii*), wet meadows, marshes, and ponds. However, dam construction in the upper basins, river channelization, and water management policies of the twentieth century have cumulatively prevented annual spring flooding in recent decades. For the middle Rio Grande, the last major floods in which large-scale cottonwood establishment occurred were in the spring of 1941 and 1942. Thus, most of the current cottonwood gallery forest reflects a legacy of flooding that occurred over half a century ago.

Structural changes in the riparian vegetation have been rapid and well documented. For example, half of the wetlands in the middle Rio Grande have been lost in just 50 years (Crawford et al., 1993). Cottonwood germination, which requires scoured sandbars and adequate moisture from high river flows, has declined substantially (Howe and Knopf, 1991). Meanwhile, invasion by exotic phreatophytic plants such as saltcedar and Russian olive has greatly altered the species composition of the riparian forests within the valley. Native cottonwood stands are in decline in many sections of the river, and the cottonwood-dominated bosque at the Nature Center in Albuquerque has experienced a 40 percent decline in cottonwood leaf litterfall over the past decade (see figure below). Without a change in water management strategies, exotic species are predicted to dominate riparian forests within the next 50–100 years.

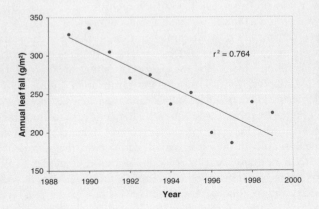

The immediate upstream effects of dam construction are obvious—the complete loss of riparian structure and functioning due to inundation, with other important changes in aquatic species, hydrology, and sediment dynamics of the inundated reaches. In particular, wildlife shifts from predominantly terrestrial species and stream-dwelling fish to predominantly lake dwelling fish. The streambank is replaced by extensive and often unstable shoreline in which floodplain vegetation is eliminated. Five percent of the total length of major rivers has been permanently inundated by large reservoirs, essentially removing their associated riparian areas (Brinson et al., 1981).

More recently, attention has been paid to the principal physical alterations of rivers downstream of dams (Rood and Mahoney, 1991). In general, dams reduce the biophysical variability (in flow, temperature, and materials transport) characteristic of rivers, which in turn reduces the biodiversity of both riparian and instream flora and fauna (Stanford et al., 1996). First, with regard to sediment dynamics, suspended sediment (clay, silt, and fine sand) and bedload sediment (coarse sand, gravel, and cobble) transported by a river settle in the slow-moving waters of a reservoir. Although their trapping effectiveness can vary somewhat, most reservoirs are effective at trapping silt-sized and larger particles. If residence times of the stored water are relatively long, large reservoirs may also be effective at trapping clay-sized particles. Over long periods, the channels below a dam can become increasingly "sediment starved," with a concurrent coarsening of sediments comprising the channel bottom. Following impoundment, a reduction in the sediment load can prevent the regular development of such geomorphologic features as point bars and islands in larger scale rivers, as was demonstrated in the Slave River Delta (English et al., 1997).

Although this is the general paradigm, actual changes depend on local conditions downstream from a dam. For example, if high flows have been suppressed by an upstream dam, sediment-laden tributaries that enter a river below the dam may cause large amounts of sediment to accumulate in the main river. In essence, a loss of river transport capacity due to flow modifications by the upstream dam encourages incoming tributary sediments to accumulate over time.

A second category of downstream alteration is related to the pattern of river flow, where the magnitude of such effects is largely dependent upon the degree of hydrologic alteration created by the dam. Dams that are used only for flood control and hydropower generation may not significantly diminish the amount of water available to downstream channels, although these structures can have a major effect on the overall flow regime (the frequency, magnitude, and temporal distribution of flows). For example, flood-control dams that store water during periods of peak runoff for later release will dampen the magnitude of high flows that would occur normally and increase the duration of moderate flows. Large flood-control dams can effectively accomplish this goal over a wide range of peak flow magnitudes (although the effectiveness of a given dam for dampening downstream peaks tends to diminish with increasingly larger precipitation events).

streambank protection methods involve such structural approaches as rip-rap, concrete, dikes, fences, asphalt, gabions, matting, and bulkheads; less than 15 percent of the information was directed towards the use of vegetation. Unlike options utilizing vegetation, structural approaches to streambank stabilization can have deleterious effects on riparian areas (Sedell and Beschta, 1991; Fischenich, 1997).

Rip-rap (large rock, pieces of concrete, or other material) remains a common solution for "hardening" a streambank or shoreline in an effort to stem erosion. It is also utilized to stabilize streambanks in the vicinity of bridge abutments, culvert installations, or other features in need of special protection from erosion during high flows. Rarely are the ecological impacts of such projects considered, either for individual projects or cumulatively where multiple projects are implemented. Rip-rap affects the riparian habitat directly by eliminating microhabitats of plant species that naturally stabilize banks. The large pore sizes typically associated with rip-rap treatments seldom contain soil and thus create poor substrates for plant establishment and growth. In addition, because many bank structures reduce the hydraulic roughness (i.e., the frictional resistance to flow) along the channel margins, flow velocities are greater along the bank during high flows, which often precludes the survival of many riparian plant species.

With the loss of riparian vegetation brought about by structural modification of a streambank, important contributions of that vegetation to the aquatic ecosystem (e.g., shading, leaf fall, structural integrity from roots, nutrient inputs) are

reduced, as are its functions as habitat for animals that commonly use streambanks and shorelines. Rip-rap can impede movement of animals that use streambanks and shorelines as migration corridors and destroy nesting areas, as has been documented for the wood turtle (Buech, 1992). Avifaunal studies along the Colorado River showed that, on average, the number of species inhabiting a rip-rapped riparian area was only about half that of an undisturbed river with intact riparian vegetation (Ohmart and Anderson, 1978).

In some cases, the use of rip-rap can have a deleterious effect on water quality. For example, runoff channels constructed of rip-rap or impervious materials can shunt water from roadways, other impermeable surfaces, or erosion-prone areas directly into nearby streams and rivers. Such warmed and often pollutant-laden water enters the river without the benefit of having been filtered by vegetation or soil of the riparian area.

Channelization

Channelization converts streams into deeper, straighter, and often wider waterbodies, making fundamental geomorphic and hydrologic transformations that would not occur under natural conditions. The most common purpose of channelization of small streams is to facilitate conveyance of water downstream so that the immediate floodplain area will not flood as long or as deeply, resulting in reduced soil water content. Channelization is widespread throughout the United

States. Prior to 1970, an estimated 200,000 miles of streams were channelized (Schoof, 1980). In Maryland, 17 percent of all stream miles have been channelized; the Pocomoke River has an estimated 81 percent of its stream miles channelized.

Channelization has the direct effect of destroying riparian vegetation by the actions of heavy equipment or by moving the stream channel to a new location where no natural riparian vegetation exists. Indirectly, channelization impacts riparian vegetation by lowering the water table (Gordon et al., 1992) and otherwise altering riparian hydrology. Because channelization reduces the frequency of overbank flow, the adjacent riparian area becomes drier and the connection between aquatic and terrestrial ecosystems is severed. In addition, channelization often creates a spoil bank next to the newly cut channel that further blocks exchanges between the now-isolated floodplain and the channel. This combination of conditions can eliminate aquatic sites within floodplains, especially high-flow channels and oxbow lakes that would otherwise serve as temporary or permanent habitat for aquatic organisms. Any riparian vegetation left intact during channelization is likely to experience drought stress and eventually be replaced by less flood tolerant species, a phenomenon similar to that which occurs in floodplains below dams (described earlier).

The increased flow capacity afforded by channelization compresses the period of water conveyance, making streams "flashier." Downstream effects include higher flood peaks and greater loading of sediment, nutrients, and contaminants. Locally, the kinetic energy of water flow is concentrated in the stream channel rather than being dissipated across the floodplain during normal overbank flows. In the absence of streambank stabilization, the channelized reach may undergo a period of accelerated erosion that can lead to additional channel incision or channel widening, or both. Channel incision is of particular concern as it often leads to a headward incision or gullying of the original channel. Thus, the effects of channelization are experienced both upstream (gullying) and downstream (increased sediment production).

Figure 3-3 illustrates the channelization scenario in western Tennessee around the 1900s in which increased channel flow caused streambank erosion and headward channel incision (Hupp and Simon, 1991). An initial stage of headward incision downcutting was initiated by channelization just downstream from the site depicted. After about 50 years, a new geomorphic surface and quasi-equilibrium condition became established. Both bank accretion and regrowth of vegetation were responsible for recovery of the riparian area.

Levees

Levees are large embankments along rivers or other waterways that are designed to not be overtopped during periods of high water and are largely employed for flood-control purposes on land along a stream, river, or other body

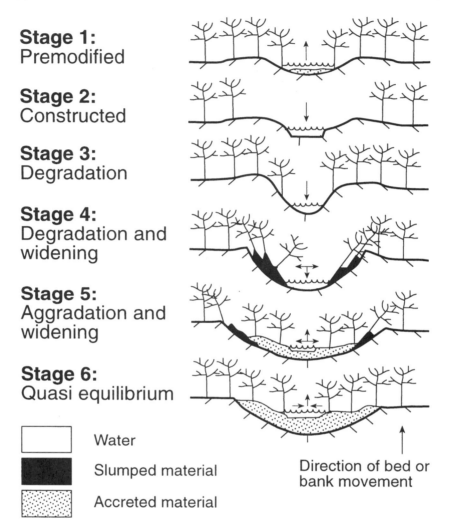

FIGURE 3-3 Stages of stream and floodplain evolution following channelization that occurred in western Tennessee streams around the 1900s. Sites depicted are just upstream from the channelization. Arrows indicate whether aggradation or degradation is taking place. SOURCE: Reprinted, with permission, from Hupp and Simon (1991). © 1991 by Elsevier Science.

of water. Like large dams, large levees are built to eliminate the occurrence of overbank flows, thus curtailing the periodic flow of water, nutrients, sediment, and organic matter between the channel and its riparian system. Except for unusually large and infrequent hydrologic events, levees are normally effective at severing the hydrologic linkages (i.e., frequency, magnitude, and duration of overbank flows) between a channel and its adjacent riparian areas. In riverine systems, levees tend to be linear features because most are constructed parallel to the river system. The U.S. Army Corps of Engineers has built over 10,500 miles of levees and floodwalls, most of which have then been assigned to non-federal sponsors for operation and maintenance (NRC, 1982). Levees are a ubiquitous feature of the United States and exist on streams of all sizes.

Levees can be broadly classified according to their use or purpose (mainline, tributary, ring, and setback levees and spur levees), according to the type of lands being protected (urban or agricultural), or according to their method of construction (compacted, subcompacted, and uncompacted) (Petersen, 1986). Mainline and tributary levees are those constructed along mainstem rivers and tributaries, respectively. Ring levees are used to completely encircle an area subject to inundation from all directions. Setback levees are built some distance landward from the channel's bank or edge of water. From an ecological perspective, setback levees have important advantages over those constructed immediately along the streambank in that they allow many of the riparian functions to still occur while still protecting most areas of concern. Spur levees project from a mainline levee toward the streambank and act to divert streamflow away from mainline levees.

Because most mainline and tributary levees are constructed close to the streambank (and may employ rip-rap, concrete, fill, or other coarse material for stabilizing the bank), they typically result in the nearly total destruction of riparian plant communities. In addition, the streamward side of the constructed levee is often maintained free of riparian vegetation or is constructed in a manner whereby riparian vegetation can no longer establish and grow. As a consequence, the area behind the levee (the landward side) becomes hydrologically disconnected from the river.

If levees can be set back (i.e., constructed some distance from the bank), particularly if they are located outside the general meander-belt of a river (see Figure 2-3), their impacts to ecological and hydrologic functions can be greatly reduced. Setback levees generally allow for natural riparian plant communities and normal floodplain dynamics by maintaining relatively frequent overbank flows, providing detention storage of flood water, and allowing for deposition of fine sediments along the entire streambank and at least a portion of the floodplain. In essence, setback levees represent a compromise between the development goals of protecting floodplain areas from overbank flows and the ecological goals of maintaining riparian and floodplain functions. Large portions of a floodplain system can be protected from overbank flows and inundation while still allowing for the maintenance of riparian and floodplain functions between the channel and the levee. By placing these levees away from the channel, construction costs are often reduced (because smaller levees may suffice), and the natural long-term adjustments in the morphology and migration of channels can occur unhindered.

Surface Water and Groundwater Withdrawals

Withdrawals, both from surface waters and groundwater, can have serious deleterious effects on riparian area functioning because of the lowering of water tables in the vicinity of riparian vegetation. Groundwater pumping for municipal and industrial water supply and agriculture throughout large areas of the West is increasingly common, as appropriate sites for dam construction on surface waters dwindle. Assessments of impacts of groundwater withdrawal rarely take riparian areas into account.

Because groundwater and surface water are generally connected in floodplains, declines in groundwater level can also come about as an indirect effect of surface water withdrawals or of regulation of surface water flow by dam construction (Rood et al., 1995). Other mechanisms that can cause groundwater declines beneath floodplains include sand and gravel mining in channels, which lowers the elevation of river channel beds (Meador and Layher, 1998), or downcutting of the channel bed either naturally or as an adjustment to the engineered straightening of channels (Bravard et al., 1997). In these situations, lowering of surface water levels in the channel produces a similar effect on the groundwater system. Initially, there is a temporary increase in hydraulic gradients and groundwater discharge to the channel. Usually the increased discharge only partly compensates for the lowering of the surface water level. Eventually, the increased discharge of groundwater to the channel lowers groundwater levels beneath the floodplain until a new equilibrium is achieved.

Decreases in groundwater levels of just one meter or more beneath riparian areas are enough to induce water stress in some riparian trees, especially in the western United States. For example, sustained declines in the water table of greater than one meter are likely to cause leaf desiccation in cottonwoods, leading to branch die-back and eventual mortality for a significant proportion of the population (Scott et al., 1993). Groundwater pumping for water supply in the West has caused a decline in the number of miles of river with perennial streamflow that can most easily support healthy riparian forests (Luckey et al., 1988). The lowering of water tables via groundwater pumping has aggravated problems caused by the invasion of exotic, drought-tolerant plants. Portions of the middle San Pedro River affected by lowering groundwater tables and reduced stream flow have seen an increase in the relative abundance of saltcedar compared to native Freemont cottonwood (Stromberg, 1998).

Phreatophyte Control and Eradication

Phreatophytic (water-loving) plants historically have been cleared from riparian areas in arid and semiarid climates because they have been viewed as competing with other users of water, particularly irrigated agriculture and municipalities. In arid climates where there is no excess precipitation (see Plate 2-3), water availability limits the species composition and productivity of riparian areas. Phreatophyte eradication has been used to supplement water availability by suppressing the amount of water that is transported from groundwater to the atmosphere via plants.[2]

The Gila River in Arizona has been the site of studies on vegetation removal beginning over 50 years ago (Turner and Skibitzki, 1952). Gatewood et al. (1950) estimated that losses of water from evapotranspiration were as much as five times greater than water loss due to evaporation from the river surface and wet sand bars. By 1963, it was acknowledged that specific conditions were necessary in order for phreatophyte removal to successfully augment stream flow (Rowe, 1963). For example, the water supply must exceed evapotranspiration after plants are removed (i.e., the stream does not go dry under normal conditions). Second, the water table must be high enough for riparian plants to reach it, or their removal will have no effect on water availability. Even where these conditions are met, phreatophyte eradication destroys nearly all ecological and geomorphic benefits provided by riparian vegetation, including stabilization of alluvial fill, shading, and provision of wood and microhabitats.

Although evapotranspiration is an inevitable consequence of the growth of riparian vegetation, it may be insignificant in comparison to other water uses. For example, reservoirs for surface water storage can lead to water losses by evaporation that may exceed those caused by evapotranspiration. For example, in the Middle Rio Grande in New Mexico, evaporation from Elephant Butte Dam is a larger component of loss before delivery to downstream users than is evapotranspiration from riparian vegetation (Figure 3-4). Increasing regulatory constraints on stream and floodplain alteration, and more limited access to public and especially federal funds, have also resulted in a decrease in or elimination of large-scale phreatophyte-eradication programs. For these reasons, current efforts are comprised of comprehensive studies assessing the role that riparian vegetation plays in ecosystem processes and even attempts to restore and enhance phreatophytes, especially in urban areas where riparian vegetation has been degraded or eliminated (J. Stromberg, Arizona State University, personal communication, 2001).

Phreatophyte control continues today for reasons such as reducing mosquito habitat. Saltcedar is the species most often targeted for control, partly because it

---

[2]Riparian vegetation has been removed for numerous reasons other than enhancing water supply, such as for the planting of crops or to create grazing areas. Although these activities have detrimental impacts on riparian areas, they are not the focus of this discussion.

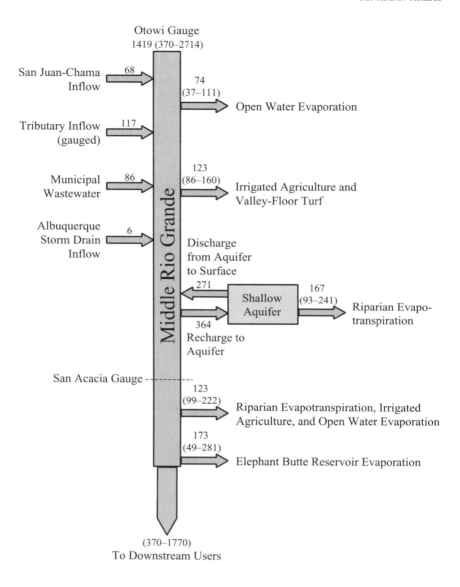

FIGURE 3-4 The Middle Rio Grande water budget with units of 10^6 m^3 yr^{-1}.

is an exotic species and because it happens to dominate in areas where attention is focused on phreatophyte issues. Saltcedar represents a particularly troublesome case because of the perceived advantages and disadvantages of removing it from riparian areas. On the one hand, saltcedar is an exotic whose removal not only could increase floodwater conveyance (although this has yet to be demonstrated), but could also provide more habitat for native plant species. At the same time, saltcedar has been proposed for protection because of its role as nesting habitat for the southwestern willow flycatcher, a federally listed endangered species (Leon, 2000). The Bureau of Reclamation has been controlling saltcedar on about 40,000 acres in New Mexico since the 1960s, but it has been unable to demonstrate positive changes in streamflow as a result, probably because of other factors such as groundwater pumping. Insects for biological control of saltcedar are now being released on a trial basis (DeLoach, 2000; J. Stromberg, Arizona State University, personal communication, 2002).

AGRICULTURE

Traditional Agricultural

Nationwide, agriculture is probably the largest contributor to the decline of riparian area quality and functioning (Dillaha et al., 1989). Because some of the most fertile soils are often located in riparian areas, there is an economic incentive for their conversion to cropland. These areas are also convenient sources of water for irrigation of adjacent cropland and, as previously discussed, excessive water withdrawal from streams lowers water tables and causes significant change to riparian area structure and functioning. In nonforested areas, there can be a tendency to encroach into the riparian area each time the field is plowed in an attempt to gain more cropland. Natural riparian areas are sometimes viewed as a potential source of plant and animal pests, a source of shade that may reduce crop yields, and competition for scarce water resources. In areas where agriculture is concentrated, such as the Midwest, these activities have converted millions of acres of native grasslands, prairie, and wetlands, including riparian areas, into croplands.

Direct effects of agricultural management practices on riparian areas are listed in Figure 3-5 and illustrated in Figure 3-6. Under natural settings, riparian vegetation protects the soil surface, and soil fauna and flora are constantly creating macropores, which maintains high infiltration and percolation rates. When land is converted to agriculture—particularly row crops—vegetative cover is reduced, which exposes soil to raindrop impact and surface sealing, thereby decreasing infiltration. Although agricultural tillage does help to maintain porosity in soil, which promotes infiltration and percolation, it does not do so to the extent achieved by undisturbed populations of soil flora and fauna. The machinery used in tilling can also compact soils. Together, these practices alter the

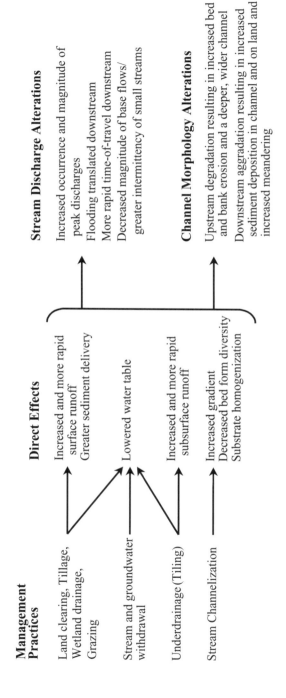

FIGURE 3-5 Alterations in stream discharge and morphology brought about by agricultural land-use practices. SOURCE: Adapted from Menzel (1983).

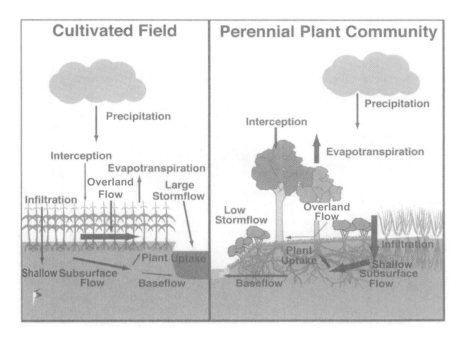

FIGURE 3-6 Differences in water movement in a non-tiled annual row-crop field and a perennial riparian forested buffer. More overland flow and less total evapotranspiration result in larger storm flow in the row-crop field while in the perennial riparian plant community, higher rates of infiltration and annual evapotranspiration reduce storm flow and increase baseflow. SOURCE: Reprinted, with permission, from Schultz et al. (2000). © 2000 by American Society of Agronomy.

hydrology by increasing overland flow volumes, peak runoff rates, and potential pollutant delivery to riparian areas. Stream channels respond to these increased runoff frequency, volumes, and peak flow rates by increasing their cross-sectional area to accommodate the higher flows—either through widening of the stream channel, downcutting of the streambed, or both—similar to what is observed during channelization (see Figure 3-3).

The altered hydrology characteristic of row-crop agriculture and some erosion control structures tends to concentrate overland flow within fields and transport it downslope in grass waterways or other ephemeral drainageways (Schultz et al., 2000). Although grass waterways are very effective in reducing gully erosion, transformation processes that could improve water quality are limited because the upland runoff enters the riparian area as concentrated flow. Also, the increased overland flow over agricultural land promotes relatively high erosion rates, such that adjacent riparian areas trap substantial amounts of sediment (Dillaha et al., 1989). Over time, the upslope portion of the riparian area evolves into a terrace or berm that, if not managed via tillage, can hinder further inflow. When this occurs, runoff flows parallel to the riparian area until a low point or drainageway is reached. The diverted overland flow enters the riparian area as concentrated flow, which again reduces its effectiveness for water-quality protection.

Agricultural chemicals (both pesticides and fertilizers) in overland flow can also negatively impact fauna and flora located in riparian areas and downstream receiving waters. Edge-of-field pesticide losses are common, with 1–10 percent of the amount of pesticide applied being entrained in overland flow (Wauchop, 1978; Baker, 1983). Similarly, fertilization can cause nutrient losses from the land to nearby streams to increase by an order of magnitude or more.

Healthy riparian areas often provide significant benefits to traditional agricultural activities. Riparian areas protect the quality of water resources used for agricultural and domestic purposes by trapping sediment, nutrients, and other pollutants. They stabilize stream channels and they promote the infiltration of overland flow. They increase groundwater resources by enhancing groundwater recharge in losing streams. They can reduce wind erosion; trap snow, thus reducing drifting; protect livestock, wildlife, and buildings from excessive wind; and reduce noise and odors associated with some agricultural activities. Riparian areas can also be a potential source of income through their use for hunting and fishing and for timber and biomass production. Unfortunately, these benefits have historically not played a role in agricultural management of riparian lands.

Drainage Tiles and Ditches

The draining of water from urban and suburban lands for the purposes of improved crop production has been practiced since the 1870s, spurred by the 1849 and 1850 Swamp Act and the subsequent organization of local drainage districts. Farmers have relied on drainage to improve soil aeration, alter soil moisture conditions to allow earlier planting and easier fall tillage, and combat disease organisms that thrive in high-moisture conditions. Without drainage, many Midwest farmlands would be significantly reduced in productivity or simply unfarmable (Fausey et al., 1995).

Drainage occurs through subsurface tiles (e.g., perforated polyethylene pipe or other older methods such as clay tiles) or by networks of ditches. In practice, surface and subsurface systems often are used together. For example, drainage tiles often intercept channelized streams or ditches created for the purpose of collecting tile outflow. Table 3-2 shows the acreage of drained land in the most heavily affected regions of the United States. Drainage impacts approximately 20.8 million hectares or 37 percent of the 55.7 million hectares of cropped farmland in the Midwest (Pavelis, 1987; Zucker and Brown, 1998). In Illinois, the

TABLE 3-2 Agricultural Drainage for the Most Heavily Drained States

State	Harvested Cropland (1,000 ha)[a]	Drained Cropland (1,000 ha)[b]	Percent of Cropland Drained	State Total Area (1,000 ha)[c]
Great Lakes and Cornbelt States				
Illinois	9,014	3,569	40	35,580
Indiana	4,742	2,782	59	22,957
Iowa	9,439	2,834	30	35,760
Ohio	4,007	2,397	60	26,209
Minnesota	7,677	1,934	25	50,954
Michigan	2,721	1,563	57	36,358
Missouri	5,038	1,202	24	44,095
Wisconsin	3,491	409	12	34,761
Mississippi Delta				
Arkansas	3,102	2,151	69	33,328
Louisiana	1,571	1,562	99	27,882
Mississippi	1,756	1,440	82	30,025
Southeast				
Florida	986	1,146	100	34,558
North Carolina	1,713	984	57	31,180
South Carolina	670	426	64	19,271
Georgia	1,523	219	14	37,068
Other States				
North Dakota	8,271	910	110	44,156
Texas	7,935	1,283	16	167,625
Tennessee	1,645	256	16	26,380
New York	1,504	333	22	30,223
Maryland	559	367	66	6,256
Delaware	189	130	69	1,251
U.S. Total	125,212			2,263,587

[a]From 1997 National Agricultural Statistics for harvested cropland, which includes land from which crops were harvested or hay was cut and land in orchards, citrus groves, Christmas trees, vineyards, nurseries, and greenhouses. NAS also reports total cropland, which includes cropland used for pasture or grazing, land in cover crops, legumes, and soil-improvement grasses, land on which all crops failed, land in cultivated summer fallow, and idle cropland.
[b]From Pavelis (1987) converted to metric units and rounded to nearest 1,000 ha.
[c]From USDA (1997).

state with the greatest amount of drained farmland acres, it is estimated that over 4 million hectares are drained by a vast network of underground drainage tiles. In some highly drained areas, such as the Embarras River watershed of east central Illinois, tiles drain 70 percent to 85 percent of the cropland (David et al., 1997). Other areas such as the Southeast (6 million hectares) and the Mississippi Delta (5 million hectares) also have significant areas of drained cropland.

Because drainage was traditionally a tool for managing soil moisture, the resulting water quality of receiving streams and other ecological factors were

rarely if ever considered. It is now known that drainage has had a dramatic impact on stream hydrology and water quality and on the functioning of riparian areas (Evans et al., 1995; David et al., 1997; Kovacic et al., 2000). By concentrating flows and circumventing the biological processes that typically occur in riparian areas, drainage tile effluent can have greater peak flows, increased concentrations of nutrients, and either increased (surface drainage) or decreased (some types of subsurface drainage) sediment load. Many of the effects of surface drainage are similar to those discussed above for channelization and traditional agriculture.

The hydrologic differences among drained cropland, non-drained cropland, and undisturbed land have been investigated by Zucker and Brown (1998). Compared to non-drained cropland, tile-drained cropland has less erosion and phosphorus runoff because of limited overland flow. However, in relation to non-cropped areas or cropped areas with various conservation practices, the environmental advantages of tile drainage are less clear or nonexistent. For example, studies in North Carolina have shown that compared to undisturbed sites, total outflow is increased by 5 percent with surface drainage and 20 percent for subsurface drainage (Evans et al., 1995). Evans et al. (1995) found that both total flow and peak outflow were increased in drained areas compared to undeveloped areas. Depending upon conditions—such as antecedent soil moisture and storm intensity—surface and subsurface drainage were found to increase peak outflow rates by four and two times, respectively (Figure 3-7). This increased outflow often results in streambank erosion, channel incision, flooding, or other impacts.

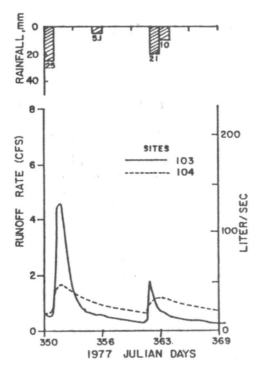

FIGURE 3-7 Increase in peak outflow rates typically associated with drainage and land conversions to agriculture. Site 104 is a natural, undrained site and Site 103 is a surface drained and developed pocosin converted to agricultural use. SOURCE: Reprinted, with permission, from Evans et al (1995). © 1995 by American Society of Civil Engineers.

Indeed, the changes in hydrology characteristic of extensively tiled areas can be so extreme that in many first- to third-order streams, flow from drainage tiles may constitute 90 percent of the baseflow during summer months (Schultz et al., 2000).

Drainage also affects the transport of particles and chemical pollutants through riparian areas. As shown schematically in Figure 3-8, subsurface drainage can expedite direct transport of chemicals (such as NO_3-N) from the soil zone to surface waters—often completely circumventing riparian areas. Thus, approximately 37 percent of the cropped land in the Midwest is not afforded the beneficial nutrient absorbing and transforming processes of riparian areas. As a result, where nutrients are added to cropland, they often are delivered to the stream systems at highly elevated levels (David et al., 1997; Kovacic, 2000). Surface drainage systems typically produce higher concentrations of phosphorus and sedi-

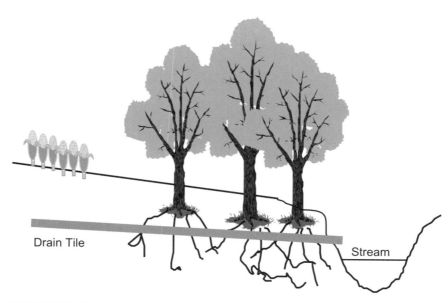

FIGURE 3-8 Short-circuiting of the riparian area by a drainage tile. Drainage tiles typically bypass the functioning of riparian areas by conveying water directly from upland areas to the stream systems. Tiles prevent riparian-related activities such as denitrification, and they enhance water conveyance, resulting in higher peak flows and greater total runoff. SOURCE: Reprinted, with permission, from Kovacic (2000).

ment than do subsurface systems, while subsurface systems typically contain higher concentrations of NO_3-N than do surface systems (Evans et al., 1991; Thomas et al., 1995). The short-circuiting of riparian areas via drainage is especially troubling in areas like the Midwest where soils are underlain by an impermeable aquiclude (Schultz et al., 2000). In such places, the riparian area may constitute the only biologically active zone through which pollutants from cropland could be transformed. The high nutrient loadings resulting from drainage networks have been implicated in the hypoxia in the Gulf of Mexico (Turner and Rabalais, 1991) as described in Box 3-2.

Grazing

Domestic Livestock

The history of grazing by domestic livestock in much of the world has been one of large-scale degradation of native plant communities (Chaney et al., 1990; Kauffman and Pyke, 2001). Although domesticated livestock have played a

> **BOX 3-2**
> **Hypoxia in the Gulf of Mexico**
>
> The hypoxic zone in the Gulf of Mexico has increased in size since the 1950s, nearly doubling in average size from 1985–1992 to 1993–1999. The area, defined by dissolved oxygen levels of less than 2 mg/L, averaged 5,500 mi^2 (14,000 km^2) in size over the 1996–2000 period, and is found off the Louisiana coast near the outflow areas of the Mississippi and Atchafalaya Rivers. These and other waters in the northern Gulf of Mexico constitute approximately 40 percent of U.S. fisheries, generating $2.8 billion annually, which makes the potential effect of hypoxia a critical issue (CENR, 2000).
>
> The hypoxic zone has been caused by a complex mix of increased nutrient loads transported by the rivers and physical changes to the basins through activities such as channelization and loss of wetlands and riparian vegetation. These factors produce a higher oxygen demand that, when coupled with water column stratification in the Gulf resulting from the freshwater–saltwater interface, can lead to hypoxic lower layers of water. It has been estimated that 90 percent of the nitrates entering the Gulf come from urban and agriculture runoff (56 percent from the Mississippi River Basin and 34 percent from the Ohio River Basin).
>
> Two primary approaches have been developed to address hypoxia (CENR, 2000; Mitsch et al., 2001). The first approach involves efforts to reduce nitrogen loads in streams and rivers in the basin through activities such as reducing fertilizer applications to recommended rates, increased use of conservation tillage systems, and improved sewage treatment. The second involves enhancing denitrification and nitrogen retention within the Mississippi–Atchafalaya River Basins through restoration of ecological systems such as riparian areas and wetlands. The stated goal of the Mississippi River/Gulf of Mexico Watershed Nutrient Task Force is to reduce by the year 2015 the average hypoxic area to 2,000 mi^2 (5,200 km^2). One of the many programmatic indicators (22 were defined) that will be used to track progress is the establishment of vegetative and forested buffers along rivers and streams in watersheds known to contribute significant quantities of nitrogen. Using an annual denitrification rate of 40 kg N/ha for riparian areas, 7.8 million acres of new riparian areas would be needed to attain a 20 percent reduction in nitrogen loads. Other estimates were also developed for wetlands acreage needed, fertilizer application options, tillage, and other possible remedies. In the final assessment, however, it was recognized that no single approach would be completely successful and that a wide variety of approaches relying upon the many existing federal, state, local, and private programs will be needed to accomplish the changes necessary to solve the Gulf hypoxia problem (EPA, 2001).

prominent and largely beneficial role in human society for thousands of years, providing food, fuel, fertilizer, transport, and clothing, they have had a dramatic negative impact on global biodiversity. As shown in Figure 3-9, primary grazing effects include the removal of vegetation, trampling of vegetation, destruction of biological soil crusts, compaction of underlying soils, redistribution of nutrients, and dispersal of exotic plant species and pathogens. Secondary effects include

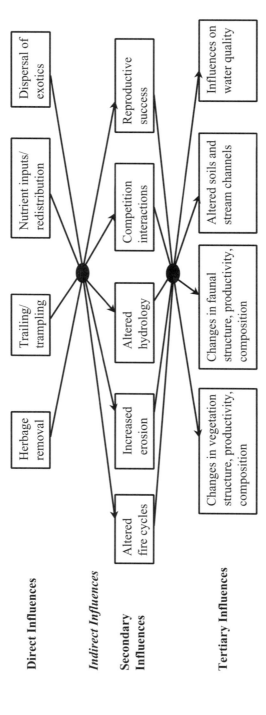

FIGURE 3-9 Direct and indirect influences of grazing. SOURCE: Reprinted, with permission, from Kauffman and Pyke (2001). © 2001 by Harcourt, Inc.

altered disturbance regimes associated with hydrology (runoff and infiltration rates and water-holding capacity) and fire (frequency and severity), accelerated erosion, altered competitive relationships among organisms, and changes in plant or animal reproductive success and/or establishment of plants. Long-term cumulative effects of domestic livestock grazing involve changes in the structure, composition, and productivity of plant and animal communities at community, ecosystem, and landscape scales. These tertiary effects often include overall declines in biotic richness or diversity of affected aquatic, riparian, and terrestrial areas.

In 1980, the U.S. Department of Agriculture estimated that vegetation on more than half of all western rangelands had deteriorated to less than 40 percent of productive potential. Although this reflects changes principally in upland conditions, there is no doubt that the impacts to western riparian areas are likely to have been much more severe, for reasons described below. Although upland range conditions reportedly have improved in many areas since 1980, extensive field observations in the late 1980s suggest that riparian areas remain in degraded condition (Chaney et al., 1990; BLM and USFS, 1994).

The disproportionate impact of livestock on riparian areas is a product of both management and animal behavior. First, until the 1960s (if not later), riparian areas were considered "sacrifice areas," used chiefly for supplying forage and water to livestock (Stoddart and Smith, 1955). Although riparian areas comprise 1 percent or less of the arid land area of the 11 western states (Belsky et al.,

1999), they nevertheless provide a substantial amount of the available forage. Roath and Krueger (1982) found that a riparian area in eastern Oregon occupied less than 2 percent of a grazing allotment's area, but it produced 21 percent of the available forage and supplied 81 percent of forage actually consumed by cattle. Second, cattle in particular congregate in riparian areas and other wet areas because of the availability of water, shade, and more succulent forage—spending from 5 to 30 times more time in these cool, productive zones than would be predicted from surface area alone (Belsky et al., 1999).

The grazing of riparian areas by domestic livestock involves the periodic removal of native streamside vegetation—particularly herbaceous plants, shrubs, or young trees. Along many streams and rivers, it has been a common practice to remove certain plants over time to create livestock pastures or hay fields or to convert the land to crop production. Grazing itself occurs over varying time periods (e.g., days, weeks, months, or seasons) and is typically repeated on an annual basis. Characteristics of the riparian plant communities, such as composition, cover, density, or other measures of plant communities, are likely to show significant changes relative to ungrazed areas (Kauffman and Pyke, 2001). In addition, a variety of effects on soils (e.g., reduced litter cover, increased bulk density, greater percentage of bare ground, decreased infiltration) and impacts on local wildlife and aquatic systems are common (Dwyer et al., 1984; Kauffman and Krueger, 1984; Howard, 1996; Ohmart, 1996). Where riparian vegetation has been suppressed or removed via grazing over long periods of time, the root biomass along channel banks and the resistance to overbank flow may become sufficiently reduced such that channels become unstable. Channel widening and gullying (as shown in Figure 3-3 for channelization) are common features of areas that have experienced the effects of season- or year-long grazing or other intensive grazing practices. Intensive grazing of the arid southwest in the late nineteenth century is thought to have played a role in the extensive arroyo cutting observed in this area, although cycles of arroyo cutting and filling prior to the introduction of domestic livestock have also been documented (Bull, 1997; McFadden and McAuliffe, 1997; Gonzalez, 2001).

Season-long grazing (commonly used throughout the West) results in major impacts to riparian areas because a large proportion of plant biomass is removed, the remaining vegetation has little opportunity to recover, and the grazing is generally repeated year after year. Grazing systems that employ rest-rotations or that result in less intensive utilization of riparian forage can potentially reduce these impacts, but these approaches have not been widely used and their potential ecological effects have received little study (Elmore and Kauffman, 1994). Of 17 grazing strategies evaluated by Platts (1991), only a few were consistently rated "well,"[3] including

---

[3]"Well" refers to a rating of 8 or higher, where 10 = highly compatible with fisheries needs and 1 = poorly compatible.

riparian pasture, corridor fencing to exclude cattle, rest-rotation with seasonal preference (sheep only), and total exclusion of sheep and cattle.

Given the many impacts of grazing described above, it is no surprise that aquatic organisms and riparian wildlife have been profoundly impacted by historical grazing practices. Two reviews have illustrated the adverse effect grazing has had on fisheries and wildlife. Over 95 percent of the studies reviewed by Platts (1991) showed "stream and riparian habitats had been degraded by livestock grazing, and that these habitats improved when grazing was prohibited." In Ohmart's (1996) view, "Unless grazing management changes are made soon it is predictable that many more species, especially neotropical birds will be placed on the endangered species list." Of the 76 federally listed plant and animal species on Bureau of Land Management (BLM) lands, for which livestock grazing was a significant factor in their decline, approximately 80 percent were dependent on or associated with riparian habitats (Horning, 1994).

Federal land management agencies have often concurred with these assessments. In 1994, BLM and the U.S. Forest Service (USFS) concluded that "watershed and water quality would improve to their maximum potential" if livestock were removed entirely from federal lands (BLM and USFS, 1994). The USFS concluded that livestock grazing is the fourth major cause of species endangerment nationwide, the second major cause of plant endangerment, and the number one cause of species endangerment in certain arid regions of the West, such as the Colorado Plateau and Arizona Basin (Flather et al., 1994). Several writers have suggested that "livestock grazing may be the major factor negatively affecting wildlife in the 11 western states" (Ohmart and Anderson, 1986; Fleischner, 1994; Ohmart, 1996). Although there is limited evidence from more humid regions, Belsky et al. (1999) suggest that environmental impacts of grazing in these regions are similar to those in drier areas.

Native Ungulates

Like livestock, native ungulates can modify riparian areas by eating plants, dispersing seeds, disturbing soil, and modifying channel morphology. Impacts on plants can include suppressed vigor, reduced reproductive output and regeneration, and increased mortality (Opperman and Merenlender, 2000). For example, successful regeneration of white cedar in winter deeryards can be virtually nonexistent because of concentrated seasonal browsing (Verme and Johnston, 1986). The effects of native ungulates depend on their populations, which fluctuate in response to predation, competition, weather, disease, and other influences (Naiman and Rogers, 1997). White-tailed, mule, and black-tailed deer, elk, and moose have drawn attention when their numbers are particularly high or when their presence is concentrated temporally. Such situations are most likely to occur as a result of human-induced changes in the landscape, or a change in predator–

prey dynamics [as exemplified by exploding deer populations (McShea et al., 1997; Hubbard et al., 2000)].

In some of the nation's parklands, native ungulates have increasingly become a riparian management issue. Elk and moose browsing have caused damage (e.g., reduced or eliminated woody species cover, limited regeneration) to cottonwoods, willows, and aspens in riparian areas and other portions of Yellowstone and Grand Teton National Parks and the National Elk Refuge in Jackson Hole, Wyoming (Kay, 1997a,b; Matson, 2000). Moose browsing on riparian willow thickets is believed to suppress both density and diversity of migrant breeding birds dependent on riparian vegetation (Berger et al., 2001). The extent and causes of this damage are controversial, though a lack of ungulate population regulation by either hunting or predation is considered at least partially to blame. In the Greater Yellowstone area, the extinction of grizzly bear and wolf populations has been linked to increases in moose density (Berger et al., 2001).

In areas supporting both livestock and wild ungulates, livestock have been observed to do greater damage to forage resources. For example, native ungulates are scattered over their summer range, making their impact on forage resources minimal to moderate, while many domestic livestock graze on aspen-covered ranges in the West during the peak of the growing season and commonly use at least 50 percent of the annual production of palatable forage (DeByle, 1985). Another study found that wild ungulate use of riparian sites in Idaho, Utah, and Nevada was "trivial" compared to livestock use of the same areas (Platts and Nelson, 1985). Long-term studies in Utah and Nevada showed that aspen fails to regenerate or regenerates only at low stem densities when it is grazed by both livestock and native ungulates (Kay and Bartos, 2000; Kay, 2001). In the absence of livestock, however, aspen regenerated successfully, provided that deer numbers were low.

In many human-modified landscapes, losses in the amount of available native habitat have concentrated herbivore pressure in an area that is already under stress. Hobbs and Norton (1996) used exclusionary fencing to show that deer were a limiting factor to the restoration of a riparian area that had been previously degraded by domestic livestock. It was suggested that the site had reached a threshold of degradation beyond which recovery was not possible without exclusionary fencing to reduce ungulate browsing. Given the high populations of deer in many areas, particularly urbanized landscapes, exclusionary fencing or targeted population control may be needed to reduce herbivore pressure and assist in riparian area recovery.

Forestry

The removal of trees by forestry operations has the potential to alter long-term composition and character of riparian forests, and thus the structure and function of these systems (Ralph et al., 1994). If selection harvest methods are employed and small amounts of timber removed, and if the frequency of harvest is separated by several decades, the effects on riparian plant communities may be relatively small. However, where large portions of the standing timber are harvested or where the period between harvest operations is short, substantial changes to the composition, structure, and function of riparian forests almost certainly will result. Figure 3-10 shows the decline in virgin forest in the United States from 1620 to 1920.

The location and construction of logging roads (e.g., temporary or permanent, loggers choice or a planned transportation system) along streams can affect

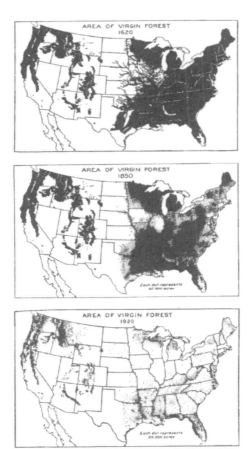

FIGURE 3-10 Virgin forest area in 1620, 1850, and 1920. This figure shows an estimate of forests have *never* been cut. It does not show the current total area of forest. SOURCE: Reprinted, with permission, from Verry et al. (2000). © 2000 by CRC Press, LLC via Copyright Clearance Center.

the long-term character of riparian forests (Adams and Ringer, 1994). Upslope roads can increase hillslope erosion rates (either surface erosion or landslides) or materially alter flow pathways, for example via the interception of shallow subsurface flow into ditches and its rerouting to locations of instability (Furniss et al., 1991).

Forest harvesting can occur in a variety of ways depending upon forest type, age, and density and upon topography, climate, and utilization standards. The impacts of forest harvest systems on riparian structure and function are much greater when forests are clearcut or harvested right up to streambanks and lake shorelines. The total harvest of riparian vegetation and adjacent terrestrial forests can increase the amount of solar radiation reaching the water surface, which can increase water temperatures and affect aquatic primary production. Temperature increases are of particular concern during summer when streams and rivers are naturally warmer. In addition, removal or alteration of the riparian vegetation changes the quantity and quality of terrestrial food resources for a stream, such as leaves, needles, and other forms of organic matter. Removal of riparian forests and repeated harvest over short rotations (e.g., 40–80 years) greatly reduce the potential for large-wood recruitment into a stream. Harvest of streamside forests also removes the vegetative cover that can slow the delivery of sediment into streams and retain nutrients, such as nitrogen and phosphorus.

As discussed in Chapter 2, riparian forests collectively provide for an array of sustainable processes and functions that make them exceptionally important for maintaining productive aquatic and terrestrial ecosystems (Johnson et al., 1985; USFS, 1993; Laursen, 1996; Verry et al., 2000). Those functions, as measured by species richness and diversity, can be impaired by forestry operations. Studies have demonstrated the habitat value of uncut riparian areas for woodpeckers (Conner et al., 1975) as well as for secondary cavity nesters such as chickadees, swallows, bluebirds, and nuthatches (Balda, 1991). The red-shouldered hawk is associated with wooded bottomlands of major rivers (Brewer et al., 1991; Robbins, 1991); work in Ontario suggests that cutting specifically in riparian woodlots may be responsible for declines in this species (Bryant, 1986).

The hydrologic effects of timber harvesting, such as increased annual water yields, increased sediment production, and altered stream chemistry, have been documented from a large number of watershed studies in forested areas (Ponce, 1983; Binkley and Brown, 1993; Adams and Ringer, 1994; Murphy, 1995). Such responses have not occurred universally and are typically dependent upon terrain conditions, the amount of timber removed, the type of logging system, post-harvest rainfall patterns, soil type, and other factors. Although increased water yields are most common when large proportions of the forest are harvested, increases in peak flows have not occurred consistently (Reiter and Beschta, 1995; Beschta et al., 1999). Increased sediment production is most likely in steep terrain where ground-based logging systems are employed or where soils are disturbed severely by post-harvest site preparation (e.g., mechanical scarification, hot slash burns) (Beschta, 1990).

Chapter 5 discusses how to diminish the potentially adverse effects of timber harvest upon aquatic and riparian resources by the use of various types of buffers or riparian reserves. Even in cases where forestry has been moderated for restoration purposes (e.g., by using partial harvest), riparian function may be impaired more than simple buffer width would indicate. Partial harvest often allows selective removal of larger or older trees, reducing ecological function more than width and targeted stem densities might reflect. Streamside buffers are generally not designed to mirror the stand composition and dynamics of desired healthy riparian forests for a given age class, especially when harvest decisions are strongly governed by social concerns about economic impacts.

Nearly 136 million acres of the nation's forestland are in the public domain, with 85 million acres being managed by the USFS, 8 million by BLM, and 43 million by state, county, and municipal governments. Private holdings amount to 347 million acres, of which 71 million are controlled by the forestry industry (Coggins et al., 2001). One of the major challenges in riparian management on public lands is the lack of a consistent scientific framework for determining widths of forested riparian areas that will sustain their desired structural and functional attributes. Differences in management between forest regions and individual national forests, between forests managed by the USFS and BLM, and among federal, state, and privately owned forests are more often based on policies, economic considerations, political pressure, and litigation than on differences in forest types, hydrologic regimes, climate, geology, physiographic provinces, or the ecological functions of riparian plant communities. Significant protection and restoration of forested riparian areas across the United States are unlikely until a common framework is developed.

INDUSTRIAL, URBAN, AND RECREATIONAL IMPACTS

Mining

Mining has historically been, and continues to be, an important land use in many portions of the United States, particularly on public lands. The General Mining Law of 1872 authorizes hardrock mineral extraction (e.g., gold, silver, nickel, copper) on all public lands that have not been specifically withdrawn from mineral development. Approximately 147 hectares (364 million acres) of public lands (constituting 80 percent and 90 percent of all lands managed by the USFS and BLM, respectively) are open to mining (NRC, 2000a).

Because only a small percentage of the U.S. land surface has been mined—less than 1 percent in recent decades according to Starnes (1983)—the effects of mineral extraction might initially be assumed slight. However, local degradation can have major downstream effects, thus affecting aquatic and riparian resources for long periods of time (Richardson and Pratt, 1980; Nelson et al., 1991; Wilkinson, 1992).

The mining of hillslopes and valley bottoms for minerals ranging from gold and silver to coal and gravel has involved a wide variety of approaches depending upon geology, topography, available technology, market value, and other factors. In hard-rock mining, the excavation of rock and soil to retrieve a mineral or ore of value to society often results in large amounts of waste rock or spoils. The extent to which such materials influence riparian areas depends on the amount of spoils deposited along stream channels; in other situations the acidity of the spoils can be a major concern. Acid mine drainage is considered to be one of the major water pollution concerns associated with many mining operations (Nelson et al., 1991). In addition, mining may introduce toxic metals such as arsenic, cadmium, chromium, copper, lead, mercury, and zinc, particularly when surface or groundwater is allowed to flow through waste piles.

Open-pit mining, where soils and rock overburden are excavated and embanked at a nearby location, is often employed when relatively low-grade ores or less valuable minerals are sought. The potential for riparian areas in or near these types of mining operations to be affected is often great. Depending upon the size and location of the mining operation, total hillsides can be excavated and their stream systems moved or buried. For example, so-called "mountaintop removal" for the mining of coal, which occurs principally in West Virginia and parts of Pennsylvania, involves placing excess spoil material into valley bottoms. This practice, which can bury and literally destroy streams, was ruled illegal in a 1999 federal district court decision. But since then, federal rule changes have been proposed that would again permit the practice under certain conditions.

When a mining operation exposes large areas of bare ground, substantial increases in overland flow and sediment production can occur during rainfall periods. Unless a well-designed and operated system of detention ponds is in place, such runoff may greatly increase sediment loading to nearby streams and rivers. Revegetation of embanked overburden and spoils is often a challenge for many open-pit mining operations.

Historically, placer mining was a common means of accessing certain types of minerals, particularly gold. Some placer operations utilized high-pressure water directed at hillslope soils or deposited alluvium—an incredibly effective method for eroding and washing large volumes of sediment into streams and riparian areas. Unable to transport the massive volumes of alluvium and hillslope sediment produced over a short time period, channels became quickly clogged. Channel aggradation, floodplain aggradation, and highly unstable channels downstream of placer mining operations were common. As might be expected, such operations have major detrimental effects on both aquatic and riparian areas and often present formidable restoration challenges (Rundquist et al., 1986; InterFluve, Inc., 1991).

In some portions of the United States (e.g., the West and Alaska), dredging of valley bottom sediments with floating dredges was a common means of mining mineral deposits, typically gold. The use of floating dredges was limited to valleys with significant floodplains so that the dredge could excavate its own flotation pond as it progressed across as well as up and down a particular valley bottom. To retrieve the gold present in valley sediments, all vegetation was removed, and the soils and underlying gravel substrates (often to depths of several meters) were mechanically dredged to the surface. Once the gold was separated on the dredge, the remaining mixture of soil and rock was dumped in arc-shaped spoil piles behind the dredge. Although most dredge mining occurred many decades ago, the resulting coarse-rock spoil piles often remain, typically unvegetated. Little effort has been made to reclaim the streams and riparian areas where dredge mining occurred.

Another form of mining practiced along many rivers and streams for extended periods of time is gravel mining (Williamson et al., 1995). Extraction of gravel, primarily for use in construction products, typically occurs along rivers and adjacent floodplains where extensive gravel deposits, often sorted by size class, are naturally found. The excavation of gravel from terraces (i.e., inactive floodplains) may have little impact on riparian systems. However, gravel excavation on active floodplains can directly reduce riparian vegetation and alter groundwater patterns. Impacts to riparian areas also can occur when gravel is mined from channels. In these situations, bar-scalping and streambed excavation can greatly influence long-term sediment transport, channel morphology, and bank stability of specific stream reaches. If large amounts of gravel are removed, channel down-cutting or incision may occur, potentially influencing local groundwater levels, the frequency of overbank flows, bank stability, and the character of riparian vegetation (Collins and Dunne, 1989; Kondolf, 1995).

Mining of heavy mineral sands for titanium-bearing minerals, zircon, and monazite is another potential threat to riparian areas. Although most heavy mineral sand is mined outside of the United States because of wetland protection laws, economic deposits of such sands are found along the Atlantic Coast and are currently mined in Florida and Virginia (Brooks, 2000). These deposits are often located in and adjacent to riparian areas. Heavy mineral sands are usually extracted using surface mining with floating dredges and concentrators after removing harvestable timber and other vegetation from the site. Topsoils are then removed and stockpiled for reclamation purposes, unless they contain high concentrations of heavy sands. The mineral concentrate produced at the mine is typically 90 percent heavy mineral, which is transported to a plant for separation into constituent minerals (Brooks, 2000). Unlike many of the other mining activities described, heavy mineral sand mining sites are amenable to restoration. Reclaimed mine sites have been successfully reestablished as wetlands, forest, pastures, and row crops.

Transportation

River Transportation and Removal of Large Wood

The rivers of the United States provided the first systematic transportation system for a developing nation. Lewis and Clark, in their exploration of the Louisiana Territory and lands west to the Pacific Coast in the early 1800s, relied primarily on rivers to transport their party across the uncharted lands. Keelboats, barges, river steamboats, canoes, and other watercraft plied the nation's rivers, moving people, farm products, and other materials over long distances.

To improve a river for transportation purposes often necessitated the removal of large wood and other obstructions. By the early 1900s, most rivers in the United States had experienced the systematic removal of large wood, or snags (Sedell et al., 1982; Maser and Sedell, 1994). Thousands of kilometers of river length were "snagged" and more than 100 snags per kilometers were often removed. Although there has been little systematic study of the effects of these snagging activities upon channel characteristics, riparian functions, and floodplain processes, the effects are likely to have been significant.

In the early years of this country, transportation of logs and timber to market was a major challenge. The downstream movement of aggregations of logs—log driving—was a relatively inexpensive means of transporting large volumes of wood over long distances. However, before a log drive could be conducted, boulders, leaning trees, sunken logs, and other forms of obstructions were blasted or otherwise removed in order to more easily float logs downstream. The number of streams affected by log drives was large. For example, by 1900 over 130 incorporated river- and stream-improvement companies were operating in Washington State. To assist the downstream movement of wood, splash dams were commonly employed to provide a surge of water (Sedell and Luchessa, 1981). Log drives that occurred on the Yukon, Chena, and Tanana Rivers in Alaska have been well documented (Sedell et al., 1991). Splash dams and log drives have also been used on rivers in the Rocky Mountains and in the eastern portion of the country. Although log drives and associated wood removal are now only a part of history, there is no doubt that the effects to channels and their riparian areas have been substantial and long-lasting.

Today, the nation's major rivers continue to be extensively utilized as major transportation routes. Ocean-going ships use the St. Lawrence Seaway and the Great Lakes for transporting a wide variety of goods and materials. Similarly, the use of barges on the Mississippi, Missouri, Ohio, and Columbia Rivers and on other waterways of the nation is an important means of transporting large amounts of cargo. Although the economic importance of these waterways is obviously great, so are the ecological effects of channelization, construction of locks, and other facets of maintaining transportation corridors along these river systems (NRC, 2001, 2002).

Roads and Railroads

Vehicular access to homes and communities, factories and production facilities, farms and ranches, recreational areas, rangelands, forests, and other locations is a characteristic feature of American society. Many of the country's road systems have had far-reaching and often permanent impacts to riparian areas, which were seldom considered during the planning of most highway and railroad systems. For example, the placement of highways along rivers and lakeshores has been a particularly common practice, the ecological effects of which have been observed to extend as much as 600 m on each side of the road (Formann and Deblinger, 2000). Significant impacts to riparian areas are likely to have occurred, particularly where narrow valleys and steep hillsides (and associated high construction costs) generally precluded the location of a road some distance from a river or shoreline. Nationally, similar impacts have occurred with railroads, though at a much-reduced scale.

The direct effects of highways and railroads within riparian areas include (1) the removal of riparian vegetation from the area occupied by the roadbed and the right-of-way, (2) the alteration of topography (extensive fills are often used to provide a roadbed foundation), and (3) local hydrologic modifications involving changes in infiltration and the rerouting of both surface and groundwater. Where sinuous streams were encountered during highway or railroad construction, portions of the channel were often filled to maintain a straight road alignment at the cost of reduced channel length. In some instances, the effects on river length have been substantial. Concurrently, riparian vegetation was often eliminated and replaced with a roadbed.

Another important feature of highway and railroad systems is that they periodically cross watercourses. A wide range of structures can be used for such purposes, but most fall into two general categories—bridges or culverts. Abutments along the bank are typically needed to provide sufficient support at each end of a bridge; in the case of a relatively wide river crossing, additional midspan piers or pilings are usually employed to provide intermediate support(s) to the bridge span. Because the abutments physically constrain the stream, future lateral adjustments by the stream are effectively eliminated. Similar effects occur at culvert locations—i.e., both vertical and lateral channel adjustments are constrained.

The construction of highway systems and urban roads outside of riparian areas can also have important indirect effects upon streams and riparian areas. In urban areas, roads and other impervious surfaces can increase peak overland flows, thus fundamentally altering the hydrologic disturbance regime for those systems. Roads can also concentrate overland flows to specific locations where channel erosion and gullying and accelerated sediment loading may be initiated. In steep mountainous terrain, an increased frequency of landslides associated with roads may alter the delivery of sediment and large wood to forested streams and riparian areas. Roads and their associated ditches also increase dispersal of exotic plant species from uplands to riparian areas (Parendes and Jones, 2000; Trombulak and Frissell, 2000). Thus, although the local effects of streamside roads are a high concern with regard to riparian processes and functions, the effects of roads immediately outside riparian areas cannot be ignored (Furniss et al., 1991; Adams and Ringer, 1994).

Urbanization

Urbanization and development have profound impacts on watershed hydrology and vegetation, and consequently on the structure and functioning of riparian areas. Among the most important impacts of urbanization are the increased frequency and magnitude of flooding and decreased baseflow that result from land-use changes typical of development (Schueler, 1987). In its natural state, vegetation intercepts a portion of precipitation, with the remainder being stored in or on the soil surface or infiltrating into the soil where it recharges groundwater or is used by plants. Typically, only a small portion of the precipitation ends up as direct overland flow. Thus, peak flows are moderated by high infiltration rates, and many streams are perennial due to groundwater flow during periods of the year when overland flow is uncommon. As urbanization increases and more of the land surface is covered with homes, buildings, roads, sidewalks, parking lots, and other structures, the imperviousness of the watershed increases. With increased imperviousness, infiltration, interflow, groundwater recharge, groundwater contributions to streams, and stream baseflows all decrease, while overland flow volumes and peak runoff rates increase, as shown in Figure 3-11. As urban-

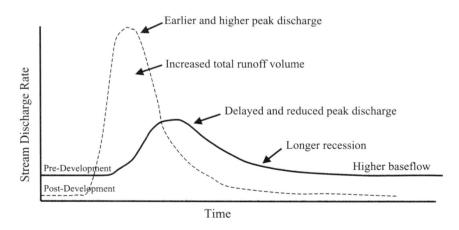

FIGURE 3-11 Effects of urbanization and development on stream flow.

ization and imperviousness increase further, the capacity of natural channels to transport the increased overland flow is exceeded, with the undesirable consequences of accelerated channel erosion and increased flooding. Downstream flooding is further exacerbated by gutters, curbs, culverts, stormwater sewers, and lined channels, which are installed to transport runoff from impervious surfaces to streams as quickly as possible.

The changing land use and hydrology of urbanizing watersheds have multiple impacts on stream channels, aquatic ecosystems, and water quality within riparian areas. As runoff frequency, volumes, and peak flow rates increase during urbanization, stream channels respond by increasing their cross-sectional area to accommodate the higher flows—either through widening of the stream channel, downcutting of the streambed, or both. This phase of channel instability, in turn, triggers a cycle of streambank erosion and habitat degradation (Schueler, 1995). Sediment loadings may increase by one to two orders of magnitude compared to pre-development conditions, such that streambeds are covered with shifting deposits of sand and mud (Schueler, 1987). Fish and aquatic insect diversity and abundance decrease because of changes in temperature, benthic substrates, dissolved oxygen levels, and pollutant loadings. Finally, increased loading of nutrients, bacteria, oxygen-demanding materials, oil, grease, salts, heavy metals, and other toxics is evident.

Depending on the location of urbanization, riparian areas may be converted to urban land uses. Riparian areas of lower-order streams are often totally eliminated, and the width of riparian areas along higher-order streams is generally reduced. In major cities, entire stream systems that have been removed during urban development have been replaced by underground culverts, pipes, and other similar structures to transport overland flow. A secondary effect of urbanization is caused by changes in how overland flow and shallow subsurface flow enter and transverse riparian areas that remain after development. Prior to urbanization, overland flow enters the more extensive riparian areas as either sheet flow from areas immediately adjacent to the riparian areas or through small ephemeral drainageways, thus allowing sediment to be deposited and other substances to be transformed. Much like traditional agriculture, development promotes the formation of concentrated flows that are less likely to be dispersed within the riparian area, greatly reducing the potential for pollutant removal (Dillaha et al., 1989). A similar paradigm holds true for shallow subsurface flow and the removal of dissolved substances. Development is marked by the construction of gutters, storm sewers, and lined channels that often pass directly through the riparian area and discharge directly into the stream channel (much like agricultural drainage tiles). Even when drainage structures are not constructed to bypass riparian areas, flow rates are generally so high that there is little opportunity for transformation processes (such as degradation and assimilation discussed in Chapter 2) to occur. Figure 3-12 illustrates how overland flow increases and infiltration decreases with imperviousness.

The site-specific effects of urbanization on stream habitat and water quality are highly variable. In general, as urbanization, population density, and imperviousness increase, water quality declines. Although somewhat controversial, a threshold in habitat quality is thought to exist at approximately 10 percent to 15 percent watershed imperviousness, beyond which urban stream habitat quality is consistently classified as poor (Shaver et al., 1994; Booth, 1991; Schueler, 1995; Booth and Jackson, 1997).[4] However, impacts of urbanization will vary with alternative development models. For example, clustered residential developments, which have the same overall population density as more traditional residential developments, can reduce disturbance of riparian areas and decrease water-quality impacts compared to traditional development. This is accomplished by devel-

---

[4]The 10 to 15 threshold reported in the literature must be used with caution because many studies do not specify whether they have measured total or effective impervious area. Total impervious area is that fraction of the watershed covered with impervious surfaces such as concrete, asphalt, and buildings. This is relatively easy to measure using areal photography and other remote sensing techniques. Effective impervious area, or directly connected impervious area, is less than total impervious area because it excludes impervious areas that drain to adjacent pervious areas; it is also more difficult to estimate. For lower density land uses, total impervious area may be twice the effective area. Approximately 10 percent is a safe impact threshold for effective impervious area and this corresponds to a total impervious area threshold of approximately 20 percent (Booth and Jackson, 1997).

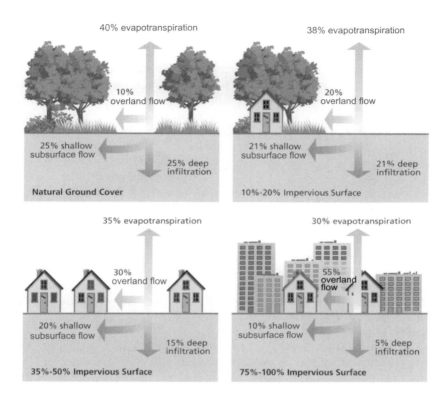

FIGURE 3-12 Relationship between impervious cover, shallow subsurface flow, deep infiltration, and overland flow. SOURCE: Modified from the Federal Interagency Stream Restoration Working Group (1998).

oping smaller lots in areas that are less hydrologically active and are outside of riparian areas. The remaining undeveloped or lightly developed green spaces (parks, trails, ball fields, etc.) are then maintained and managed for recreational and environmental benefits. If properly designed and combined with urban stormwater best management practices (BMPs), cluster and other green development approaches can promote properly functioning riparian areas and the environmental services they provide. (Appropriate BMPs might include infiltration systems, detention ponds, minimization of impervious surfaces, and dispersion of concentrated flow from the high-density areas into the green areas.) Protection of riparian areas is much more difficult to accomplish with traditional dispersed residential development of the same overall site density, which preserves much less common green space and has a higher degree of imperviousness because more roads are required per household served. The more extensive road system, particularly if curbed, may promote more rapid movement of runoff through the drainage network, bypassing remaining riparian areas.

Urbanization and development permanently reduce the extent and functioning of riparian areas through land-use conversion and the creation of hydrologic conditions that reduce aquatic habitat quality and negate the effectiveness of the remaining riparian areas for water-quality protection. Some specific considerations related to lakeshore development, a rapidly growing phenomenon, are discussed in Box 3-3.

Recreation

River corridors have been found to draw recreationists (hikers, cyclists, horse-back riders) and other visitors more frequently and from a wider area than other types of parks and open space (Green and Tunstall, 1992; Cole, 1993). Boat landings, fishing access points, portages, parks, golf courses, campgrounds, and trails are all recreational enhancements commonly placed within riparian areas, usually without a careful assessment of their potential impacts. Negative effects on riparian areas from recreational activities and facilities stem in part from a lack of environmental assessment before plans are implemented, a dearth of sound ecological design to mitigate impacts, and an absence of ongoing monitoring to detect problems.

Recreational uses and their impacts are not peculiar to riparian areas. What sets recreation in riparian areas apart, however, is the concentration of human activities in what are often narrow strips of land and the potential of those activities to affect both aquatic and terrestrial ecosystems. Effects of recreational use are roughly grouped into impacts on water, soils, vegetation, and animals (Cole, 1989, 1993).

Recreational activities in riparian areas can introduce sediment, nutrients, bacteria, petrochemicals, pesticides, and refuse to adjacent waterbodies (Andereck, 1995). Conversely, motorized boats and personal watercraft contribute water and noise pollution and cause erosion and disturbance of aquatic and riparian animals through the creation of wakes. Outboard motors have been shown to resuspend sediments in the littoral zone and negatively affect plant growth in that portion of the riparian area (Garrison and Wakeman, 2000). Effects on riparian soils include trampling by foot, animal, or vehicle traffic, which leads to compaction, destruction of soil biota, and increased erosion. Damage to vegetation can be incidental, as through trampling, or deliberate, as in its removal for the construction of recreational facilities or collection of firewood. Impaired riparian vegetation translates into both a depauperate terrestrial community and an impaired aquatic community, as shading benefits and inputs of woody material and organic nutrients to the waterbody decrease. A study in Colorado showed dramatic improvements in stream structure and trout populations in sections of streams that were fenced to protect against recreational use and grazing; projected fishing opportunities in the protected stretches were nearly double those in the unfenced stream (Stuber, 1985).

Animal life can be affected negatively by recreation in riparian areas in ways that include direct disturbance, modification, or destruction of habitat (Cole, 1993; Knight and Cole, 1995); pollution; direct exploitation (hunting and fishing); or introduction of pathogens, often through recreationally motivated introductions of animals (Cunningham, 1996). Responses of animals to disturbance can range from an immediate effect such as a heightened physiological response ("fight or flight") to a long-term effect like population decreases due to increased mortality or lowered reproductive rates (Knight and Gutzwiller, 1995). Negative impacts of lakeshore development and recreational activity on nesting common loons have been demonstrated in several studies (Robertson and Flood, 1980; Heimberger et al., 1983; Meyer et al., 1997). In coastal riparian areas, beach-nesting birds are particularly vulnerable to almost all types of recreational use, with common outcomes being disruption of incubation, increased mortality of young, and wholesale abandonment of nesting sites (Burger, 1995). Sea turtles are particularly vulnerable to mortality from recreational vehicles, beach lighting that confounds hatchlings, ingestion of plastic and other debris, and destruction of nesting habitat by beach stabilization or replenishment activities (NRC, 1990).

Standard access sites (e.g., boat landings and fishing docks) rarely have an ecological perspective incorporated in their planning. Instead, the emphasis is usually on creating tidy, safe, and accessible places for the public to reach the water. The riparian area is frequently seen as an impediment and obstacle on the way to the water. Pavement for vehicle access creates a sloped surface that funnels sediment and polluted water into the water body. Mowed lawns, a common feature of boat landings, further increase runoff and nutrient and pesticide loadings as they replace riparian vegetation. Boat landings are frequent dispersal nodes for unwanted exotic plant species such as purple loosestrife and Eurasian milfoil in the eastern United States, as evidenced by natural resource agency signage at such locations.

Golf course construction and maintenance can impact first- and second-order streams that flow through them and their adjacent riparian areas via the removal of natural vegetation and increased loadings of pesticides and fertilizers. Furthermore, many golf courses are heavily landscaped; natural drainages and streams are often destroyed or largely reconfigured. Removal of water for irrigation of greens can also negatively affect aquatic ecosystems and riparian areas.

As discussed above, roads that provide public access to riparian recreational areas can affect the structure and functioning of those areas (Findlay and Bourdages, 2000; Jones et al., 2000). Roads, trails, and other structures can increase the rate of introduction of exotic plant species and modify microclimates required by particular plant species, rendering the habitat unsuitable for sensitive species (Green, 1998). A particularly destructive recreational force is all-terrain vehicles, which can cause environmental degradation through destruction of veg-

BOX 3-3
Lakeshore Development in Wisconsin

Riparian areas of lakes face a unique threat from development. Cottages, resorts, and second homes are usually firmly situated in riparian areas, with attendant modifications of vegetation and additions of impervious surfaces. This phenomenon is particularly prevalent in Wisconsin, a state purported to have the third-largest concentration of fresh water glacial lakes, totaling 15,000 miles of shoreline. The number of dwellings per mile has risen on Wisconsin lakes of every size, with an average increase in density of almost 60 percent between 1960 and 1995 (Daulton and Hanna, 1997). Undeveloped lakes that are not completely in public ownership are now rare (Korth and Cunningham, 1999). There have been attempts in some regions to regulate development via local or state zoning restrictions that dictate lot size, setbacks, and limits to vegetation modification.

Historically, most development on lakes in the Great Lakes region consisted of small, seasonal dwellings, usually with surrounding natural vegetation left fairly intact. In recent years, these small 1940s-style cottages have given way to large, year-round dwellings. Computer modeling was recently conducted to compare runoff volume of water, sediment, and phosphorus to a lake from (1) an undeveloped wooded lot, (2) a small, 1940s-style cottage (700 ft^2) with grass path to the lake, and (3) a 1990s-style dwelling (over 3,000 ft^2) with lawn and paved driveway (J. Panuska, unpublished data, cited in Korth and Cunningham, 1999). The 1940s-style development had a fourfold increase in sediment input compared to an undeveloped lot, but runoff volume and phosphorus were virtually unchanged from the undeveloped condition. In contrast, the 1990s-style development showed a nearly sevenfold potential increase in phosphorus input, an 18-fold increase in sediment, and five times the volume of runoff water compared to the undeveloped lot.

Paleolimnological data (Garrison and Wakeman, 2000) has been used to document the long-term effects of development on lakes in Wisconsin. Redox-sensitive elements were used to address changes in hypolimnetic oxygen levels, and changes in the diatom community were used to assess impacts on lakes' trophic statuses. Historically, the greatest shift in the diatom community, from species inhabiting clear water to species tolerant of higher phosphorus loading, appears to correspond with the period of development from 1950 to 1970. More specifically, on all lakes the researchers found the highest input of sediment occurring during the construction or reconstruction phases of development. The U.S. Geological Survey (USGS) is currently undertaking actual monitoring of runoff from developed and undeveloped lots.

In a related study, the Wisconsin Department of Natural Resources conducted an inventory of shoreline plants, frogs, and birds on developed and undeveloped shoreline (Meyer et al., 1997). Of particular interest was the effectiveness of the current statewide Wisconsin Shoreland Zoning Program in achieving one of its stated goals, that of protecting aquatic life. Not surprisingly, sharp decreases in plant abundance and diversity along developed shorelines were discovered. Green frog numbers showed an inverse relationship to number of homes per mile. Songbirds showed a shift from forest birds, including thrushes, vireos, and warblers, to more common and cosmopolitan "suburban" birds, such as blue jay, American goldfinch, and black-capped chickadee. Based on regression models, the study concluded that the current statewide zoning guidelines are inadequate for the long-term protection of the riparian community and could lead to the elimination of many species on developed lakes. Other work has demonstrated a similar degradation of lakeshore fish habitat and decrease in species diversity that follows shoreline development (Bryan and Scarnecchia, 1992; Christensen et al., 1996; Jennings et al., 1996).

Wisconsin has instituted a program to assist counties in classifying their lakes based on physical characteristics such as depth and areal extent, presence of wetlands, presence of developed and undeveloped shoreline, and predictions of use (Korth and Cunningham, 1999) (a program that has met with resistance from some local governments and people employed in development-related enterprises). The intent is to use two- or three-tiered classifications to design management plans for particular lake types. In such a scheme, development and motorboat use might be strictly limited on a "pristine" lakeshore. A "low density" lakeshore might be amenable to more development subject to a requirement for large setbacks of homes to minimize disruption of the riparian area. On "high development" lakeshores, the focus might be shoreline restoration (Korth and Cunningham, 1999). Lakes also differ in vulnerability to impacts based on their water residence times. For example, seepage lakes, with no inflow or outflow streams, have the lowest capacity for flushing out pollutants and thus the greatest risk of rapid degradation of water quality.

Although some water-quality problems in lakes are best addressed through a watershed approach—such as addressing agricultural or forestry runoff—the best hope for protecting *lake riparian* areas lies in modifying activities of individual homeowners at the construction phase and during the management of their properties. This requires ongoing education of builders, real estate professionals, buyers, and lakeshore residents. Lake associations provide effective ways to disperse such information. In addition, publications available to homeowners in the Midwest and Northeast provide useful suggestions for preserving or creating buffers of native vegetation, limiting application of lawn chemicals, minimizing impervious surfaces, and reducing modification of vegetation on the building site (Moore, 1994; Dresen and Korth, 1995; Henderson et al., 1998). These sources emphasize an approach that seeks to "naturalize" the lakeshore and thus maintain its functions as habitat for a diversity of plants and animals as well as its functions as a filter for the lake.

etation, soil erosion, and disturbance of wildlife (Sheridan, 1979; Luckenbach and Bury, 1983; Webb and Wilshire, 1983; Bleich, 1988).

Exotic Species Invasion

Exotic or nonindigenous species have sometimes been intentionally introduced to accomplish specific objectives, such as the use of reed canary grass for erosion control. Unfortunately, some of these species can come to dominate the native plant or animal community and spread to off-site locations. Exotic plant species have been introduced to native communities around the world in such numbers that they now constitute a large proportion of the total number of plant species in many regions. Within the United States, an estimated 23 percent of the approximately 22,000 species of plants are exotic (Heywood, 1989). The proportion of plant species in riparian areas that are exotic can be even higher. For example, along the Rio Grande in New Mexico, exotic species represent over 25 percent of herbaceous plant species and over 40 percent of tree species (Muldavin et al., 2000).

The most common concern about exotic organisms is their displacement of native species and the subsequent alteration of ecosystem properties. Because they have been moved to areas outside their native range, exotic species are usually faced with fewer population-control mechanisms, especially biological agents such as predators, parasites, and pathogens. As a consequence, populations of exotic species can grow explosively and may dominate large areas of the landscape in the process. Generally, they replace indigenous species with a more homogenous community that supports lower wildlife diversity. Exotic plant species may create health or safety problems when they include toxic fruits or allergens or when they promote wildfires. Within the United States, exotic species are the primary cause for the decline of approximately 42 percent of those native species now listed by the federal government as threatened or endangered (Stein and Flack, 1996; Wilcove et al., 1998).

Several of the most aggressive exotic plant species in the United States are invaders of riparian areas. Indeed, it has been suggested that the disturbance regimes characteristic of riparian areas (e.g., from flooding) may make riparian communities vulnerable to invasion by non-native plant species (Stohlgren et al., 1998). Of the exotic weeds listed as candidates for the worst weeds in North America, as many as a third are found in riparian areas or wetlands (Stein and Flack, 1996; Plant Conservation Alliance, 2000; The Nature Conservancy, 2001). Prominent examples include saltcedar (*Tamarix*), which has replaced cottonwood and other native riparian plants throughout much of the southwestern United States. Invasion by saltcedar and its subsequent competition with native species is exacerbated by a reduction in flood flows caused by dams and by the lowering of water tables caused by water withdrawal and consumption. Other exotic plants that have become abundant in riparian communities include reed canary grass, buckthorns, scotch broom, blackberry, and kudzu. Table 3-3 lists 15 of the more prominent exotic plants currently invading the riparian areas of the United States. Those listed in the table represent the most serious current threats to riparian diversity and function, because all have a demonstrated capacity to spread rapidly and form large, dense stands of high biomass. Box 3-4 highlights some of the technical and financial problems associated with invasions of riparian areas by exotic plants.

Many exotic species of fish, amphibians, and invertebrates have been introduced either intentionally or accidentally into streams and rivers. These species frequently alter the abundance and distribution of native species through competition or predation, and they sometimes lead to local extirpations or extinctions. Introduction of predators (e.g., bullfrogs, largemouth and small mouth bass, brown trout) have been linked to declines of fish and amphibians, leading to the listing of some as threatened or endangered (Taylor et al., 1984; Crossman, 1991). Rainbow trout from the western United States introduced into East Coast streams has reduced populations of native brook trout (Larson and Moore, 1985). Although native fish species often have competitive advantages within their native habitats (Baltz and Moyle, 1993), habitat degradation can shift that advantage to introduced species from regions with habitats more similar to the degraded conditions. Use of waterways for navigation also causes extensive introduction of exotic species through disposal of bilge water or attachment to vessel hulls. Introduction of exotic species closely related to native species can cause genetic degradation through hybridization (e.g., brook trout introduced into the range of bull trout).

In the face of human population growth, introductions of nonindigenous species are likely to increase in riparian networks throughout North America. Exotic plant species continue to be used in some restoration efforts. However, most efforts that intentionally use non-native plants are designed to provide short-term functions and little if any long-term survival. Currently, there are no long-term monitoring systems for tracking the extent of intact riparian plant communities, the composition of riparian communities, or the distribution of exotic species. As a result, it is unlikely that riparian areas will be adequately protected.

Global Climate Change

Although predictions concerning global warming are uncertain, there is wide agreement that human activities will cause average global temperatures to continue to rise over the next century (NRC, 2000b,c). The latest Intergovernmental Panel on Climate Change (IPCC) estimates of global mean temperature and rising sea level by the end of the twenty-first century are 1.4–5.8 °C and 0.09–0.88 m, respectively (IPCC, 2001). The expected changes in temperature, precipitation, oceanic and atmospheric circulation, and frequency and severity of

TABLE 3-3 Exotic Plant Species Currently Recognized as Threats to Riparian Areas Across the United States

Exotic Riparian Plants[a]	Pacific Region	Great Basin	Arid and Semiarid Southwest	Rocky Mtns.	Great Plains	Cool Temperate East	Warm Temperate East	Southeast
Amur Peppervine *Ampelopsis brevipedunculata*						X	X	x
Chinese Privet *Ligustrum sinense*	x						X	X
Fig Buttercup *Ranunculus ficaria*	X					X	X	
Garlic Mustard *Alliaria petiolaria*				x	X	X		x
Giant Reed *Arundo donax*	X		x				x	X
Japanese Knotweed *Polygonum cuspidatum*	X			x		X	X	x
Japanese Meadowsweet *Spiraea japonica*					x	X	X	X
Nepalese Browntop *Microstegium vimineum*						X	X	X
Princesstree *Paulownia tomentosa*					x	x	X	X
Purple Loosestrife *Lythrum salicaria*	X	x	x	X	x	X	X	x
Russian Olive *Elaeagnus angustifolia*	x	X	X	X	X	x	x	
Saltcedar *Tamarix ramosissima*	X	X	X	X	X			x
Silktree *Albizia julibrissin*	x		x		x	x	X	X
Tallowtree *Triadica sebifera*					X		X	X
Tree of Heaven *Ailanthus altissima*	X	x	X	x	x	x	x	x

NOTE: Upper case X indicates regions of greatest current impact. Darkest shading indicates species of the greatest threat.
[a] Plant names follow USDA NRCS (2001b).

storms will probably vary regionally and locally. Increased summer maximum temperatures, fewer cold days and less frost over inland areas, reduced diurnal temperature changes over land, more intense precipitation events, and increased summer continental drying and drought are likely during the twenty-first century (IPCC, 2001). Changes in the frequency and intensity of extreme events, such as hurricanes, may be more ecologically significant than moderate changes in the mean values of environmental factors (Michener et al., 1997). All these changes include the climate variables most likely to influence riparian communities. The sections below consider possible changes to riparian structure and functioning that may occur as a result of global change. Nonetheless, as significant as climate changes are likely to be, land- and water-use changes have had and will continue to have the greatest effect on riparian areas in the near and medium term (Graf, 1999).

Riverine Settings

Changes in precipitation brought about by global climate change are likely to have substantial ecological consequences (Poff et al., 1997). The magnitude of such changes, however, is difficult to estimate. In general, it is expected that less variation will occur in the eastern United States, where precipitation is generally high enough to sustain perennial flow in most rivers.

As indicated in Chapter 2, snowmelt is an important contributor to runoff in a number of geographic regions. One major consequence of global warming might be a shift from spring peak flows to late-winter peaks in snowmelt-dominated regions. A shift to higher winter flows associated with rain or rain-on-snow events may scour streambeds and destroy overwintering eggs of some fish species (Montgomery et al., 1999). Floodplain wetlands along rivers with a snowmelt hydrology may also be altered.

With a warmer climate, streamflows in snowmelt systems would decline earlier in the summer and corresponding water tables would drop, with consequences for invertebrates and fish in addition to the lack of a high water table for riparian plants (Scott et al., 1999; Stromberg et al., 1991). A transition to lower flows under drier conditions would be particularly stressful for the aquatic and riparian areas that are already water-limited (Grimm and Fisher, 1989). For more humid regions, Poff et al. (1997) predicted a transition from perennial to intermittent flow. Analogues of this effect are seen below dams on impounded rivers, where reservoirs store flood flows and thus "shave" downstream flood peaks, transforming floodplain forests to communities adapted to drier conditions (Johnson et al., 1976). Under a drying climate, the effects of groundwater pumping, irrigated agriculture, and grazing can be expected to intensify, resulting in greater competition for water, space, and food.

An increase in "storminess," expressed as increasing concentration of rainfall on fewer days, has also been projected to be a consequence of global climate

BOX 3-4
Three Case Histories of Exotic Plant Species in Riparian Areas

Chinese Privet

Chinese privet, *Ligustrum sinense*, was introduced into the United States from China in 1852 and has since been planted as an ornamental, mainly as a hedge. Herbarium collections in Georgia include specimens collected from the wild as early as 1900 (USDA NRCS, 2001a). Farther west, the species escaped cultivation in Louisiana by the 1930s and continued to spread throughout the southeastern and eastern United States during the middle of the twentieth century. For example, after a fairly moderate rate of initial spread in Oklahoma beginning in 1900, Chinese privet populations increased rapidly from 1960 to 2000 (Taylor et al., 1996). Though Chinese privet can grow in a wide variety of conditions, it thrives in areas with moist soil, conditions that are often found in riparian areas.

Chinese privet disperses through its production of abundant seeds, which are spread by birds. Once established in a new location, it can form dense and nearly impenetrable thickets through vegetative reproduction. Its threat to native species is mainly the result of its ability to form these dense thickets that exclude other plants. In addition, Chinese privet fruits are toxic to humans, and where it flowers in abundance it can induce respiratory problems.

Control of Chinese privet is difficult because the plant resprouts following fires and has no known effective biological control agents. The best results have been obtained through mechanical removal, herbicidal applications, or a mix of the two. However, care needs to be taken during mechanical removal because plant fragments left on the site have the potential to resprout. In addition any disturbance to the soil during mechanical control efforts offers an opportunity for recolonization.

Saltcedar

Several species of saltcedar (*Tamarix*) have been introduced into the United States during the past two centuries as ornamentals or sources of wood or for erosion control. Of these species, *Tamarix ramosissima* has emerged as a very serious threat to riparian areas. This species has quickly spread beyond the areas where it was planted, and it now dominates approximately one million acres in the western United States, particularly riparian areas.

Saltcedar's invasion of riparian areas in the Southwest has produced large changes in ecosystem structure and function. One damaging ecosystem change in this arid region has been an increase in the rates of water use by riparian forests where dense stands of saltcedar dominate. Where saltcedar has invaded moist soils around desert springs, it has often dried up these critical water sources. In addition, saltcedar has been shown to raise the salinity of surface soils, which has also been implicated in its successful out-competition of natives (Busch and Smith, 1995; DiTomaso, 1998). Another major effect of saltcedar is related to the frequency and intensity of fires. Because saltcedar readily sprouts following a fire, compared to the native cottonwoods and willows, higher fire frequency appears to increase the rate at which salt cedar dominates southwestern riparian areas. Saltcedar stands support reduced diversity of several important biological taxa (understory vegetation, butterflies, and cavity-nesting birds).

The arid Southwest is particularly vulnerable to saltcedar because the tree is more tolerant of higher soil salinity and alkalinity, and it produces seed during a longer period than native cottonwood and willow. Saltcedar is also capable of developing very deep tap roots and higher leaf area within stands compared to native riparian trees, and it grows rapidly, up to 10 feet per year. Human activities such as dam building and cattle raising appear to promote the spread of saltcedar. Cattle feed preferentially on native cottonwood and willow trees, adding to saltcedar's competitive advantage. Dams alter the natural flow regime to which native cottonwood and willow are adapted. Relative to native Fremont cottonwood (*Populus fremontii*), leaf litter of saltcedar decomposes more rapidly, which is associated with at two-fold decrease in macroinvertebrate richness and a four-fold decrease in macroinvertebrate abundance (Bailey et al., 2001).

Several methods have been used or are being developed to control saltcedar, including mechanical, chemical, and biological control. The most effective management of saltcedar infestations appears to be a combination of several methods. Mechanical techniques range from simple hand removal of young saltcedar in small infestations to bulldozing and root plowing for the control of large infestations of mature saltcedar. The saltcedar that resprouts after mechanical control may be controlled with herbicides. Fire is of limited use as a control agent, because saltcedar resprouts readily after fire. Flooding kills saltcedar only if the root crowns remain submerged for at least three months. Several insects that attack saltcedar in its native range are now being investigated as biological control agents.

Purple Loosestrife

Purple loosestrife (*Lythrum salicaria*) is a serious exotic invader of both wetlands and riparian areas. It was introduced from Europe into New England in the early 1800s either accidentally or as an ornamental. Although it is a potential threat in every state, purple loosestrife is most problematic in the wetlands of the northeastern United States and the Great Lakes region. Since its introduction, several cultivated varieties have been developed, and though reported as sterile, these cultivated varieties have been shown to hybridize readily with wild populations to produce viable seed. Though its sale has been banned in many areas, purple loosestrife continues to be marketed in many regions. Its commercial promotion may have contributed to the rapid spread of the species.

Since 1930, purple loosestrife has spread rapidly, being recorded throughout southern Canada and in every state of the United States except Alaska and Florida. Estimates vary widely, but its current rate of spread is over 280,000 acres per year. Once established, purple loosestrife can create large nearly monospecific stands that may reach thousands of acres in size. These homogenous patches provide little of value for most wildlife species, and they displace native vegetation. The spread of purple loosestrife is thought to have been aided by increased rates of land disturbance from agricultural activity and the construction or development of transportation corridors such as roads and canals. Enrichment of wetland soils by fertilizers from runoff from agricultural lands may also increase the spread of purple loosestrife.

Like other successful invaders, purple loosestrife reproduces at a prodigious rate. A single mature plant produces nearly 3 million seeds, approximately 80 percent of which remain viable after 3 years of submergence. Seed densities of over 400,000 per square meter have been recorded in the upper 5 cm of wetland soils in Minnesota. Though vegetative reproduction is not a major contributor to the spread of this species, the biomass of individual plants increases substantially as they mature, contributing to the overall increase in exotic biomass in established stands.

Control of purple loosestrife infestations is difficult and costly. Mowing, burning, and flooding are generally ineffective, and flooding may contribute to the spread of purple loosestrife seeds and to the establishment of new populations. Small infestations can be removed by hand, but care must be taken to remove all root fragments as these can resprout. Herbicides can be successful if used with care. However, the greatest potential for control appears to be the introduction of several insects that attack purple loose-

continues

> **BOX 3-4**
> **Continued**
>
> strife in its native range. Five beetle species and one gall midge appear to hold the greatest promise as control agents and may over the long term reduce the cover of purple loosestrife to levels more similar to that in its native range where its cover ranges from 1 percent to 4 percent. The cumulative costs of controlling purple loosestrife as of 1999 amounted to approximately $45 million within the United States alone (Pimentel et al., 1999). This is only one example of the severity of the ecological and economic burden caused by exotic species. These mounting costs prompted President Bill Clinton to issue an executive order in 1999 to minimize the ecological, economic, and health damage caused by exotic species.

change, with important implications for streams and rivers (Michener et al., 1997). Where this would lead to an increased frequency or severity of flooding in riparian areas, shifts in the distribution of vegetation on floodplains are likely. As these new conditions become established, changes toward more flood-tolerant and disturbance-dependent species might be expected.

Lake Settings

Lake-fringe riparian areas typically respond both to seasonal variation in lake water levels and to interannual variation that causes the position of wetlands to migrate back and forth over longer periods of time (Keough et al., 1999). Susceptibility of riparian systems to climate change will depend largely on shoreline morphology. For example, deepening of water under a wetter climate would eliminate some riparian species, especially in areas where the shoreline is too steep to allow plants to establish. In contrast, lower water levels would require that plants establish further toward the lake, but this could happen only if protective barrier beaches form along the shoreline to reduce wave energy (Kowalski and Wilcox, 1999). This situation may become an issue for the Great Lakes in particular, where lake levels are expected to drop rapidly over the next several decades (Chao, 1999).

Marine and Estuarine Coastal Settings

The positive relationship between sea surface temperature and frequency of Atlantic hurricanes has led to speculations of greater hurricane activity with global warming. However, increases in hurricane intensity predicted by models fall within the range of natural interannual variability and of uncertainties of current studies (Henderson-Sellers et al., 1998).

A more likely consequence of global climate change is a rise in sea level, which continues to cause salinity intrusion and wetland loss in a number of coastal areas (Warren and Niering, 1993; Williams et al., 1999). The effects of rising sea level can be viewed in two dimensions: vertical and horizontal. The vertical dimension means that the soil surface of coastal riparian systems must keep pace with rising water levels (Brinson et al., 1995; Cahoon et al., 1995). Given that sea level is projected to rise by 2–9 mm/year over the next 100 years, sediment accretion will have to occur at a rate two- to nine-times that observed over the last century. The exact nature of vertical accretion is quite complicated because it is dependent on inorganic sediments mostly from continental sources, as well as on subsidence. For example, the sediment supply to the Mississippi Delta has decreased by 70 percent since 1860, largely because of the building of Missouri River dams. Increases in freshwater inputs to coastal zones, if accompanied by greater sediment supplies, could compensate for dam-induced decreases.

As for the horizontal component, Titus et al. (1991) predicted that a 1-m rise in sea levels would cause the loss of 36,000 square kilometers (14,000 square miles) of land in the contiguous United States, half of which would be wetlands in coastal riparian areas. Allowing riparian areas to migrate landward where sea level is rising could alleviate some of this loss, especially along coastlines with very low elevations. However, the presence of highways, cities, and other valued obstructions in many places will prevent this migration and will result in the compression, or loss, of riparian areas.

General Predictions

Riparian areas respond to changes in climate largely through characteristics of the terrestrial watersheds that supply their water. Riverine, lake, and estuarine-marine riparian areas occur in virtually all climates, so we should be able to predict change by arranging sites along a moisture continuum. In other words, transformation to a warming and drier (or wetter) climate in a particular region will produce conditions that already exist for riparian areas in a similar climate at a different geographic location (Figure 3-13) (Michener et al., 1997). For example, cypress–tupelo swamps, currently limited to the Southeast and up the Mississippi Valley, could replace silver maple–ash–elm forests of the warm temperate east. Mangroves would be expected at higher latitudes as the frequency of frost decreases. In each case, the composition of riparian plants and animals would be determined by additions of species that migrate to their correspondingly more favorable climatic conditions elsewhere and subtractions of species that become locally extinct because of less favorable environmental conditions.

Although species migration is likely to occur over long periods, there are nevertheless many formidable barriers to species migration, not only natural and human-influenced upland barriers, but also dams, dikes, and drainage systems. In addition, species vary greatly in their capacity to disperse, so the ability of plants

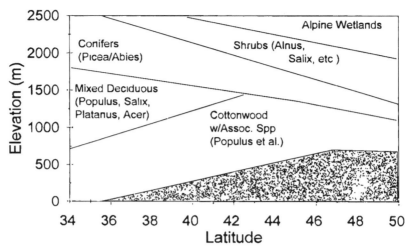

FIGURE 3-13 Distribution of riparian vegetation along elevational and latitudinal gradients. Increases in global temperature would shift distributions to higher elevations, making alpine wetland species locally extinct. SOURCE: Reprinted, with permission, from Patten (1998). © 1998 by Dr. Douglas A. Wilcox, Editor-in-Chief, Wetlands.

and animals to find suitable habitats under a changing climate will be highly variable. Recently introduced, non-native species, such as saltcedar, Russian olive, and Chinese privet, may provide insight into traits necessary for the dispersal of native species. These invaders may be expected, however, to also interfere with the gradual spread of indigenous riparian plants because exotics may be able to monopolize available space and nutrients before indigenous plants can arrive (Galatowitsch et al., 1999). On one hand, resident species, by co-opting space, may inhibit colonization by potential invaders. On the other hand, colonizers are facilitated when resident species become extinct or become stressed from human activities unrelated to climate change such as altered hydrology, nutrient loading, and sedimentation. There is some evidence that species-rich riparian communities are no more resistant to invasion by aliens than are those communities of lower species diversity (Planty-Tabacchi et al., 1996).

Just because a given species is effective in migrating and colonizing another geographic region does not mean, however, that it will also become extinct when the climate changes locally. For example, mangroves will not disappear from lower latitudes just because a warming climate allows them to expand into recently transformed frost-free areas. Populations and communities seldom respond in a monolithic fashion, and changes in hydrology seldom occur without corresponding changes in water quality, which itself can influence riparian productivity and species composition.

CURRENT STATUS OF RIPARIAN LANDS IN THE UNITED STATES

Determining the extent and condition of the nation's riparian areas is fundamental to managing them for multiple purposes. Although snapshots in time of riparian areas are useful for knowing their present status, the true utility of acreage and condition information lies in observing trends in these data over time and in understanding the factors causing such trends. Trends information is critical for relating riparian conditions (e.g., wildlife populations and vegetation) to other factors such as human population growth and water use. Such information can be used to predict future conditions in the presence or absence of restoration activities. And it can motivate decision-makers to take action before riparian areas are irreversibly impacted or destroyed. Surprisingly, there have been very few assessments of riparian acreage across the United States and only a handful of comprehensive studies on the condition of riparian lands.

Riparian Acreage

The amount of total land classified as "riparian" obviously depends on one's definition of that term. Indeed, variable definitions partially account for the inconsistent data found in reports of riparian acreage across the country. Many reports measure riparian areas in stream miles rather than acres, making direct comparisons difficult. Figure 3-14 shows the distribution of estimated stream miles and riparian acreage across the United States. National Resources Inventory and U.S. Environmental Protection Agency (EPA) estimates of current riparian acreage—which assume that the riparian area extends 50 ft from the edge of waterbodies—are 62 million and 38 million acres, respectively (excluding Alaska). Brinson et al. (1981) estimates a liberal upper limit of 121 million riparian acres, which includes all land in the 48 contiguous states that is within the 100-year floodplain and is thus potentially able to support riparian vegetation. This estimate was refined by Swift (1984) to those areas within the 100-year floodplains of streams and rivers that have certain vegetative characteristics. Swift estimated that there were at least 67 million acres of riparian land in the United States prior to European settlement, with about 23 million acres remaining. Reasons for the decrease in acreage include removal of vegetation along streambanks, channel straightening to remove meanders, and flooding of riparian areas upstream of impoundments. For example, an estimated 70 percent of the original floodplain forests have been converted to agricultural and urban land uses. Brinson et al. (1981) estimated that impoundments alone had inundated more than 24,000 km of streams, while the downstream effects of modified streamflow on riparian functions have been seldom documented. Case histories show that in some areas loss of natural riparian vegetation is as much as 95 percent—indicating that riparian areas are some of the most severely altered landscapes in the United States (Brinson et al., 1981).

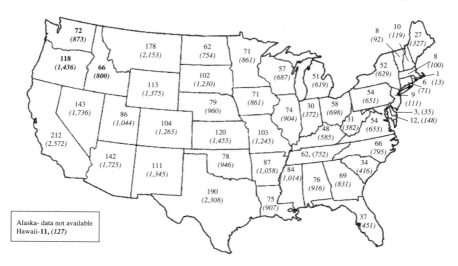

FIGURE 3-14 Stream miles and riparian acreage (in parentheses) in 1,000s. Acreage values assume 50-ft buffers on either side of streams. SOURCE: EPA (1999), except for bolded numbers, which are from the National Resources Inventory.

A more limited source of information on riparian acreage is public land data, which identify 23 million acres of combined riparian areas and wetlands on BLM-administered lands in the United States, although it is not known what portion is riparian (BLM, 1998). BLM and USFS (1994) suggests that total riparian acreage in the public domain is only 3.2 million acres. Because BLM statistics exclude lands bordering intermittent and ephemeral streams, these data are likely to underestimate the amount of existing riparian acreage by two to ten times. However, what is clear from all these sources is that riparian areas constitute a small fraction of total land area in the United States, probably less than 5 percent (Swift, 1984).

Riparian Condition

Determining the condition of riparian lands can be accomplished in multiple ways, for example, by measuring plant community composition or certain functions of riparian areas. As discussed in Chapter 5, several methods have been developed to assess riparian condition and functioning that cast their results in qualitative terms (e.g., Prichard et al., 1998). For example, in 1999, 40 percent of riparian areas administered by the BLM (excluding Alaska) were rated healthy ("proper functioning condition"), 41 percent were rated "at risk," 10 percent were found to be in poor health ("not functioning"), and 9 percent had not been assessed (BLM, 1999). A similar methodology used on USFS lands rated 78 percent of their riparian areas to be healthy and 22 percent as "not meeting objectives" (BLM and USFS, 1994). The variability of these data is extremely high, as evidenced by a USFS report that less than 10 percent of the riparian areas associated with 36 miles of perennial streams on one allotment in Lincoln National Forest, New Mexico, are in satisfactory condition (USFS, 1999). In addition, these statistics reveal little about the condition of riparian areas in the eastern states because of the small percentage of public lands found in the East.

Information on water quality can also be used to make inferences about riparian conditions, given the proximity and interconnections of streams and riparian areas. Data from EPA's 305(b) program indicate that there are at least 300,000 miles of streams (10 percent of the total) and more than 5 million acres of lakes that are not meeting water-quality standards (EPA, 2000). Given that many states' lists of impaired waters have been found to underestimate the true number of affected waters (NRC, 2001b), these numbers are likely to be conservative. Indeed, in many regions of the country, the percentage of impaired stream segments is greater than 25 percent. Sediment, nutrients, and pathogens are the top three pollutants responsible for water-quality impairment, followed by anoxia, metals contamination, and habitat destruction. The fact that properly functioning riparian areas can reduce levels of these contaminants in nearby streams and rivers (see Chapter 2) suggests that riparian areas adjacent to impaired streams are also suffering from quality degradation.

The health of riparian areas has also been assessed simply by observing or estimating trends in riparian acreage over time. Swift (1984) estimates that 66 percent of all riparian areas in the United States have been destroyed, reflected as losses of vegetation and conversion to some other land use, predominantly agriculture. These declines have been most severe in the Mississippi Delta, the agricultural Midwest, California, and the arid Southwest. In particular, an estimated 85 percent to 95 percent of Arizona and New Mexico riparian forests have been lost to grazing and other land uses, with almost 100 percent modification along stretches of the lower Colorado River, lower Gila River, lower Salt River, and the Rio Grande (Ohmart and Anderson, 1986; Noss et al., 1995; Mac et al., 1998). Indeed, an Arizona Executive Order, Order 91-6 (1991), stated that "over 90 percent of the native riparian areas along our major desert watercourses have been lost, altered, or degraded." Similarly dire statistics exist for California, where only 5 percent to 10 percent of the original riparian habitat remains (Mac et al., 1998). A recent report card on Oregon's riparian areas found that riparian forests along the Willamette River have been reduced by more than 85 percent since the 1850s (Gregory, 2000). Even in the absence of more comprehensive historical data, it is apparent that riparian areas across the nation have declined in both acreage and condition.

Historical trends for wetlands have received considerably greater attention, and such data may provide clues about trends in riparian acreage, given that these areas sometimes overlap. Between 1780 and 1980, every state experienced declines in wetland acreage, with greater than 50 percent loss in 22 states (Table 3-4). Wetlands in California, Ohio, Iowa, Missouri, Indiana, Illinois, and Kentucky were subject to the most intensive alteration. Between 1986 and 1997, the estimated total loss of wetlands was 644,000 acres (Dahl, 2000). It should be noted that in some states, particularly in the agricultural Midwest, widespread drainage of wetlands, wet prairies, and other areas has resulted in the creation of extensive networks of drainage ditches, underground drainage tiles, and other types of channels to convey agricultural drainage water to river systems. These new channels may be bordered by riparian areas that previously did not exist, and thus might be thought of as increasing overall riparian acreage. However, because such riparian areas are often hydrologically disconnected from the constructed channel, they are generally in poor condition and possess limited functioning compared to native riparian areas (see earlier discussion on drainage networks). In fact, about 50 percent of the land along streams and drainage ditches in Iowa is cultivated to the bank, and another 30 percent is in pasture, most of which is overgrazed (Schultz et al., 2000).

The Emergency Wetland Resources Act requires that wetlands in the United States be inventoried on a regular (10-year) basis to observe ongoing changes, but the statute does not extend to riparian areas. Given the profound lack of information on riparian land status and trends, a comprehensive and rigorous assessment of riparian acreage similar to the National Wetlands Inventory is greatly needed. State interest in such an inventory is great, as evidenced by the creation of individual programs (e.g., Colorado's) in lieu of a national program. The satellite and spectral technology needed to conduct an inventory of riparian areas exists today, as described below.

Remote Sensing of Riparian Condition

Although there is no comprehensive or methodologically consistent monitoring of trends in the nation's riparian areas, recent technology makes landscape assessment of riparian community composition and distribution possible and cost effective. Land use/land cover can be used as an indicator to evaluate the quality and spatial extent of riparian areas along streams, rivers, lakes, and wetlands. Remotely sensed information (aerial photos, satellite spectral data) can be used to predict the composition of vegetation or major classes of land uses. Land cover/land use information, gathered systematically at regular intervals, can inform many community-based decisions, especially those concerning riparian areas. Fundamental goals for human activity—preservation of agricultural soils, preservation of wetlands, control of rural residential sprawl, limitation of impervious surfaces in urban areas—can be assessed and quantified through patterns of vegetation derived from remotely sensed data.

TABLE 3-4 Wetland Loss By State From the 1780s to the 1980s

State	Total surface area (acres)	1780 Wetlands estimates (acres)	1980 Wetlands estimates (acres)	% Surface area that is wetlands in 1980	% Wetlands lost
AL	33,029,760	7,567,600	3,783,800	11.5	−50
AK	375,303,680	170,200,000	170,000,000	45.3	−0.1
AZ	72,901,760	931,000	600,000	0.8	−36
AR	33,986,560	9,848,600	2,763,600	8.1	−72
CA	101,563,520	5,000,000	454,000	0.4	−91
CO	66,718,720	2,000,000	1,000,000	1.5	−50
CT	3,205,760	670,000	172,500	5.4	−74
DE	1,316,480	479,785	223,000	16.9	−54
FL	37,478,400	20,325,013	11,038,300	29.5	−46
GA	37,680,640	6,843,200	5,298,200	14.1	−23
HI	4,115,200	58,800	51,800	1.3	−12
ID	53,470,080	877,000	385,700	0.7	−56
IL	36,096,000	8,212,000	1,254,500	3.5	−85
IN	23,226,240	5,600,000	750,633	3.2	−87
IA	36,025,600	4,000,000	421,900	1.2	−89
KS	52,648,960	841,000	435,400	0.8	−48
KY	25,852,800	1,566,000	300,000	1.2	−81
LA	31,054,720	16,194,500	8,784,200	28.3	−46
ME	21,257,600	6,460,000	5,199,200	24.5	−20
MD	6,769,280	1,650,000	440,000	6.5	−73
MA	5,284,480	818,000	588,486	11.1	−28
MI	37,258,240	11,200,000	5,583,400	15.0	−50
MN	53,803,520	15,070,000	8,700,000	16.2	−42
MS	30,538,240	9,872,000	4,067,000	13.3	−59
MO	44,599,040	4,844,000	643,000	1.4	−87
MT	94,168,320	1,147,000	840,300	0.9	−27
NE	49,425,280	2,910,500	1,905,500	3.9	−35
NV	70,745,600	487,350	236,350	0.3	−52
NH	5,954,560	220,000	200,000	3.4	−9
NJ	5,015,040	1,500,000	915,960	18.3	−39
NM	77,866,240	720,000	481,900	0.6	−33
NY	31,728,640	2,562,000	1,025,000	3.2	−60
NC	33,655,040	11,089,500	5,689,500	16.9	−49
ND	45,225,600	4,927,500	2,490,000	5.5	−49
OH	26,382,080	5,000,000	482,800	1.8	−90
OK	44,748,160	2,842,600	949,700	2.1	−67
OR	62,067,840	2,262,000	1,393,900	2.2	−38
PA	29,013,120	1,127,000	499,014	1.7	−56
RI	776,960	102,690	65,154	8.4	−37
SC	19,875,200	6,414,000	4,659,000	23.4	−27
SD	49,310,080	2,735,100	1,780,000	3.6	−35
TN	27,036,160	1,937,000	787,000	2.9	−59
TX	171,096,960	15,999,700	7,612,412	4.4	−52
UT	54,346,240	802,000	558,000	1.0	−30
VA	26,122,880	1,849,000	1,074,613	4.1	−42
VT	6,149,760	341,000	220,000	3.6	−35
WA	43,642,880	1,350,000	938,000	2.1	−31
WV	15,475,840	134,000	102,000	0.7	−24
WI	35,938,560	9,800,000	5,331,392	14.8	−46
WY	62,664,960	2,000,000	1,250,000	2.0	−38

SOURCE: Dahl (1990).

Land use refers to "human activities on the land which are directly related to the land" (Clawson and Stewart, 1965, in Anderson et al., 1976); land cover describes "the vegetational and artificial constructions covering the land surface" (Burley, 1961). On any patch of land, there can be both a land use and a land cover. In some cases, land use and land cover can be consistent in that one infers the other. For example, a land cover of agricultural row crop is consistent with a land use of agriculture. In other cases, forest land cover may be coincident with a land use of urban, agriculture, residential, and forestland use. Nonetheless, these two types of data allow interpretation of human use and of potential habitat conditions and ecological processes.

Recently, land use/land cover has been used as an indicator of the status of various resources. Models of populations, communities, and habitats can be driven by information on either land cover or land use. Ecological models allow quantification of current, historical, or future biological responses to land-use patterns.

Prior to the development of satellite remote sensing, interpretation of aerial photographs was the primary method of acquiring land cover data over large areas. Today, spectral data from satellites and aerial photography are both commonly used. The ability to differentiate between areas of different land use/land cover depends on the spatial resolution of the sensor and the complexity of the information contained in the spectral data or photographic image. For example, images acquired below 10,000 ft can generally provide more detail than can images obtained at 100,000 ft. However, images taken at higher altitudes cover more area. Finer spatial resolution requires larger amounts of data per unit area and more extensive computer resources for the manipulation and storage of those data. Thus, there is a trade-off between spatial resolution of the data and the cost and time required for analysis.

For changes in measured land use/land cover data to reflect changes in resource quality, the spatial and temporal resolution of those data should ideally match the spatial and temporal scales over which the physical mechanisms that determine resource quality operate. If sufficient resolution is lacking, important trends will not be observable. The resolution of the sensor and the analysis technique are particularly crucial for the accurate measurement of thin, linear features such as streams and riparian corridors. The masking of streams by overhanging vegetation as well as seasonal changes may preclude accurate measurements of stream areal or lineal extent via land use/land cover data.

How Land Use/Land Cover Data Might Be Used

Riparian areas and land cover can be mapped using remotely sensed data from different sources (e.g., satellites, high- and low-elevation aircraft, and balloons) and that provide different spectral information (e.g., multispectral, infrared, visible, ultraviolet, videography, or digital photographs). These methods are variable in their spatial resolution (e.g., less than one meter for low-elevation videography to 80 m for satellite Multispectral Scanner sensors) and temporal representation (e.g., seasonal analysis of phenology as in Oetter et al., 2001, vs. a representation from a single image at a standardized time of year). Like all scientific measurements, the techniques and sources of information must be carefully matched to the specific questions or objectives.

Although a growing array of remotely sensed information offers many options for assessment of riparian conditions, land use/land cover data for large basins or regions are most efficiently obtained by the processing and analysis of remotely sensed images acquired by satellite. The USGS through the EROS Data Center allows federal government and affiliated users to access Landsat data from both the Thematic Mapper (TM) and the Multispectral Scanner (MSS) sensors. The spatial resolution of TM data is 30 m, while MSS data are of 80-m resolution. Data from other sensors are also commercially available. These sensors are carried on satellites that have been launched by either commercial organizations such as the Space Imaging Corporation or other countries; resolution of multispectral data from these satellites begins at 4 m (IKONOS 1). TM scenes are available from 1982 to the present, while MSS data of good quality are available from 1975 to 1992. The same area of the satellite's footprint is revisited every 16 days for the Landsat TM scenes; the revisit time for other satellites varies from a few days to in excess of a month.

Several studies have acquired land use/land cover data over relatively large land areas. For example, as a part of a recurrent forest inventory, changes in area of forest, agriculture, low-density urban, and urban land use between 1971–74 and for 1982 were mapped from aerial photographs for western Oregon on a countywide basis by the Forest Inventory and Analysis Unit of the USFS Pacific Northwest Research Station (Gedney and Hiserote, 1989). The Natural Resources Conservation Service (NRCS) of the U.S. Department of Agriculture provides an assessment of resources on national and state scales at 5-year intervals as part of the National Resources Inventory (NRI) (NRCS, 1998). The NRI is the one major national source of information on land cover, providing "updated information on the status, condition, and trends of land, soil, water and related resources on the Nation's nonfederal land" (Florida Center for Public Management, 1998).

Spatial configuration of habitat across the landscape is an essential component of conservation strategies for wildlife, and indices of landscape pattern have been linked in many studies to ecological function (Schumaker, 1996). Given a land use/land cover database of appropriate resolution and extent, metrics describing landscape patterns (patch size, shape, connectivity, distribution, interior size, edge length, edge-to-area ratio, etc.) can be extracted (Turner, 1989). If the land use/land cover classes adequately define the habitat type of critical species, biological diversity indicators can make use of these metrics. In the Netherlands, for example, wood lot size was found to be the best single indicator of bird species richness (Van Dorp and Opdam, 1987). In the Pacific Northwest, the abundance of spotted owls was found to vary with the proportion of old growth in the forest, with home ranges increasing in partially harvested forests as fragmentation of the old growth areas increased (Carey, 1985). Thus, land use/land cover can serve as an indicator that attempts to locate watersheds or perhaps river reaches where aquatic and terrestrial species are exposed to greater risk because of the changes in the landscape.

Recent Land Use/Land Cover Studies in Riparian Areas

Recent watershed studies have used multivariate statistics and geographical information systems (GIS) to examine how terrestrial ecosystems, human activities, climate, and geology affect nutrient concentrations in streams (e.g., Biggs et al., 1990; Richards and Host, 1994; Richards et al., 1996). Johnson et al. (1997) used land use/land cover, topography, hydrology, and geology of the Saginaw Bay watershed of central Michigan (4.03 million acres) and 62 water quality sampling sites to investigate the relationships between these landscape factors and nutrient and sediment concentrations in streams. Derived land-use metrics included the percent non-row-crop agriculture, the percent urban, the percent forested wetland, and patch density. Non-anthropogenic metrics included catchment area, mean slope, surficial geology, and the percent coarse till. Land-use mapping resolution was about 2.5 acres (1 hectare), based on aerial photography, with six classes of land use defined; a 100-foot (30-meter) digital elevation model was used. Although complex relationships between the seasons, chemicals, and landscape factors were determined, about 50 percent of the variation in chemical concentrations was not explained by any of the landscape factors studied. The scale (resolution) of the land cover data was suggested as part of the problem: forested riparian buffers were thought to be underrepresented. Also, the temporal scales of the processes controlling the fluxes of concentration (e.g., storms) are such that there may have been undersampling. This study and other studies show that land use/land cover alone cannot act as an indicator of water quality; quantities that describe the mechanisms by which the chemical inputs to streams are linked to land use must also be considered. The study also highlights the need to understand the limitations caused by the spatial resolution of the land use/land cover data (here, ignorance of the full extent of riparian areas) and either to develop an adequate measurement technique for poorly sampled components or to develop proxies to represent these elements.

Fish and Wildlife Service Riparian Mapping System

Detailed mapping of riparian plant communities has been initiated by the U.S. Fish and Wildlife Service for western areas of the United States (FWS, 1998). The riparian mapping effort is an outgrowth of the National Wetlands Inventory and thus has all of the strengths of that program, including field validation and development of standardized protocols. The approach, based on aerial photography, provides detailed descriptions of the composition and distribution of riparian plant communities, breaking down riparian areas by plant type (useful for detailed modeling of wildlife).

Some weaknesses of the method are that it is limited in spatial extent because of the time and resources required for the fine-scale mapping. Because it uses aerial photography, it is only applicable to the western two-thirds of the country where vegetative differences between riparian areas and uplands are obvious. The extensive effort required for this type of assessment prevents its use as a tool for monitoring broad trends in riparian conditions across large regions over long periods of time. Although some groups are more comfortable with the finer spatial resolution and additional information provided by aerial photography (or related techniques such as airborne videography and digital imaging), the expense and analysis time required to process these forms of land cover data make them computationally and financially unsuitable for assessment of the entire United States on a frequently recurring basis. These finer resolution sources could be used by states and local resource management agencies to augment a broader national assessment and provide local detail on spatial extent and composition of riparian vegetation. FWS should help develop a uniform national program for mapping riparian areas that relies on measurements of land cover and land use from broadly available remotely sensed data, such as satellite multispectral data. As described in Box 3-5, Illinois and Oregon have mapped land use and land cover via satellite remote sensing and are using those data in land planning and resource management.

CONCLUSIONS AND RECOMMENDATIONS

Historical and current land-use practices across the United States significantly affect the hydrologic, geomorphic, and biological structure and functioning of riparian areas. Land-use practices that directly remove native vegetation such as row-crop agriculture, grazing, timber harvesting, urban development, and mining have altered the character of riparian systems. Changes in hydrologic regimes as a result of water resources development across the nation have been particularly widespread and effective in degrading riparian areas. Other effects have been more indirect in that the management of upslope areas has brought about changes in adjacent riparian areas (e.g., accelerated erosion and pollutant transport from upslope areas following development or flow modification through tile drainage). The degraded conditions caused by some land and water uses are reversible over the short term—for example, by implementing agricultural best management practices or restricting grazing. The effects of other activities, such as large dams and levees and the extensive modification of hydrology in many agricultural areas, can only partially be ameliorated.

BOX 3-5
Remote Sensing of Land Cover/Land Use in Illinois and Oregon

Illinois

Characterization of riparian lands in Illinois has been attempted using two approaches. In the early 1980s, with the development of the Illinois Streams Information System (ISIS) and subsequent revisions (Johnston et al., 1999), aerial photograph slides from the Agricultural Stabilization and Conservation Service were projected onto USGS 1:24,000 topographic maps, and land cover was interpreted for 0.1-mile segments of streams. ISIS includes riparian cover descriptions for land immediately adjacent to the stream and for the dominant cover in a 300-m-wide strip adjacent to the stream channel. This was a laborious process and will be difficult to duplicate on any regular schedule.

The challenging task of rapidly and frequently quantifying riparian areas, or at least describing land cover within a corridor that reasonably reflects the riparian area, can be accomplished much more easily through use of satellite imagery. One frequently used source of imagery is Landsat Thematic Mapper (TM), which provided Illinois with a dataset that, when interpreted, described the land cover in 23 categories at a ground resolution of 28.5 x 28.5 meters (93.5 x 93.5 feet or about 0.2 acres). These categories include urban and rural cover types with such broad definitions as row crop, small grains, orchards/nurseries, urban grassland (parks, residential lawns, golf courses, cemeteries, and other open space), rural grassland (pastureland, grassland, waterways, buffer strips, Conservation Reserve Program land, etc.), deciduous (two types), coniferous, and wetland (five types).

The original data used were TM satellite imagery from Landsat 4. Imagery from two dates was used for each area of the state, and all imagery was taken from the period 1991–1995. Since the original collection, the two state agencies and the National Agricultural Statistics Service have entered into agreements that will result in agricultural land being reassessed each year and non-agricultural land being assessed on at least a five-year basis. These data can be used to characterize land cover within any predetermined area. For example, if a riparian area was determined from field elevation data or through simple width definitions, land cover within this area could easily be determined. However, limitations of resolution do prevent accurate description of narrow riparian areas, and the generalities of land cover classification can be problematic. Recently available satellite information will provide resolution at the 5-meter pixel size and will be much more useful where riparian areas have been drastically reduced in width or are naturally narrow.

Oregon

The Willamette River basin encompasses 30,000 km^2, entering the Columbia River roughly 90 km upstream of its mouth at the Pacific Ocean. The basin contains several major urban centers, residential areas, agricultural lands, and commercial and federal forests. Human population in the Willamette Valley is expected to double in the next 50 years, placing tremendous demands on limited land and water resources. Much of this future human population growth will be focused on the riparian corridors throughout the lowland portions of the valley. Effective environmental management policies will require explicit analysis of landscape features and integration of appropriate management practices. This requires a scientifically credible assessment of riparian conditions.

The Pacific Northwest Ecosystem Research Consortium developed a land use/land cover map of the basin based on classification of more than 50 vegetation cover classes from satellite spectral data. Riparian systems were evaluated in 120-m widths on both sides of all perennial streams in the basin. The results revealed that in the uplands of the Willamette Basin, conifer forests comprise only 52 percent of the riparian systems along these streams and rivers. Agriculture and development make up a small portion of the riparian systems along small streams, accounting for less than 2 percent of the area. Conifer and hardwood forests each account for less than 10 percent of the riparian areas along small streams in the lowlands. Agricultural crops and urban development along small streams occupy roughly 40 percent and 10 percent of riparian areas, respectively. Riparian conditions along small rivers in the basin exhibit similar patterns to those observed for small streams, though hardwood forests are more abundant than conifer forests. The 120-m band along the mainstem Willamette River is a small fraction of the floodplain, but within this 120-m band, hardwood forests make up 14 percent of the riparian area and conifer forests make up less than 5 percent of the riparian area. Agriculture now occupies more than a third of the riparian areas, and development occupies roughly a quarter of the riparian lands. This represents a loss of 88 percent of the floodplain forests that were present in 1850, based on surveys from the General Land Office. Development has modified riparian forests along the Willamette River to a greater extent than in the smaller streams and tributaries.

This analysis of a large basin with diverse topography and land use illustrates the potential application of satellite remote sensing for riparian assessment. Though the accuracy of this classification at the scale of a few meters from the stream and the resolution of plant types is relatively limited, accuracy of classification across the landscape is relatively high (65–80 percent accurate) and provides a large-scale perspective unavailable with any other source of information. In addition, the analytical process can be applied to data sets from future years with relatively minor costs as compared to the costs of developing the original relationships and analytical methods.

Most human development and land-use activities have cumulative impacts in riparian areas that are rarely considered during planning or management. Riparian systems typically cut through diverse landscapes, crossing political, social, cultural, and land-use boundaries. Coordinated efforts at land-use planning in riparian areas would benefit from the early identification of the multiple and often cumulative impacts associated with various human activities that can occur at both local and basin scales. There is a critical need to implement, nationwide, land-use practices that are "riparian friendly" and that are effective at eliminating or significantly reducing many of the potentially adverse effects of existing and future land uses.

The majority of riparian areas in the United States have been converted or degraded. Although landscape-scale studies assessing the extent and condition of riparian areas have been limited, results indicate that conversion and degradation have been common. The spatial extent of riparian forests has been substantially reduced, plant communities on floodplains have been converted to other land uses or have been replaced with developments, and the area of both woody and non-woody riparian communities has decreased. The ecological functions of these riparian systems are greatly diminished in comparison to their historical range of ecological condition.

Current assessments of the status of riparian areas are incomplete, cover a small fraction of perennial streams and almost no intermittent and ephemeral streams, and are not operationally consistent. It has only been relatively recently that assessments of the areal extent and condition of riparian systems have been undertaken. Unfortunately, these efforts have been limited in scope, are difficult to compare because of differing methodologies, and provide only a fragmented view of the nation's riparian areas. The few existing studies of riparian condition tend to be ground-based assessments from field studies or aerial photogrammetry, and they focus only on relatively small areas or stream lengths. Although such studies provide detailed information for managing local resources, they do not address changes in riparian conditions at regional or national scales or measure rates of change from repeated measures or assessments.

There is no comprehensive or methodologically consistent monitoring of trends in the nation's riparian areas, although current technology makes landscape assessment of riparian community composition and distribution possible and cost effective. National assessment of wetlands is far more advanced than riparian assessment. The riparian mapping effort of the FWS could promote development of a uniform national program for mapping riparian areas. Such a program should incorporate broadly available remotely sensed data, such as satellite multispectral data, which could be used to classify and map land cover and land use information in each of the states.

Remotely sensed land cover/land use information from satellite spectral data offers the greatest potential for monitoring riparian conditions consistently across the United States on a frequently recurring basis. Satellite data extending back to the early 1970s provide a 20- to 30-year record of changes in riparian resources. Future development of analytical techniques and refinement of classifications can be reapplied to historical satellite data in order to take advantage of future advancements in remote sensing. Increasing the availability of remotely sensed information on riparian conditions would allow citizens and management authorities to assess environmental status and track changes in this critical resource.

Although land-use changes have had and will continue to have the greatest effect on riparian areas in the near and medium term, global climate change is likely to exacerbate stressors on riparian areas rather than counteract them. Thus, land owners and managers should continue to strive for land uses that are consistent with protecting and restoring riparian areas in the absence of definitive information about how climate changes may be influencing those systems. This includes reducing stressors from localized human activities such as water withdrawals, flow regulation, continued land drainage, excessive sedimentation, nutrient loading, excessive grazing, and introduction and spread of exotic species.

REFERENCES

Adams, P. W., and J. O. Ringer. 1994. The effects of timber harvesting and forest roads on water quantity and quality in the Pacific Northwest: summary and annotated bibliography. Corvallis, OR: Forest Engineering Department, Oregon State University. 147 pp.

Akashi, Y. 1988. Riparian vegetation dynamics along the Bighorn River, Wyoming. M. Sc. Thesis, University of Wyoming, Laramie. 245 pp.

Andereck, K. L. 1995. Environmental consequences of tourism: a review of recent research. Pp. 77–81 In: Linking tourism, the environment, and sustainability. Gen. Tech. Report INT-GTR-323. Ogden, UT: USDA Forest Service.

Anderson, J. R, E. E. Hardy, J. T. Roach, and R. E. Witmer. 1976. A land use and land cover classification system for use with remote sensor data. Geological Survey Professional Paper 964, Washington DC: U.S. Geological Survey.

Bailey, J. K., J. A. Schwietzer, and T. G. Whitham. 2001. Salt cedar negatively affects biodiversity of aquatic macroinvertebrates. Wetlands 21:442–447.

Baker, J. L. 1983. Agricultural areas as nonpoint sources of pollution. Pp. 275–310 In: Environmental impacts of nonpoint source pollution. M. R. Overcash and J. M. Davidson (eds.). Ann Arbor, MI: Ann Arbor Sci. Publ., Inc.

Balda, R. P. 1991. The relationship of secondary cavity nesters to snag densities in western coniferous forests. Wildlife Habitat Technical Bulletin No. 1. Albuquerque, NM: USDA.

Baltz, D. M., and P. B. Moyle. 1993. Invasion resistance to introduced species by a native assemblage of California stream fishes. Ecological Applications 3:246–255.

Behan, M. 1981. The Missouri's stately cottonwoods: how can we save them? Montana Magazine, September 76–77.

Belsky, A. J., A. Matzke, and S. Uselman. 1999. Survey of livestock influences on stream and riparian ecosystems in the western United States. Journal of Soil and Water Conservation 54: 419–431.

Benke, A. C. 1990. A perspective on America's vanishing streams. Journal of the North American Benthological Society 9:77–88.

Berger, J., P. B. Stacey, L. Bellis, and M. P. Johnson. 2001. A mammalian predator-prey imbalance: grizzly bear and wolf extinction affect avian neotropical migrants. Ecological Applications 11:947–960.

Beschta, R. L. 1990. Effects of fire on water quantity and quality. Pp. 219–232 In: Natural and prescribed fire in Pacific Northwest forests. J. D. Walstad, S. R. Radosevich, and D. V. Stromberg (eds.). Corvallis, OR: Oregon State University Press.

Beschta, R. L., M. Pyles, A. Skaugset, and C. Surfleet. 1999. Peakflow responses to forest practices in the western cascades of Oregon, USA. Journal of Hydrology 233:102–120.

Biggs, B. J., M. J. Duncan, I. G. Jowett, J. M. Quinn, C. W. Hickey, R. J. Davies-Colley, and M. E. Close. 1990. Ecological characterization, classification and modelling of New Zealand rivers: an introduction and synthesis. New Zealand Journal of Marine and Freshwater Research 24:277–304.

Binkley, D., and T. C. Brown. 1993. Management impacts on water quality of forests and rangelands. Gen. Tech. Report RM-239. USDA Forest Service. 114 pp.

Bleich, J. L. 1988. Chrome on the range: off-road vehicles on public lands. Ecology L. Q. 15:159–189.

Booth, D. 1991. Urbanization and the natural drainage system-impacts, solutions and prognoses. Northwest Environmental Journal 7(1):93–118.

Booth, D. B., and C. R. Jackson. 1997. Urbanization of aquatic systems: degradation thresholds, stormwater detention, and the limits of mitigation. J. American Water Resources Assoc. 33(5):1077–1090.

Bradley, C., and D. Smith. 1986. Plains cottonwood recruitment and survival on a prairie meandering river floodplain, Milk River, southern Alberta and northern Montana. Canadian Journal of Botany 64:1433–1442.

Bravard, J., C. Amoras, G. Pautou, G. Bornette, M. Bournard, C. Des Chatelliers, J. Gibert, J. Peiry, J. Perrin, and H. Tachet. 1997. River incision in southeast France: morphological phenomena and ecological effects. Regulated Rivers: Research and Management 13:75–90.

Brewer, R., G. A. McPeek, and R. J. Adams, Jr. 1991. The atlas of breeding birds of Michigan. East Lansing, MI: Michigan State University Press. 594 pp.

Brinson, M. M., B. L. Swift, R. C. Plantico, and J. S. Barclay. 1981. Riparian ecosystems: their ecology and status. FWS/OBS-81/17. Kearneysville, WV: U.S. Fish and Wildlife Service.

Brinson, M. M., R. R. Christian, and L. K. Blum. 1995. Multiple states in the sea-level induced transition from terrestrial forest to estuary. Estuaries 18:648–659.

Brooks, D. R. 2000. Reclamation of lands disturbed by mining of heavy minerals. Pp. 725–754 In: Reclamation of drastically disturbed lands. Agronomy Monograph No. 41. Madison, WI: American Society of Agronomy.

Brothers, T. S. 1984. Historical vegetation change in the Owens River riparian woodland. Pp. 75–84: In: California riparian systems: ecology, conservation, and productive management. R. Warner and C. Hendricks (eds.). Berkeley, CA: University of California Press.

Brown, D. E., C. H. Lowe, and J. F. Hausler. 1977. Southwestern riparian communities: their biotic importance and management in Arizona. Pp. 201–211 In: Importance, preservation, and management of riparian habitat: a symposium. Tucson, Arizona, July 9, 1977. R. R. Johnson and D. A. Jones (eds.).

Bryan, M. D., and D. L. Scarnecchia. 1992. Species richness, composition, and abundance of fish larvae and juveniles inhabiting natural and developed shorelines of a glacial Iowa lake. Environ. Biol. of Fishes 35:329–341.

Bryant, A. A. 1986. Influence of selective logging on red-shouldered hawks, *Buteo lineatus*, in Waterloo region, Ontario, 1953-1978. The Canadian Field-Naturalist 100:525.

Buech, R. R. 1992. Streambank stabilization can impact wood turtle nesting areas. 54th Midwest Fish and Wildlife Conference Proceedings. Toronto, Ontario. December 6–9, 1992.

Bull, W. B. 1997. Discontinuous ephemeral streams. Geomorphology 19:227–276.

Bureau of Land Management (BLM). 1998. Public Land Statistics. Volumes 182, 183. BLM/BC/ST-99/001+1165.

BLM. 1999. Public Land Statistics. www.blm.gov.

Bureau of Reclamation. 1977. Design of small dams. Water Resources Technical Publication. Washington, DC: U.S. Department of the Interior. 816 pp.

Burger, J. 1995. Beach recreation and nesting birds. Pp. 281–295 In: Wildlife and recreationists. coexistence through management and research. R. L. Knight and K. J. Gutzwiller (eds.). Washington, DC: Island Press. 372 pp.

Burley, T. M. 1961. Land use or land utilization. Professional Geographer 13(6):18–20.

Busch, D. E., and S. D. Smith. 1995. Mechanisms associated with decline of woody species in riparian ecosystems of the southwestern U.S. Ecological Monographs 65:347–370.

Cahoon, D. R., D. J. Reed, and J. W. Day, Jr. 1995. Estimating shallow subsidence in microtidal salt marshes of the southeastern United States: Kaye and Barghoorn revisited. Marine Geology 128:1–9.

Carey, A. B. 1985. A summary of the scientific basis for spotted owl management. In: Ecology and management of the spotted owl in the Pacific Northwest. General Technical Report PNW-185. R. J. Gutierrez and A. B. Cary (eds.). Portland, OR: USDA Forest Service, Pacific Northwest Forest and Range Experiment Station.

Chaney, E., W. Elmore, and W. S. Platts. 1990. Livestock grazing on western riparian areas. Eagle, ID: Northwest Resource Information Center, Inc. 45 pp.

Chao, P. 1999. Great Lakes water resources: climate change impact analysis with transient GCM scenarios. J. Amer. Water Resources Assoc. 35(6):1499–1507.

Christensen, D. L., B. J. Herwig, D. E. Schindler, and S. R. Carpenter. 1996. Impacts of lakeshore residential development on coarse woody debris in north temperate lakes. Ecol. Application 6:1143–1149.

Clawson, M., and C. L. Stewart. 1965. Land use information: a critical survey of U.S. statistics including possibilities for greater uniformity. Baltimore, MD: The John Hopkins Press for Resources for the Future, Inc. 402 pp.

Coggins, G. C., C. F. Wilkinson, and J. D. Leshy. 2001. Federal public land and resources law, 4th ed. New York: Foundation Press.

Cole, D. N. 1989. Areas of vegetation loss: a new index of campsite impact. USDA/FS Research Note INT-389.

Cole, D. N. 1993. Minimizing conflict between recreation and nature conservation. Pp. 105–122. In: Ecology of greenways: design and function of linear conservation areas. D. S. Smith and P. C. Hellmund (eds.). Minneapolis, MN: University of Minnesota Press.

Collins, B. D., and T. Dunne. 1989. Gravel transport, gravel harvesting, and channel-bed degradation in rivers draining the southern Olympic Mountains, Washington, USA. Environmental Geology and Water Sciences 13:213–224.

Committee on Environment and Natural Resources (CENR). 2000. Integrated assessment of hypoxia in the Northern Gulf of Mexico. Washington, DC: National Science and Technology Council Committee on Environment and Natural Resources.

Conner, R. N., R. G. Hooper, H. S. Crawford, and H. S. Mosby. 1975. Woodpecker nesting habitat in cut and uncut woodlands in Virginia. J. Wildl. Manage. 39:144–150.

Crawford, C. S., A. C. Culley, R. Leutheuser, M. S. Sifuentes, L. H. White, and J. P. Wilber. 1993. Middle Rio Grande ecosystem: Bosque biological management plan. Albuquerque, NM: U.S. Fish and Wildlife Service, District 2. 291 pp.

Crossman, E. J. 1991. Introduced freshwater fishes: A review of the North American perspective with emphasis on Canada. Canadian Journal of Fisheries and Aquatic Sciences 48:46–57.

Crouch, G. 1979. Changes in the vegetation complex of a cottonwood ecosystem on the South Platte River. Great Plains Agricultural Council Publication 91:19–22.

Cunningham, A. A. 1996. Disease risks of wildlife translocations. Conservation Biology 10:349–353.

Dahl, T. E. 1990. Wetlands losses in the United States 1780s to 1980s. Washington, DC: U.S. FWS.

Dahl, T. E. 2000. Status and Trends of Wetlands in the Conterminous United States 1986 to 1997. Washington, DC: U.S. FWS.

Daulton, T., and C. Hanna. 1997. Protecting Inland Lakeshores of the North. Robert E. Matteson Workshop. Sigurd Olson Environmental Institute. Northland College. Ashland, WI.

David, M. B., L. E. Gentry, D. A. Kovacic, and K. M. Smith. 1997. Nitrogen balance in and export from an agricultural watershed. Journal of Environmental Quality 26:1038–1048.

DeByle, N. V. 1985. Animal impacts. Pp. 115–23 In: Aspen: ecology and management in the western United States. Gen. Tech. Rept. RM-119. USDA Forest Service, Rocky Mountain Forest and Range Experiment Station.

DeLoach, C. J. 2000. Saltcedar biological control: methodology, exploration, laboratory trials, proposals for field releases, and expected environmental effects. http://refuges.fws.gov/nwrsfiles/HabitatMgmt/PestMgmt/SaltcedarWorkshopSep96/deloach.html.

Dillaha, T. A., R. B. Reneau, S. Mostaghimi, and D. Lee. 1989. Vegetative filter strips for agricultural non-point source pollution control. Transactions of the American Society of Agricultural Engineers 3:513–519.

DiTomaso, J. M. 1998. Impact, biology, and ecology of saltcedar (*Tamarix* ssp.) in the southwestern United States. Weed Technology 12:326–336.

Dresen, M., and R. Korth. 1995. Life on the edge...owning waterfront property. Stevens Point, WI: University of Wisconsin-Extension Lakes Partnership, University of Wisconsin. 95 pp.

Dwyer, D. D., J. C. Buckhouse, and W. S. Huey. 1984. Impacts of grazing intensity and specialized grazing systems on the use and value of rangeland: summary and recommendations. In: Developing strategies for rangeland management. Boulder, CO: Westview Press.

Dykaar, B. B., and P. J. Wiggington, Jr. 2000. Floodplain formation and cottonwood colonization patterns on the Willamette River, Oregon, USA. Environmental Management 25(1):87–104.

Dynesius, M., and C. Nilsson. 1994. Fragmentation and flow regulation of river systems in the northern third of the world. Science 266:753–762.

Elmore, W., and J. B. Kauffman. 1994. Riparian and watershed systems: degradation and restoration. Pp. 213–231 In: Ecological implications of livestock herbivory in the West. M. Vavra, W. A. Laycock, and R. D. Pieper (eds.). Denver, CO: Society of Range Management.

English, M. C., R. B. Hill, M. A. Stone, and R. Ormson. 1997. Geomorphological and botanical change on the Outer Slave River Delta, NWT, before and after impoundment of the Peace River. Hydrological Processes 11(13):1707–1724.

Environmental Protection Agency (EPA). 1999. Riparian acreage and stream miles. Provided by Joe Williams at the first meeting of the NRC Committee on Riparian Zone Functioning and Strategies for Management, October 19–20, 1999. From EPA Reach File 3.

EPA. 2000. Atlas of America's polluted waters. EPA 840-B-00-002. Washington, DC: U.S. Environmental Protection Agency, Office of Water. 53 pp.

EPA. 2001. Action plan for reducing, mitigating, and controlling hypoxia in the Northern Gulf of Mexico. Washington, DC: EPA Mississippi River/Gulf of Mexico Watershed Nutrient Task Force. http://www.epa.gov/msbasin/actionplanintro.htm.

Evans, R. O., R. W. Skaggs, and J. W. Gilliam. 1991. A field experiment to evaluate the water quality impacts of agricultural drainage and production practices. Proceedings of the National Conference on Irrigation and Drainage Engineering. New York, NY: ASCE.

Evans, R. O., R. W. Skaggs, and J. W. Gilliam. 1995. Controlled versus conventional drainage effects on water quality. Journal of Irrigation and Drainage Engineering 121(4):271–276.

Fausey, N. R., L. C. Brown, H. W. Belcher, and R. S. Kanwar. 1995. Drainage and water quality in the Great Lakes and Cornbelt states. Journal of Irrigation and Drainage Engineering 121(4):283–288.

Federal Interagency Working Group. 1998. Stream corridor restoration. principles, processes, and practices. Washington, DC: National Technical Information Service, U.S. Department of Commerce.

Fenner, P., W. Brady, and D. Patton. 1985. Effects of regulated water flows on regeneration of Fremont cottonwood. Journal of Range Management 38(2):135–138.

Findlay, C. S., and J. Bourdages. 2000. Response time of wetland biodiversity to road construction on adjacent lands. Conservation Biology 14:86–94.

Fischenich, J. C. 1997. Hydraulic impacts of riparian vegetation; summary of the literature. Technical Report EL-97-9. Washington, DC: U.S. Army Corps of Engineers. 53 pp.

Fish and Wildlife Service (FWS). 1998. A system for mapping riparian areas in the western U.S. Washington, DC: U.S. Fish and Wildlife Service.

Flather, C. H., L. A. Joyce, and C. A. Bloomgarden. 1994. Species endangerment patterns in the United States. Gen. Tech. Rep. RM-241. USDA Forest Service, Rocky Mountain Forest and Range Experiment Station.

Fleischner, T. L. 1994. Ecological costs of livestock grazing in western North America. Conservation Biology 8:629–44.

Florida Center for Public Management. 1998. Environmental indicator technical assistance series. Volume 3: State Indicators of National Scope. Land Use/Land Cover. http://www.fse.edu/~cpm/segip/catalog/volume3.html. 18 Nov. 1998.

Formann, R. T. T., and R. D. Deblinger. 2000. The ecological road-effect zone of a Massachusetts (U.S.A.) suburban highway. Conservation Biology 14:36–46.

Friedman, J. M., W. R. Osterkamp, M. L. Scott, and G. T. Auble. 1998. Downstream effects of dams on channel geometry and bottomland vegetation: regional patterns in the Great Plains. Wetlands 18:619–633.

Furniss, M. J., T. D. Roelofs, and C. S. Yee. 1991. Road construction and maintenance. Pp. 297–323 In: Influences of forest and rangeland management on salmonid fishes and their habitats. W. R. Meehan (ed.). Special Pub. 19. Bethesda, MD: Amer. Fish. Soc. 751 pp.

Galatowitsch, S. M., N. O. Anderson, and P. D. Ascher. 1999. Invasiveness in wetland plants in temperate North America. Wetlands 19:733–755.

Garrison, P. J., and R. S. Wakeman. 2000. Use of paleolimnology to document the effect of lake shoreland development on water quality. Journal of Paleolimnology 24(4):369–393.

Gatewood, J. S., J. W. Robinson, B. R. Colby, J. D. Hem, and L. C. Halpenny. 1950. Use of water by bottomland vegetation in the Lower Safford Valley, Arizona. U.S. Geological Survey Water-Supply Paper 1103.

Gedney, D. R., and B. A. Hiserote. 1989. Changes in land use in Western Oregon between 1971–74 and 1982. Resource Bulletin PNW-RB-165. Portland, OR: USDA Forest Service, Pacific Northwest Research Station. 21 pp.

Gonzalez, M. A. 2001. Recent formation of arroyos in the Little Missouri Badlands of southwestern North Dakota. Geomorphology 38:63–84.

Gordon, N. D., T. A. McMahon, and B. L. Finlayson. 1992. Stream hydrology: an introduction for ecologists. Chichester, England: John Wiley and Sons.

Graf, W. L. 1999. Dam nation: A geographic census of American dams and their large-scale hydrologic impacts. Water Resources Research 35(4):1305–1311.

Green, C. H., and S. M. Tunstall. 1992. The amenity and environmental value of river corridors in Britain. In: River conservation and management. New York: John Wiley & Sons.

Green, D. M. 1998. Recreational impacts on erosion and runoff in a central Arizona riparian area. Journal of Soil and Water Conservation 53:38–42.

Gregory, S. V. 2000. Summary of current status and health of Oregon's riparian area. In: Health of Natural Systems and Resources.

Grimm, N. B., and S. G. Fisher. 1989. Stability of periphyton and macroinvertebrates to disturbance by flash floods in a desert stream. Journal of the North American Benthological Society 8:293–307.

Heimberger, M., D. Euler, and J. Barr. 1983. The impact of cottage development on common loon reproductive success in central Ontario. Wilson Bull. 95:431–439.

Henderson, C. L., C. J. Dindorf, and F. J. Rozumalski. 1998. Lakescaping for wildlife and water quality. Minneapolis, MN: Minnesota Department of Natural Resources Nongame Wildlife Program, Section of Wildlife.

Henderson-Sellers, A., H. Ahang, G. Berz, K. Emanuel, W. Gray, C. Landsea, G. Holland, J. Lighthill, S. L. Shieh, P. Webster, and K. McGuffie. 1998. Tropical cyclones and global climate change: a post-IPCC assessment. Bulletin of the American Meteorological Society 79:19–38.

Heywood, V. H. 1989. Patterns, extents, and modes of invasions by terrestrial plants. In: Biological invasions: a global perspective, SCOPE 37. J. A. Drake et al., (eds.). Chichester, UK: John Wiley & Sons Ltd.

Hobbs, R. J., and D. A. Norton. 1996. Towards a conceptual framework for restoration ecology. Restoration Ecology 4:93–110.

Horning, J. 1994. Grazing to extinction: endangered, threatened and candidate species imperiled by livestock grazing on Western public lands. Washington, DC: National Wildlife Federation.

Howard, W. E. 1996. Damage to rangeland resources. Pp. 383–394 In: Rangeland Wildlife. P. R. Krausman (ed.). Denver, CO: Society of Range Management.

Howe, W. H., and F. L. Knopf. 1991. On the imminent decline of the Rio Grande cottonwoods in central New Mexico. Southwestern Naturalist 36:218–224.

Hubbard, M. W., Danielson, B. J. and R. A. Schmitz. 2000. Factors influencing the location of deer-vehicle accidents in Iowa. Journal of Wildlife Management 64:707–713.

Hupp, C. R., and A. Simon. 1991. Bank accretion and the development of vegetated depositional surface along modified alluvial channels. Geomorphology 4:111–124.

Illinois Department of Natural Resources. 1996. Illinois land cover, an atlas. IDNR/EEA-96/05. Springfield, IL: Illinois Department of Natural Resources.

Inter-Fluve, Inc. 1991. Handbook for reclamation of placer mined stream environments in western Montana. Bozeman, MT: Inter-Fluve, Inc. 340 pp.

Intergovernmental Panel on Climate Change (IPCC). 2001. IPCC third assessment report: summary for policy makers (United States). Geneva, Switzerland: IPCC Secretariat, c/o World Meteorological Association.

Jennings, M. J., K. Johnson, and M. Staggs. 1996. Shoreline protection study: a report to the Wisconsin State Legislature. PUBL-RS-921-96. Madison, WI: Wisconsin Department of Natural Resources.

Johnson, L. B., C. Richards, G. E. Host, and J. W. Arthur. 1997. Landscape influences on water chemistry in Midwestern stream ecosystems. Freshwater Biology 37:193–208.

Johnson, R. R., C. D. Ziebell, D. R. Patton, P. F. Folliott, and R. H. Hamre. 1985. Riparian Ecosystems and their management: reconciling conflicting uses. USDA Forest Service General Technical Report RM-120. Fort Collins, CO:USFS. 523 pp.

Johnson, W. C., R. L. Burgess, and W. R. Keammerer. 1976. Forest overstory vegetation on the Missouri River floodplain in North Dakota. Ecological Monographs 46:59–84.

Johnston, D. M., D. L. Szafoni, and A. Srivastava. 1999. Illinois streams information system: users manual. Urbana, IL: Department of Landscape Architecture, Geographic Modeling Systems Laboratory, University of Illinois.

Jones, J. A. F. J. Swanson, B. C. Wemple, and K. U. Snyder. 2000. Effects of roads on hydrology, geomorphology, and disturbance patches in stream networks. Conservation Biology 53:76–85.
Kauffman, J. B., and D. A. Pyke. 2001. Range ecology, global livestock influences. Pp. 33–52 In: Encyclopedia of Biodiversity, Volume 5. S. Levin et al. (eds). San Diego, CA: Academic Press.
Kauffman, J. G., and W. C. Krueger. 1984. Livestock impacts on riparian ecosystems and streamside management implications. Journal of Range Management 37:430–438.
Kay, C. E. 1997a. Is Aspen Doomed? Journal of Forestry 95:4-11.
Kay, C. E. 1997b. Viewpoint: Ungulate herbivory, willows, and political ecology in Yellowstone. J. Range Mgmt. 50:139–45.
Kay, C. E., and D. L. Bartos. 2000. Ungulate herbivory on Utah aspen: Assessment of long-term exclosures. J. Range Mgmt. 53:145–53.
Kay, C. E. 2001. The condition and trend of aspen communities on BLM administered lands in central Nevada—with recommendations for management. Battle Mountain, NV: BLM. 152 pp.
Keough, J. R., T. A. Thompson, G. R. Guntenspergen, and D. A. Wilcox. 1999. Hydrogeomorphic factors and ecosystem responses in coastal wetlands of the Great Lakes. Wetlands 19:821–834.
Keown, M. P., N. R. Oswalt, E. B. Perry, and E. A. Dardeau, Jr. 1977. Literature survey and preliminary evaluation of streambank protection methods. Technical Report H-77-9. Vicksburg, MS: U.S. Army Corps of Engineers, Waterway Experiment Station.
Knight, R. L., and K. J. Gutzwiller, eds. 1995. Wildlife and recreationists: coexistence through management and research. Washington, DC: Island Press. 372 pp.
Knight, R. L., and D. N. Cole. 1995. Wildlife responses to recreationists. Pp. 51–69 In: Wildlife and recreationists: coexistence through management and research. R. L. Knight and K. J. Gutzwiller (eds.). Washington, DC: Island Press. 372 pp.
Kondolf, G. M. 1995. Managing bedload sediment in regulated rivers: examples from California, U.S.A. Pp. 165–176 In: Natural and anthropogenic influences in fluvial geomorphology. J. E. Costa, A. J. Miller, K. W. Potter, and P. R. Wilcock (eds.). Geophysical Monograph 89. Washington, DC: American Geophysical Union. 239 pp.
Korth, R., and P. Cunningham. 1999. Margin of error? Human influence on Wisconsin shores. Stevens Point, WI: Wisconsin Lakes Partnership (Wisconsin Association of Lakes, Wisconsin Department of Natural Resources, and University of Wisconsin-Extension).
Kovacic, D. A. 2000. Presentation to the NRC Committee on Riparian Zone Functioning and Strategies for Management. June 3–4, 2000, Ames, IA.
Kovacic, D. A., M. B. David, L. E. Gentry, K. M. Starks, and R. A. Cooke. 2000. Effectiveness of constructed wetlands in reducing nitrogen and phosphorus export from agricultural tile drainage. Journal of Environmental Quality 29:1262–1274.
Kowalski, K. P., and D. A. Wilcox. 1999. Use of historical and geospatial data to guide the restoration of a Lake Erie coastal marsh. Wetlands 19:858–868.
Larson, G. L., and S. E. Moore. 1985. Encroachment of exotic rainbow trout into stream populations of native brook trout in the southern Appalachian mountains. Transactions of the American Fisheries Society 114:195–203.
Laursen, S. B. 1996. At the water's edge: the science of riparian forestry. BU-6637-S. St. Paul, MN: University of Minnesota. 160 pp.
Leon, S. C. 2000. Southwestern willow flycatcher. http://ifw2es.fws.gov/swwf/.
Lorang, M. S., P. D. Komar, and J. A. Stanford. 1993a. Lake level regulation and shoreline erosion in Flathead Lake, Montana: a response to the redistribution of annual wave energy. J. Coastal Research 9(2):494–508.
Lorang, M. S., J. A. Stanford, and P. D. Komar. 1993b. Dissipative and reflective beaches in a large lake and the physical effects of lake level regulation. Ocean and Coastal Manage. 19:263–287.
Lorang, M. S., and J. A. Stanford. 1993. Variability of shoreline erosion and accretion within a beach compartment of Flathead Lake, Montana. Limnology and Oceanography 38(8):1783–1795.

Luckenbach, R. A., and R. B. Bury. 1983. Effects of off-road vehicles on the biota of the Algodoens Dunes, Imperial County, California. Journal of Applied Ecology 20:265–286.

Luckey, R. R., E. D. Gutentag, F. J. Heimes, and J. B. Weeks. 1988. Effects of future ground-water pumpage on the High Plains aquifer in parts of Colorado, Kansas, Nebraska, New Mexico, Oklahoma, South Dakota, Texas, and Wyoming. U.S. Geological Survey Professional Paper 1400-E. Washington DC: U.S. Geological Survey. 44 pp.

Mac, M. J., P. A. Opler, C. E. Puckett Haecker, and P. D. Doran. 1998. Status and trends of the nation's biological resources. Reston, VA: U. S. Geological Survey.

Maser, C., and J. R. Sedell. 1994. From the forest to the sea: the ecology of wood in streams, rivers, estuaries, and oceans. Delray Beach, FL: St. Lucie Press.

Matson, N. P. 2000. Biodiversity and its management on the National Elk Refuge, Wyoming. Pp. 101–38 In: Developing Sustainable Management Policy for the National Elk Refuge, Wyoming. Bulletin Series No. 104. T. W. Clark et al. (eds.). New Haven. CT: Yale School of Forestry and Environmental Studies.

McFadden, L. D., and J. R. McAuliffe. 1997. Lithologically influenced geomorphic responses to Holocene climatic changes in the southern Colorado Plateau, Arizona: a soil-geomorphic and ecologic perspective. Geomorphology 19:303–332.

McShea, W. J., H. B. Underwood, and J. H. Rappole. 1997. The science of overabundance: deer ecology and population management. Washington, DC: Smithsonian Press.

Meador, M. R., and A. O. Layher. 1998. Instream sand and gravel mining: environmental issues and regulatory process. Fisheries 23(11):6–13.

Menzel, B. W. 1983. Agricultural management practices and the integrity of instream biological habitat. Pp. 305–329 In: Agricultural management and water quality. F. W. Schaller and G. W. Bailey (eds.). Ames, IA: Iowa Sate University Press.

Meyer, J. L. 1996. Beyond gloom and doom: ecology for the future. Bulletin of the Ecological Society of America 77:785–788.

Meyer, M., J. Woodford, S. Gillum, and T. Daulton. 1997. Shoreland zoning regulations do not adequately protect wildlife habitat in northern Wisconsin. Final Report, USFWS State Partnership Grant, P-1-W, Segment 17. Ashland, WI: Wisconsin Department of Natural Resources, Rhinelander Wisconsin and Sigurd Olson Environmental Institute. 73 pp.

Michener, W. K., E. R. Blood, K. L. Bildstein, M. M. Brinson, and L. R. Gardner. 1997. Climate change, hurricanes and tropical storms, and rising sea level in coastal wetlands. Ecological Applications 7:770–801.

Mitsch, W. J., J. W. Day, Jr., J. W. Gilliam, P. M. Groffman, D. L. Hey, G. W. Randall, and N. Wang. 2001. Reducing nitrogen loading to the Gulf of Mexico from the Mississippi River basin: strategies to counter a persistent ecological problem. BioScience 51(5):373–388.

Molles, M. C., Jr., C. S. Crawford, L. M. Ellis, H. M. Valett, and C. N. Dahm. 1998. Managed flooding for riparian ecosystem restoration. BioScience 48:749–756.

Montgomery, D. R., E. M. Beamer, G. R. Pess, and T. P. Quinn. 1999. Channel type and salmonid spawning distribution and abundance. Canadian Journal of Fisheries and Aquatic Sciences 56:377–387.

Moore, J. L. 1994. A special place: New Hampshire's Lakes. Wolfeboro Falls, NH: Lake Winnipesaukee Association. 32 pp.

Muldavin, E., P. Durkin, M. Bradley, M. Stuever, and P. Melhop. 2000. Handbook of wetland vegetation communities of New Mexico. Albuquerque, NM: New Mexico Heritage Program..

Murphy, M. L. 1995. Forestry impacts on freshwater habitat of anadromous salmonids in the Pacific Northwest and Alaska—requirements for protection and restoration. Decision Analysis Series No. 7. Silver Spring, MD: U.S. Department of Commerce, National Oceanic and Atmospheric Administration, Coastal Ocean Program. 156 pp.

Naiman, R. J., and H. Décamps (eds). 1990. The ecology and management of aquatic-terrestrial ecotones. Paris: The Parthenon Publishing Group, UNESCO. 316 pp.

Naiman, R. J., and K. H. Rogers. 1997. Large animals and system-level characteristics in river corridors. BioScience 47:521–529.
National Research Council (NRC). 1982. A levee policy for the National Flood Insurance Program. Washington, DC: National Academy Press.
NRC. 1990. Decline of sea turtles, causes and prevention. Washington, DC: National Academy Press.
NRC. 1996. Upstream: salmon and society in the Pacific Northwest. Washington, DC: National Academy Press. 452 pp.
NRC. 2000a. Hardrock mining of federal lands. Washington, DC: National Academy Press.
NRC. 2000b. Reconciling observations of global temperature change. Washington, DC: National Academy Press. 85 pp.
NRC. 2000c. Global change ecosystems research. Washington, DC: National Academy Press.
NRC. 2001a. Inland navigation system planning: the upper Mississippi River–Illinois waterway. Washington, DC: National Academy Press.
NRC. 2001b. Assessing the TMDL approach to water quality management. Washington, DC: National Academy Press.
NRC. 2002. The Missouri River ecosystem: exploring the prospects for recovery. Washington, DC: National Academy Press.
Nelson, R. L., M. L. McHenry, and W. S. Platts. 1991. Mining. Pp. 425–457 In: Influences of forest and rangeland management on salmonid fishes and their habitats. W. R. Meehan (ed.). Special Pub. 19. Bethesda, MD: Amer. Fish. Soc. 751 pp.
Nilsson, C., R. Jansson, and U. Zinko. 1997. Long-term responses of river-margin vegetation to water-level regulation. Science 276:798–800.
Noss, R. F., E. T. LaRoe, and J. M. Scott. 1995. Endangered ecosystems of the United States: a preliminary assessment of loss and degradation. Biological Report 28. Washington, DC: National Biological Service.
Oetter, D. R., W. B. Cohen, M. Berterretche, T. K. Maiersperger, and R. E. Kennedy. 2001. Land cover mapping in an agricultural setting using multiseasonal Thematic Mapper data. Remote Sensing of Environment 76(2):139–156.
Ohmart, R. D., W. O. Deason, and C. Burke. 1977. A riparian case history: the Colorado River. Pp. 35–47 In: Importance, preservation, and management of riparian habitat: a symposium. R. R. Johnson and D. A. Jones (eds.). Tucson, Arizona, July 9, 1977.
Ohmart, R. D. 1996. Historical and present impacts of livestock grazing on fish and wildlife resources in western riparian habitats. Pp. 245–279 In: Rangeland Wildlife. P. R. Krausman (ed.). Denver, CO: Society for Range Management.
Ohmart, R. D., and B. W. Anderson. 1978. Wildlife use values of wetlands in the arid southwestern United States. Pp. 278–295 In: Wetland functions and values: the state of our understanding. Proceedings of the National Symposium on Wetlands. P. E. Greeson, J. R. Clark, J. E. Clark (eds.). Minneapolis, MN: American Water Resources Association.
Ohmart, R. D., and B. W. Anderson. 1986. Riparian habitat. Pp. 164–199 In: B. S. Cooperider (ed.). Inventorying and monitoring of wildlife habitat. Denver, CO: U.S. Bureau of Land Management.
Opperman, J. J., and A. M. Merenlender. 2000. Deer herbivory as an ecological constraint to restoration of degraded riparian corridors. Restoration Ecology 8:41–47.
Parendes, L. A., and J. A. Jones. 2000. Role of light availability and dispersal mechanisms in invasion of exotic plants along roads and streams in the H. J. Andrews Experimental Forest, Oregon. Conservation Biology 14:64–75.
Patten, D. T. 1998. Riparian ecosystems of semi-arid North America: diversity and human impacts. Wetlands 18:498–512.
Pavelis, G. A. 1987. Farm drainage in the United States: history, status, and prospects. Publication Number 1455. Washington, DC: U.S. Department of Agriculture.

Petersen, M. 1986. Levees and associated flood control works. Pp. 422–438 In: River engineering. Englewood Cliffs, NJ: Prentice-Hall.

Pimentel, D., L. Lach, R. Zuniga, and D. Morrison. 1999. Environmental and economic costs associated with non-indigenous species in the United States. Ithaca, NY: Cornell University College of Agriculture and Life Sciences.

Plant Conservation Alliance. 2000. Alien plant invaders of natural areas: weeds gone wild. (http://www.nps.gov/plants/alien/fact.htm). Washington, DC: Bureau of Land Management.

Planty-Tabacchi, A-M., E. Tabacchi, R. J. Naiman, C. Deferrari, and H. Décamps. 1996. Invasibility of species-rich communities in riparian zones. Conservation Biology 10:598–607.

Platts, W. S. 1991. Livestock grazing. Pp. 389–423 In: Influences of forest and rangeland management on salmonid fishes and their habitats. W. R. Meehan (ed.). Special Pub. 19. Bethesda, MD: Amer. Fish. Soc. 751 pp.

Platts, W. S., and R. L. Nelson. 1985. Will the riparian pasture build good streams? Rangelands 7:7–10.

Poff, N. L., J. D. Allan, M. B. Bain, J. R. Karr, K. L. Prestegaard, B. D. Richter, R. E. Sparks, and J. C. Stromberg. 1997. The natural flow regime. BioScience 47:769–784.

Ponce, S. L. (ed.). 1983. The potential for water yield augmentation through forest and range management. Water Resources Bull. 19(3):351–419.

Postel, S. L., G. C. Daily, and P. R. Ehrlich. 1996. Human appropriation of renewable freshwater. Science 271:785–788.

Prichard, D., et al. 1998. A user guide to assessing proper functioning condition and supporting science for lotic areas. Technical reference 1737-15. Denver, CO: Bureau of Land Management, National Applied Resource Science Center. 126 pp.

Ralph, S. C., G. C. Poole, L. L. Conquest, and R. J. Naiman. 1994. Stream channel morphology and woody debris in logged and unlogged basins of western Washington. Canadian Journal of Fisheries and Aquatic Sciences 51:37–51.

Reiter, M. L., and R. L. Beschta. 1995. Effects of forest practices on water. In: Cumulative effects of forest practices in Oregon: literature and synthesis. Salem, OR: Oregon Department of Forestry.

Richards, C., and G. E. Host. 1994. Examining land use influences on stream habitats and macroinvertebrates: a GIS approach. Water Resources Bulletin 30:729–738.

Richards, C., L. B. Johnson, and G. E. Host. 1996. Landscape scale influences on stream habitats and biota. Canadian Journal of Fisheries and Aquatic Sciences 53(1):295–311.

Richardson, B. Z., and M. M. Pratt. 1980. Environmental effects of surface mining of minerals other than coal: annotated bibliography and summary report. General Technical Report INT-95. Ogden, UT: USDA Forest Service. 145 pp.

Roath, L. R., and W. C. Krueger. 1982. Cattle grazing and influence on a forested range. J. Range Mgmt. 35:332–338.

Robbins, S. D., Jr. 1991. Wisconsin birdlife. Population and distribution, past and present. Madison, WI: University of Wisconsin Press. 702 pp.

Robertson, R. J., and N. J. Flood. 1980. Effects of recreational use of shorelines on breeding bird populations. Canadian Field Naturalist 94:131–138.

Rood, S. B., and S. Heinze-Milne. 1989. Abrupt riparian forest decline following river damming in Southern Alberta. Canadian Journal of Botany 67:1744–1749.

Rood, S. B., and J. M. Mahoney. 1990. The collapse of river valley forests downstream from dams in the western prairies: probable causes and prospects for mitigation. Environmental Management 14:451–464.

Rood, S. B., and J. M. Mahoney. 1991. Impacts of the Oldman River dam on riparian cottonwood forests downstream. Department of Biological Sciences, University of Lethbridge, Alberta, Canada. 34 pp.

Rood, S. B., J. M. Mahoney, D. E. Reid, and L. Zilm. 1995. Instream flows and the decline of riparian cottonwoods along the St. Mary River, Alberta. Can. J. Bot. 73:1250–1260.

Rood, S. B., K. Taboulchanas, D. E. Bradley, and A. R. Kalischuk. 1999. Influence of flow regulation on channel dynamics and riparian cottonwoods along the Bow River, Alberta. Rivers 7(1):33–48.

Rowe, P. B. 1963. Streamflow increases after removing woodland–riparian vegetation from a southern California watershed. Journal of Forestry 61:365–370.

Rundquist, L. A., N. E. Bradley, J. E. Baldrige, P. D. Hampton, T. R. Jennings, and M. R. Joyce. 1986. Best management practices for placer mining. Juneau, AK: Entrix, Inc. 250 pp.

Schoof, R. 1980. Environmental impact of channel modification. Water Resources Bulletin 16(4):697–701.

Schueler, T. R. 1987. Controlling urban runoff: a practical manual for planning and designing urban BMPs. Washington, DC: Metropolitan Washington Council of Governments.

Schueler, T. R. 1995. The importance of imperviousness. Watershed Protection Techniques 1(3):100–112.

Schultz, R. C., J. P. Colletti, T. M. Isenhart, C. O. Marquez, WE. W. Simpkins, and C. J. Ball. 2000. Riparian forest buffer practices. Pp. 189–281 In: North American agroforestry: an integrated science and practice. Madison, WI: American Society of Agronomy.

Schumaker, N. H. 1996. Using landscape indices to predict habitat connectivity. Ecology 77(4):1210–1225.

Scott, M. L., M. A. Wondzell, and G. T. Auble. 1993. Hydrograph characteristics relevant to the establishment and growth of western riparian vegetation. Pp. 237–246 In: Proceedings of the 13th Annual American Geophysical Union Hydrology Days. H. J. Morel-Seytoax (ed.). Atherton, CA: Hydrology Days Publications.

Scott, M. L., P. B. Shafroth, and G. T. Auble. 1999. Responses of riparian cottonwoods to alluvial water table declines. Environmental Management 23(3):347–358.

Sedell, J. R., and Froggatt. 1984. Importance of streamside forests to large rivers: the isolation of the Willamette River, Oregon, U.S.A., from its floodplain by snagging and streamside forest removal. International Association of Theoretical and Applied Limnology 22:1828–1834.

Sedell, J. R., and K. J. Luchessa. 1981. Using the historical record as an aid to salmonid habitat enhancement. Pp. 210–223 In: Acquisition and utilization of aquatic habitat inventory information. N. B. Armantrout (ed.). Bethesda, MD: American Fisheries Society. 376 pp.

Sedell, J. R., and R. L. Beschta. 1991. Bringing back the "bio" in bioengineering. American Fisheries Society Symposium 10:160–175.

Sedell, J. R., F. H. Everest, and F. J. Swanson. 1982. Fish habitat and streamside management: past and present. Pp. 245–255 In: Proceedings of the Society of American Foresters Annual Meeting, Bethesda, MD.

Sedell, J. R., F. N. Leone, and W. S. Duval. 1991. Water transportation and storage of logs. Pp. 325–368 In: Influences of forest and rangeland management on salmonid fishes and their habitats. W. R. Meehan (ed.). Special Publication 19. Bethesda, MD: American Fisheries Society. 751 pp.

Shaver, E., J. Maxted, G. Curtis and D. Carter. 1994. Watershed protection using an integrated approach. In: Stormwater NPDES related monitoring needs: engineering foundation. August 7–12, 1994. Crested Butte, CO: American Society of Civil Engineers.

Shaw, D. W., and D. M. Finch, tech. coords. 1996. Desired future conditions for southwestern riparian ecosystems: bring interests and concerns together. General Technical Report RM-GTR-272. Fort Collins, CO: USDA Forest Service. 359 pp.

Sheridan, D. 1979. Off-road vehicles on public lands. Washington, DC: Council on Environmental Quality.

Slack, J. R., A. M. Lamb, and J. M. Landwehr. 1993. Hydro-climatic data network (HCDN) streamflow data set, 1874–1988. Water Resources Investigations Report 93-4076. Washington, DC: U.S. Geological Survey.

Smith, S. D., A. B. Wellington, J. L. Nachlinger, and C. A. Fox. 1991. Functional responses of riparian vegetation to streamflow diversion in the Eastern Sierra Nevada. Ecological Application 1:89–97.

Snyder, W. D., and G. C. and Miller. 1991. Changes in plains cottonwoods along the Arkansas and South Platte Rivers—Eastern Colorado. Prairie Naturalist 23:165–176.

Stanford, J. A., J. V. Ward, W. J. Liss, C. A. Frissel, R. N. Williams, J. A. Lichatovich, and C. C. Coutant. 1996. A general protocol for restoration of regulated rivers. Regulated Rivers 12: 391–413.

Starnes, L. B. 1983. Effects of surface mining on aquatic resources in North America. Fisheries 8:2–4.

Stein, B. A., and S. R. Flack. 1996. America's least wanted: alien species invasions of U.S. ecosystems. Arlington, VA: The Nature Conservancy.

Stine, S., D. Gaines, and P. Vorster. 1984. Destruction of riparian systems due to water development in the Mono Lake watershed. Pp. 528-533 In: California riparian systems: ecology, conservation, and productive management. R. Warner and C. Hendricks (eds.). Berkeley, CA: University of California Press..

Stoddart, L. A., and A. Smith. 1955. Range management, 2nd edition. New York: McGraw-Hill.

Stohlgren, T. J., K. A. Bull, Y. Otsuki, C. A. Villa and M. Lee. 1998. Riparian zones as havens for exotic plant species in the central grasslands. Plant Ecol. 138:113–125.

Strahan, J. 1984. Regeneration of riparian forests of the Central Valley. In: California riparian systems: ecology, conservation, and productive management. R. Warner and C. Hendricks (eds.). Berkeley, CA: University of California Press. Pp. 58-67.

Stromberg, J. C., and D. T. Patten. 1990. Riparian vegetation instream flow requirements: a case study from a diverted stream in the eastern Sierra Nevada, California. Environmental Management 14(2):185–194.

Stromberg, J. C., and D. T. Patten. 1991. Instream flow requirement for cottonwoods at Bishop Creek, Inyo, CA. Trout Unlimited 2:1–11.

Stromberg, J. 1998. Dynamics of Freemont cottonwood (*Populus fremontii*) and saltcedar (*Tamarix chinesis*) populations along the San Pedro River, Arizona. Journal of Arid Environments 40:133–155.

Stromberg, J. C., D. T. Patten, and B. D. Richter. 1991. Flood flows and dynamics of Sonoran riparian forests. Rivers 2:221–235.

Stuber, R. J. 1985. Trout habitat, abundance, and fishing opportunities in fenced vs. unfenced riparian habitat along Sheep Creek, Colorado. Pp. 310–314 In: Riparian ecosystems and their management: reconciling conflicting uses. Gen. Tech. Bull. RM-120. USDA Forest Service.

Swift, B. L. 1984. Status of riparian ecosystems in the United States. Water Resources Bulletin 20(2):223–228.

Taylor, C. E., K. L. Magrath, P. Folley, P. Buck, and S. Carpenter. 1996. Oklahoma vascular plants: additions and distributional comments. Proceedings of the Oklahoma Academy of Science 76:31–34.

Taylor, J. N., W. R. Courtenay, Jr., and J. A. McCann. 1984. Known impacts of exotic fishes in the continental United States. Pp. 322–373 In: Distribution, biology and management of exotic fishes. W. R. Courtenay, Jr and J. R. Stauffer, Jr. (eds.). Baltimore, MD: The John Hopkins University Press.

The Nature Conservancy. 2001. Weeds on the web: the worst invaders. (http://tncweeds.ucdavis.edu/worst.html)

Thomas, D. L., D. C. Perry, R. O. Evans, F. T. Uzuno, K. C. Stone, and J. W. Gilliam. 1995. Agricultural drainage effects on water quality in Southeastern U.S. Journal of Irrigation and Drainage Engineering 121(4):277-282.
Titus, J. G., et al. 1991. Greenhouse effect and sea level rise: the cost of holding back the sea. Coastal Management 19:171–210.
Townsend, P. A. 2001. Relationships between vegetation patterns and hydroperiod on the Roanoke River floodplain, North Carolina. Plant Ecology 156:43–58.
Trombulak, S. C., and C. A. Frissell. 2000. Review of ecological effects of roads on terrestrial and aquatic communities. Conservation Biology 14:18–30.
Turner, M. G. 1989. Landscape ecology: the effect of pattern on process. Ann. Rev. Ecol. Syst. 20:171–197.
Turner, R. E., and N. N. Rabalais. 1991. Changes in Mississippi River water quality this century. BioScience 41:140–147.
Turner, S. F., and H. E. Skibitzke. 1952. Use of water by phreatophytes in a 2000-foot channel between Granite Reef and Gillespie Dams, Maricopa County, Arizona. Transactions of the American Geophysical Union 33:66–72.
U.S. Department of Agriculture (USDA). 1997. Census of agriculture, volume 1: part 51, chapter 2. Washington, DC: USDA National Agricultural Statistics Service.
USDA Natural Resources Conservation Service (USDA NRCS). 1998. State of the Land. Land Cover/Use. http://www.nhq.nrcs.usda.gov/land/index/cover_use.html, 18 Nov. 1998.
USDA, NRCS. 2001a. Chinese privet, *Ligustrum sinense*. Plants Database (http://plants.usda.gov/plants/). Baton Rouge, LA: National Plant Data Center.
USDA, NRCS. 2001b. Plants Database (http://plants.usda.gov/plants/). Baton Rouge, LA: National Plant Data Center.
USDA Forest Service (USFS). 1993. Forest ecosystem management: an ecological, economic and social assessment. 1993-793-071. Washington, DC: U.S. Government Printing Office.
USDA Forest Service. 1999. Notice: Authorization of livestock grazing activities on the Sacramento grazing allotment, Sacramento Ranger District, Lincoln National Forest, Otero County, NM. Federal Register 64:24132.
U.S. Department of the Interior Bureau of Land Management (BLM) and USDA Forest Service (USFS). 1994. Rangeland reform '94 draft environmental impact statement. Washington, DC: U.S. Department of the Interior Bureau of Land Management and USDA Forest Service.
Van Dorp, D., and P. F. M. Opdam. 1987. Effects of patch size, isolation and regional abundance on forest bird communities. Landscape Ecology 1:59–73.
Verme, L. J., and W. F. Johnston. 1986. Regeneration of northern white cedar deeryards in upper Michigan. J. Wildl. Manage. 50:307–313.
Verry, E. S., J. S. Hornbeck, and D. A. Dolloff. 2000. Riparian management in forests of the eastern United States. New York: Lewis Publishers. 402 pp.
Warren, R. S., and W. A. Niering. 1993. Vegetation change on a northeast tidal marsh: interaction of sea-level rise and marsh accretion. Ecology 74:96–103.
Wauchope, R. D. 1978. The pesticide content of surface water draining from agricultural fields – a review. J. Environmental Quality 7:459–472.
Wear, D. N., M. G. Turner, and R. O. Flamm. 1996. Ecosystem management with multiple owners: landscape dynamics in a southern Appalachian watershed. Ecological Applications 64(4):1173–1188.
Webb, R. H., and H. G. Wilshire, eds. 1983. Environmental effects of off-road vehicles: impacts and management in arid regions. New York: Springer-Verlag.
Wilcove, D. S., D. Rothstein, J. Bubow, A. Phillips, and E. Losos. 1998. Quantifying threats to imperiled species in the United States. BioScience 48:607–615.

Wilkinson, C. F. 1992. The miner's law. Pp. 28–74 In: Crossing the next meridian. Washington, DC: Island Press.

Williams, K. K., C. Ewel, R. P. Stumpf, F. E. Putz, and T. W. Workman. 1999. Sea-level rise and coastal forest retreat on the west coast of Florida, USA. Ecology 80(6):2045–2063.

Williamson, K. J., D. A. Bella, R. L. Beschta, G. Grant, P. C. Klingeman, H. W. Li, and P. O. Nelson. 1995. Gravel disturbance impacts on salmon habitat and stream health, Volume 1: Summary report. Corvallis, OR: Oregon Water Resources Research Institute Oregon State University, 52 pp.

World Commission on Dams. 2000. Dams and development: a new framework for decision-making. London and Sterling, VA: Earthscan Publications Ltd.

Zucker, L. A., and L. C. Brown (Eds.). 1998. Agricultural drainage: water quality impacts and subsurface drainage studies in the midwest. Ohio State University Extension Bulletin 871. Columbus, OH: The Ohio State University.

4

Existing Legal Strategies for Riparian Area Protection

Public recognition of the importance of wetlands resulted in the 1970s in a national-level regulatory system for their protection (NRC, 1995). Not until the 1990s, however, have riparian areas begun to receive legal recognition as places requiring special attention. During the last decade, a patchwork of federal, state, and local laws and programs has developed that, directly and indirectly, begins to acknowledge the importance of riparian areas and to require or encourage special management to restore or protect their essential functions. The degree of protection, the focus, and the spatial coverage of these laws and programs are highly variable at federal, state, and local levels. Although riparian areas perform many valuable functions, it is their importance to stream water quality and fisheries that prompted most of the laws and programs that afford them protection.

A key differential in the level of protection given to riparian areas is their ownership status. There are approximately 2.3 billion acres of land in the United States, including Alaska and Hawaii. Of this total, 550 million acres are owned by the federal government—about 24 percent. The proportion of land in the public domain varies from state to state, from less than 0.2 percent in Iowa to 77 percent in Nevada. The largest federal government holdings are in the 12 western states (AK, AZ, CA, CO, ID, MT, NM, NV, OR, UT, WA, and WY) with much of this land being managed by the U.S. Forest Service (USFS) and the Bureau of Land Management (BLM).

Ownership may be shared, as where a stream forms the boundary between adjoining tracts, or along federally defined navigable waters where the state owns the bed and banks, but the adjacent lands above the mean high water are owned by another entity. An additional property interest is ownership of the water itself or of a right to its use. Thus, legal protections for riparian areas and any recommendations for changes in their management must account for both the property interests in the relevant waters and lands and the fact that most riparian areas are linear features that cross ownership/jurisdictional boundaries. An example of this complexity is the Interior Columbia River Basin, which contains 74 separate federal land units, including 35 national forests and 17 BLM districts, as well as significant private, state, and tribal holdings—all of which must be taken into account when formulating a joint, comprehensive management plan.

Protection of riparian areas has been approached in a variety of ways (Table 4-1). One approach, exemplified by the National Environmental Policy Act (NEPA)—and comparable laws in some states—is to require identification and analysis of adverse environmental effects that would be caused by federal actions, along with consideration of less environmentally damaging alternatives. Such an approach is not specific to riparian areas, nor does it require their protection, but it does ensure attention to their environmental values if they would be potentially affected by a proposed federal action. Examples in which environmental impact statements have focused on riparian values are discussed in this chapter.

A second approach is to place special limitations on activities in riparian areas on publicly owned lands. For example, in the Pacific Northwest logging and other activities are restricted in riparian reserves that have been established on federal lands in order to protect salmon. Many of the benefits provided by riparian areas—wildlife habitat, water quality protection, channel stability, and maintenance of fisheries—are public in nature.

A third approach is to regulate activities on private riparian areas. Such regulation must necessarily protect the legal rights of property owners while limiting those land uses deemed unacceptably harmful to public interests. Protec-

TABLE 4-1 General Approaches for Riparian Area Protection

Approach	Example
1. Required impact identification	National Environmental Policy Act and State Environmental Policy Acts
2. Special management areas on public lands	Northwest Forest Plan
3. Private land development/use regulation	State and local stream buffer requirements
4. Financial incentives, technical assistance, education	Farm Bill programs
5. Public/nonprofit purchase of private riparian lands or interests in lands	Greenway programs, conservation easements

tion of riparian habitat essential for federally protected endangered species is one example of this approach. Other examples are found in statewide programs that restrict certain types of activities on lands adjacent to waterbodies, such as the Massachusetts Rivers Protection Act and the New Hampshire Comprehensive Shoreland Protection Act. Many states restrict timber harvesting on private lands adjacent to streams. In addition, many local communities use their land-use authority to limit new construction in streamside areas.

Fourth, incentives such as cost-sharing, low-cost loans, or tax reductions may be used to encourage good practices on private riparian areas, and special technical assistance and education may be used as well. At the national level, several Farm Bill programs provide incentives for moving intensive agricultural practices away from streams; several states have similar programs.

Fifth, privately owned riparian lands can be purchased—either in fee or by easement—for public management. The desirability of riparian areas for recreational use has prompted urban areas to acquire riparian lands for greenways (Smith and Hellmund, 1993). The remarkable growth in the number of land trusts in recent years has provided another vehicle for protection of private lands utilizing conservation easements. Federal—and an increasing number of state—tax laws provide incentives for landowners to donate such easements.

This chapter sets out the general legal and management frameworks that now apply to the protection of riparian areas. First reviewed are the federal, state, and local laws and programs directed, at least in part, to protect and restore essential functions and values of privately owned riparian lands. Both regulatory and nonregulatory approaches are discussed. Then, federal laws and policies applying to publicly owned riparian areas are presented, organized by category of federal public land. Next, laws governing the use of water resources are considered in relation to supporting riparian areas. Two federal programs that have significant potential to expand protection of riparian areas are given in-depth consideration. A final section considers the efficacy of the existing framework and evaluates the need for additions and changes.

PROTECTION OF PRIVATELY OWNED RIPARIAN AREAS

Most riparian lands are in private ownership, especially in the eastern portion of the United States. The value of riparian lands to a private landowner most often is measured in terms of their economic benefits rather than their ecological functions. Private owners of riparian lands typically have only limited motivation to use these areas in a manner protective of their functions. In the absence of improved education about riparian functioning, legal strategies for protecting the ecological values of privately owned riparian lands must be based either on implementing regulatory requirements or on providing special incentives. Alternatively, such areas may be purchased for public ownership and management.

Regulatory Approaches

Regulatory approaches are especially well suited to situations in which private gain from the development and use of land and natural resources causes unacceptable public loss, with the negative consequences of the action falling largely or entirely on someone other than the developer or user. Typically, land and resource developers may be required through a permit process to alter or restrict the manner of development in order to reduce its negative effects. Even-handed application of the requirements imposes the same burden on all similarly situated land and resource owners and developers. Regulatory approaches must further a legitimate public purpose and cannot deprive a property owner of all economically beneficial use of the property (unless the government pays compensation).

Federal Programs

Except for wetlands, there is no national regulatory program that attempts to manage ecologically harmful activities within riparian areas. Although the link with water, and hence commerce, has provided a legal basis for federal control of dredge-and-fill activities in wetlands, private land-use regulation generally is within the province of states. Nevertheless, there are federal programs that apply to certain activities in riparian areas.

National Environmental Policy Act. The National Environmental Policy Act (NEPA) requires federal agencies to examine the potential adverse environmental effects of proposed major actions that would significantly affect the quality of the human environment. Alternatives to the proposed action must also be considered. NEPA is not itself a regulatory law in that no particular result is required by this statute, but the environmental analysis serves to disclose both the existence of environmental problems and less environmentally damaging approaches. Although NEPA applies only to proposed federal actions, it often extends to private activities requiring some form of federal approval or receiving federal financing.

Litigation under NEPA involving riparian areas has chiefly involved claims that environmental impacts were not adequately addressed or mitigated. *Friends of the Payette v. Horseshoe Bend Hydroelectric Co.*, 988 F.2d 989 (9th Cir. 1993), for example, considered whether likely impacts of a proposed hydroelectric power facility on a riparian area and its bald eagle habitat were adequately evaluated. The need to prepare a full environmental impact statement rather than simply an environmental assessment has been the subject of other litigation. For example, in *Sierra Club v. Babbitt*, 69 F. Supp. 2d 1202 (E.D. Cal. 1999), a federal district court concluded that a National Park Service environmental assessment of a proposed highway reconstruction project provided insufficient details to assess the likely project impacts on the Merced River or its riparian corridor.

Clean Water Act. The Clean Water Act (CWA) has the stated goal of restoring and maintaining the chemical, physical, and biological integrity of the nation's waters. It focuses jointly on human and aquatic ecosystem health by establishing a water-quality goal of "fishable and swimmable" and "zero pollution" for all bodies of water. Although many sections of the act indirectly address riparian areas, the section most relevant to their protection (other than Section 404) is Section 303(d), which requires states and the U.S. Environmental Protection Agency (EPA) to identify waters not meeting state water-quality standards and to develop Total Maximum Daily Loads (TMDLs). A TMDL is the maximum amount of a pollutant that a water-body can receive and still be in compliance with state water-quality standards. After determining TMDLs for impaired waters, states are required to identify all point and nonpoint sources of pollution in a watershed that are contributing to the impairment and allocate reductions to each source in order to meet the state standards. Although TMDLs have been required under the Clean Water Act since 1972, their development did not begin in earnest until forced by widespread litigation during the 1990s (Houck, 2000; NRC, 2001).

Although it has been a matter of debate for some time, recent court rulings have confirmed that the TMDL program applies to both point and nonpoint sources of pollution (*Pronsolino v. Marcus*, 1999). Thus, it is likely that implementation plans to achieve water quality standards in impaired water-bodies will involve a variety of management strategies in riparian areas. The potential for such strategies, and for application of the TMDL program in general, to protect and restore riparian areas is considered in depth later in this chapter.

Wetlands Regulation. Section 404 of the Clean Water Act provides authority for a national program supervising the discharge of dredged and fill materials into "waters of the United States," defined by regulation to include at least some wetlands.[1] Under this regulatory program, wetlands that meet the jurisdictional definition cannot be dredged or filled without a permit from the U.S. Army Corps of Engineers (Corps). The Corps subjects permit applications to a review that considers a wide range of factors, including whether there are reasonable alternative locations. If no reasonable alternative can be identified and the need for the activity is demonstrated, consideration is given to mitigation measures and, as a last resort, to compensation for lost wetlands.

Although some wetlands occur within riparian areas, many riparian areas do not meet the jurisdictional definition of wetlands. Thus, activities occurring in

---

[1] The Supreme Court case known as SWANCC, decided in January 2001, raises some doubts as to whether all wetlands are within the Corps' regulatory jurisdiction. A federal judge in Virginia is currently considering whether the Corps has authority to require a 404 permit to fill a wetland that is not adjacent to a navigable stream.

such areas are not subject to the permit requirements of Section 404. Moreover, Congress and the Corps have exempted from 404 requirements normal farming, silviculture, and ranching activities as well as maintenance of structures such as dikes, levees, and dams and construction and maintenance of irrigation and drainage ditches.

Floodplain Regulation. The federal government spends several billion dollars a year on flood control and related water management projects in an attempt to reduce the roughly $4 billion per year of flood losses that occur in this country (Federal Interagency Floodplain Management Task Force, 1992). Encouraged by the National Flood Insurance Act of 1968, all states now have legislation authorizing local governments to adopt regulations restricting certain types of development within floodplains. Several states have adopted statewide floodplain management regulations. To be eligible for federal flood insurance, state and local programs must delineate the "regulatory floodway"—an area capable of passing a 100-year flood without increasing the water surface elevation by more than one foot. "Encroachments" such as buildings that would increase this elevation more than one foot may not be permitted.

Traditionally, floodplain management has focused on human safety and protection of investments. Riparian protection has not been a stated objective of such management. In fact, many of the structural responses to flood control, such as construction of levees and straightening of stream channels, have been harmful to riparian areas. The Interagency Floodplain Management Review Committee (the Galloway Committee) recommended more explicit recognition of the environmental values of floodplains in its 1994 report *Sharing the Challenge: Floodplain Management into the 21st Century*. In particular, the committee recommended a better-focused and more coordinated federal effort under the U.S. Department of the Interior to purchase either fee or conservation easement interests in frequently flooded lands with environmental values. In addition, it urged the commitment of ongoing federal funding following construction of federal flood-control projects to protect associated environmental values. It encouraged expanded use of the authority now given to the Corps to mitigate the environmental losses associated with already-constructed flood-control projects. In response, Congress expanded the 1996 Water Resources Development Act, Section 1135, program to allow for small environmental restoration projects when it is found that a Corps project has contributed to environmental degradation.

Endangered Species Act. The federal Endangered Species Act (ESA) has served as authority to regulate the development and use of land in riparian areas that provide essential habitat for a listed threatened or endangered plant or animal species. Under this law, federal agencies are prohibited from taking any action likely to jeopardize the continued existence of protected species, including destroying or adversely modifying their designated critical habitat. Moreover, the

ESA makes it unlawful for any person to "take" a protected animal species, defined to include harming, harassing, or killing such species.

Because riparian areas provide habitat for an abundance of plant and animal species, especially in the more arid western states, they have been the focus of the ESA is some cases. For example, federal land management agencies in the Pacific Northwest have established extensive networks of riparian reserves along streams in national forests and other federal public lands to afford protection to the Northern spotted owl and anadromous fish. These riparian reserves are expected to also provide habitat protection to a wide range of other aquatic and terrestrial wildlife species.

The prohibition against "taking" a protected species is motivating habitat protection on private lands as well, sometimes in riparian areas. Authorized under Section 10 of the ESA, habitat conservation plans provide a means whereby otherwise lawful activities that might incidentally cause take of a protected species can go forward in return for implementation of conservation measures. For example, as a condition of undertaking development that would destroy the riparian habitat of a protected species, a habitat conservation plan could provide for protection of similar habitat on some other private land. Such an approach is now under development in the Front Range of Colorado to provide protection for the Preble's meadow jumping mouse, a listed threatened species found only within riparian areas of foothills streams (see Box 4-1).

Riparian areas are entitled to *affirmative* protection under the ESA if (1) they occur on federal lands and provide habitat to any listed species or any species proposed for listing or (2) if they are within designated critical habitat. Riparian habitat has been included in the critical habitat designations for numerous fish species or stocks (e.g., coho salmon, steelhead, winter-run chinook, desert pupfish, Sonoran chub, Railroad Valley springfish), mammals (riparian brush rabbit and riparian woodrat), birds (least Bell's vireo and southwestern willow flycatcher), and reptiles (concho watersnake) (50 C.F.R. §§ 17.11, 17.95, 226.10, 226.12, 226.204).

Surface Mining Control and Reclamation Act. The Surface Mining Control and Reclamation Act (SMCRA) sets permitting requirements, environmental protection performance standards, and reclamation requirements for surface coal mines on private and public lands. The regulatory structure can afford protection of riparian areas. For example, applicants for permits are required to submit site-specific information about fish and wildlife resources when the permit area or adjacent area is likely to include "habitats of unusually high value for fish and wildlife such as important streams, wetlands, [and] riparian areas" (30 C.F.R. § 780.16). Applications must also include a protection and enhancement plan, which includes "protective measures that will be used during the active mining phase," such as establishing buffer zones and monitoring surface water quality and quantity. The plan must also include "enhancement measures that will be used during

BOX 4-1
Preble's Meadow Jumping Mouse

On May 12, 1998, the U.S. Fish and Wildlife Service (FWS) listed the Preble's meadow jumping mouse as a threatened species under the Endangered Species Act. The Preble's mouse, an 8- to 10-inch-long mouse with a tail that accounts for at least 60 percent of its length and with long hind feet adapted for jumping, lives only in well-vegetated riparian areas along the foothills and adjacent plains of the Front Range of Colorado and Wyoming. Sites with willows are particularly favored. Preble's mice are nocturnal creatures, and they hibernate for a good portion of the year. Always considered rare, the Preble's mouse has been declining in numbers in recent years. Fragmentation and loss of its riparian habitat from human use and development have been identified as the primary factors causing this decline.

The Endangered Species Act attempts to protect listed species in two primary ways. First, it prohibits a federal agency from taking an action that might adversely affect the continued existence of the species, including modification of its designated critical habitat. Second, it prohibits any person subject to the jurisdiction of the United States from "taking" a listed species. "Taking" is defined to include harassing, harming, or killing a species, as well as destroying its habitat.

Counties along the rapidly growing Colorado Front Range are attempting to develop habitat conservation plans that will provide for long-term protection of riparian habitat needed by the Preble's mouse. Likely conservation strategies include precluding or minimizing new development within known or likely Preble's habitat, requiring new activities within riparian areas to use best management practices to minimize impacts, and developing habitat "banks" of preserved or restored riparian areas to compensate for habitat unavoidably lost to new development.

the reclamation and postmining phase of operation to develop aquatic and terrestrial habitat," such as stream and wetland restoration.

Mine operators are directed to use the best technology available to minimize disturbances and adverse impacts on fish, wildlife, and related environmental values and to "achieve enhancement of such resources where practicable" (30 C.F.R. § 816.97a). Operators must further "avoid disturbances to, enhance where practicable, restore, or replace wetlands, and riparian vegetation along rivers and streams and bordering ponds and lakes," as well as "habitats of unusually high value for fish and wildlife" (30 C.F.R. § 816.97f). Generally, mining is not to occur within 100 feet of a stream.

Coastal Zone Management Act. Originally enacted in 1972 and significantly amended by the Coastal Zone Act Reauthorization Amendments of 1990, the Coastal Zone Management Act (CZMA) authorizes significant federal financial and technical assistance to states that establish a satisfactory Coastal Management Plan. All federal actions occurring within or affecting the coastal zone are to be "consistent" with the state Coastal Management Plan. Minimum plan requirements include identification of permissible land uses within the coastal zone; designation of areas of particular concern; identification of means for controlling land uses; and establishment of planning processes for providing public access to beaches and other high-value areas, for preventing erosion, and for siting of energy facilities.

The 1990 amendments required states to develop a coastal nonpoint source pollution control program and to submit it to the National Oceanic and Atmospheric Administration (NOAA) and EPA for approval. NOAA and EPA must evaluate whether the state's coastal zone boundary extends inland sufficiently far to control land and water uses significantly impacting coastal waters. EPA guidance for program compliance endorses many familiar best management practices for controlling nonpoint source pollution, including "streamside special management areas" to protect streams from logging and measures for controlling grazing in erosion-sensitive areas such as riparian areas and wetlands (EPA, 1993). The guidance recognizes both the pollution-abatement functions of riparian areas as well as their potential to become sources of nonpoint pollution if degraded. Because the CZMA is designed to protect water quality, and riparian areas are the last line of defense between receiving waters and upland sources of pollution, most state CZMA programs require riparian area protection as a means of meeting the goals of the CZMA. This is being done through protection of functioning riparian areas and restoration of nonfunctioning riparian areas where possible.

Federal Power Act. Under the Federal Power Act, the Federal Energy Regulatory Commission (FERC) regulates essentially all nonfederal hydroelectric power facilities. In 1986, Congress amended the Federal Power Act to require FERC to give "equal consideration" to energy conservation, protection of fish and wildlife, protection of recreational opportunities, and preservation of general environmental quality, along with the power generation potential of a river, in its licensing and relicensing process. FERC's "Manual of Standard Special Articles" requires license applicants to submit a wetland mitigation plan and a wildlife mitigation plan that will be included in their license (FERC, 1992). In addition, for projects with a reservoir, the applicant must provide a management plan providing for a shoreline buffer zone.

The relicensing process on the Deerfield River in Vermont and Massachusetts in the early 1990s provides an example of the act's potential for protecting riparian areas. New England Power Company (NEP) operates eight dams along the river with 15 generating units, all covered by a single FERC license that expired in 1991. As part of its license-renewal process, NEP worked out a comprehensive settlement with 15 parties that included commitments to maintain flows below each of its dams at levels sufficient to protect fisheries and to make scheduled white-water releases for boaters. In addition, NEP committed to spending $200,000 to improve waterfowl and wildlife habitat and to permanently protect (with conservation easements) over 18,000 acres of riparian and watershed lands owned by NEP—primarily as shoreline buffers around its reservoirs (Kimball, 1997).

State and Local Regulatory Programs

States can regulate land use in the exercise of their sovereign police power. Traditionally, states have delegated this authority to local government. For matters determined to be of statewide importance, however, states may exercise this authority directly. The importance of protecting riparian areas has prompted several states to establish state-level regulatory programs beyond those authorized for floodplain regulation. The most common form of regulation is to establish buffer zones (setbacks) adjacent to waterways in which development is precluded or limited. It should be noted that state and local setback regulations on private land have been or are likely to be ruled constitutional, as discussed in Box 4-2.

Statewide Shoreline or Riverfront Protection. Several states have made riparian areas a subject of special attention. A comprehensive approach has been taken by Massachusetts. The 1996 Rivers Protection Act established a state-level permit system for development activities within a "riverfront" area (Rivers Protection Act, MGL chapter 131). A riverfront area is defined as a corridor 200 feet wide (or 25 feet on each side in large municipalities and in densely populated areas) along all perennial rivers and streams. Proposed development in riverfront areas must demonstrate no significant adverse effects on water supplies, wildlife habitat, fisheries, shellfish, groundwater, and flood and pollution prevention.

Moreover, there must be no practicable economic alternatives to the development for which effects would be less adverse.

New Hampshire enacted the Comprehensive Shoreland Protection Act in 1994. Under this law, shorelands within 250 feet of public waters are designated for special protection. Public waters are defined to include all fresh waterbodies, natural or impounded; coastal waters; and rivers of fourth-order size or greater. Certain types of activities, such as solid or hazardous waste facilities and automobile junkyards, are prohibited within protected shorelands. Statewide minimum standards, which relate to such things as location of septic systems, sediment controls, tree cutting, and minimum lot size, are established to govern all development within protected shorelands (North Country Resource Conservation and Development Area Inc., 1995).

Wisconsin has a "shoreline" zoning program regulating property uses within 1,000 feet of a lake or 300 feet of a stream or its floodplain (Wis. Admin. Code ch. NR 115). This program establishes minimum lot sizes within a shoreline area, requires a 75-foot setback for buildings, and restricts clearcutting activities.

The Montana Natural Streambed and Land Preservation Act requires any person or entity proposing to do work that would physically alter the bed or banks of a perennial stream on public or private land to obtain a permit from the Board of Supervisors of the local Conservation District.

Forest Practices Acts. Forestry practices on private riparian forestlands are prescribed by the individual states. Oregon enacted the first legislation for private forest practices in 1972. Since that time, 40 states and U.S. territories have established either mandatory forest practices or best management practices (BMPs) (Figure 4-1). Oregon, Washington, California, Idaho, Montana, Alaska, and Minnesota have established regulations for forest practices on private lands that generally specify widths of riparian management zones (RMZs) and the amount of partial timber harvest allowed within the RMZs. RMZs on private lands generally apply to riparian areas within 100 feet or less of perennial streams. Additional rules address road building, road crossings, yarding systems, replanting, leave trees, and harvest unit dimensions. Thirty-three (33) states and territories have used voluntary programs based either on best management practice guidelines or on achieving water-quality standards. These programs rely on training and education programs and on voluntary compliance by forest operators. Almost a quarter of the states have no explicit guidelines or legislation for private forest practices. Table 4-2 lists the riparian management approaches required by state forest practice acts in states and territories and the agencies responsible for their enforcement (if necessary).

In general, riparian buffers on public lands are often more extensive than those on private lands. On public lands, buffer widths range from less than 25 ft (7.5 m) to more than 500 ft (150 m), while widths of riparian buffers on private

BOX 4-2
Takings

Land-use regulation to protect riparian areas may be hindered by private property rights-related concerns that stem in part from "takings" law. Takings law derives from the Fifth Amendment to the U.S. Constitution, which states "nor shall private property be taken for public use without just compensation." In other words, governments are not prohibited from "taking" property; they are only forbidden to appropriate property for a nonpublic purpose or to appropriate it without paying for it. A taking can be direct, physical occupation or confiscation of property, or it may result from a regulation that restricts property use. Physical invasions by government are more readily ruled takings than are restrictions of use, known as "regulatory takings." Only rarely have courts ruled that land-use regulation has "taken" property in violation of the Constitution.

Defining an unconstitutional taking is difficult. The U.S. Supreme Court's approach involves ad hoc, factual inquiries rather than developing a set formula for determining when "justice and fairness" require that the economic injuries caused by public action be compensated by the government (*Penn Central Transportation Co. v. City of New York*, 1978). Key considerations, however, include the economic impact of the regulation on the claimant and, particularly, the extent to which the regulation interferes with distinct, investment-backed expectations, as well as "the character of the government action" (*Ibid*).

The Court has often said mere diminution in the value of property is insufficient to demonstrate a taking. For instance, the Court upheld a Los Angeles zoning ordinance, which reduced the value of property by 92 percent by forbidding the continuation of brick-making (*Hadacheck v. Sebastian*, 1915). Comprehensive zoning regulations, which can sharply reduce property values, have long been upheld (*Euclid v. Ambler Realty Co.*, 1926).

One example of a regulation having gone too far was established in *Lucas v. South Carolina Coastal Council* (1992). The Court held that a regulation that "deprives land of all economically beneficial use" effects a taking, unless the "proscribed use interests were not part of the [landowner's] title to begin with." Courts have long recognized that private property rights are not absolute; for instance, they are subject to the state's power to abate "nuisances." The *Lucas* Court explained that if the government prohibits some use (here, the right to build on beachfront property) not traditionally prohibited or deemed a nuisance under state law, and thus destroys the economic value of private property interests that pre-date the regulatory prohibition, it must pay for the property taken.

One difficulty with takings law is identifying the relevant parcel or unit of property—i.e., the "denominator." For example, if a landowner challenges a regulation that prevents her from building on two riparian acres of an undivided ten-acre tract, a court should examine not the reduced value of the two acres, but the impact of the prohibition on the entire tract. However, not all courts have concurred. In *Loveladies Harbor, Inc. v. U.S.* (1994), the Federal Circuit Court of Appeals was asked to decide whether the

Corps' refusal to issue a permit to fill wetlands "took" the plaintiff's property. Before wetlands regulation, Loveladies had developed and sold the majority of a 250-acre tract. In the 1980s, Loveladies was denied a federal permit to develop the remaining 12.5 acres. The court decided that 12.5 acres was the proper denominator. The fair market value of the 12.5-acre tract fell from $2,658,000 to $12,500 after permit denial. The court concluded that this was a taking.

Another wetlands case, *Florida Rock Industries v. United States*, took up a different issue—whether a regulation that "deprives the owner of a substantial part but not essentially all of the economic use or value of the property, constitute[s] a partial taking, and is it compensable as such." According to the Federal Circuit, the line between a compensable partial taking and a noncompensable "mere diminution" is unresolved. This issue, it said, necessitates a balancing of the benefits to the public (of which Florida Rock is a part) and the burden on the individual owner. Compensation is due, the court said, when application of the *Penn Central* tests "indicates that this plaintiff was singled out to bear a burden which ought to be paid for by society as a whole." The claims court concluded that Florida Rock was such a plaintiff because its investment-backed expectation had been entirely frustrated and the government had entirely taken the company's common-law right to remove minerals from its property.

A more recent Supreme Court takings case examined not a denial of development rights, but the conditions placed on development. *Dolan v. City of Tigard* (1994) involved the expansion of a retail store. By a 5-4 vote, the Court struck down building permit conditions requiring the property owners to dedicate to the city about ten percent of their property, within the floodplain of a creek, for use as a public greenway and bicycle path. The Court said the city had not justified why the Dolans must dedicate property to the city, as opposed to simply leaving it undeveloped, nor had the city documented the bike path's effect on traffic congestion. The majority did not discuss the fact that the regulation would affect only ten percent of the property, nor that the property's value would be *increased* as a result of the conditional permit. The majority did express concern that the permit condition would deny the owner "one of the most essential sticks in the bundle of rights that are commonly characterized as property"—the "right to exclude" the public.

Thus, several takings issues remain unresolved: (1) whether and how other courts will distinguish between compensable partial takings and noncompensable mere diminution in property values, (2) how to determine the denominator in takings analyses, (3) whether *Dolan* applies to all conditions on development permits or only to required dedications of land, and (4) how courts should weigh benefits of a land-use regulation to the public against its burden on an individual property owner. Despite this uncertainty, certain predictions concerning riparian area regulations are possible. First, protecting or restoring riparian areas is undoubtedly a public purpose. Thus, a regulation that restricts or conditions use of property to promote that purpose would *not* likely be deemed to "take" property unless it (1) left the owner no economically viable use of the entire parcel, (2) authorized public access, and/or (3) imposed a burden on the landowner that is disproportionate to the impact of the desired property use.

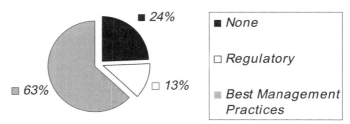

FIGURE 4-1 Proportion of states and U.S. territories with regulations on forest practices, with best management practice guidelines (includes use of BMPs or water quality standards), and with no legislated guidelines.

commercial forest lands, if required, generally range from 25 to 100 ft (10–30 m). Although an increasing number of states have incorporated regulatory buffers into forest practice rules, no states have regulatory buffer requirements for agricultural or grazing practices.

Special Area Protection. Some state legislatures have identified areas in which protection of riparian areas is particularly important and have established a special program for this purpose. For example, in the Delta Protection Act of 1992, California established special planning requirements for activities on lands within the delta of the Sacramento and San Joaquin Rivers—a 738,000-acre area that drains 40 percent of the state's water. One of the requirements is to preserve and protect riparian and wetlands habitat. In 1999, the North Carolina Environmental Management Commission adopted rules to protect 50-ft-wide riparian buffers along waterways in the Neuse and Tar-Pamlico river basins. Although existing uses are exempt, new activities and land uses are prohibited within 50 feet of waterbodies, unless approved by the state. The primary motivation for establishing these buffers was to reduce nutrient loadings.

On a larger scale, the federal and state governments have combined to establish the Chesapeake Bay Program to restore and protect this major estuary. Parties to the 1983 Chesapeake Bay Agreement include Maryland, Pennsylvania, Virginia, the District of Columbia, the Chesapeake Bay Commission (a tri-state legislative body), EPA, and others. As part of its ongoing efforts to reduce water pollution, the Chesapeake Executive Council adopted a directive in 1994 to protect and restore riparian buffer forests along tributaries to the Chesapeake Bay.

The Chesapeake Bay states themselves have established special programs. Virginia's 1988 Chesapeake Bay Preservation Act defines "resource protection areas" as "sensitive lands at or near the shoreline that have an intrinsic water quality value due to the ecological and biological processes they perform or are sensitive to impacts which may cause significant degradation to the quality of state waters" (Section 9VAC10-20-80 A). The act designates a riparian buffer of not less than 100 ft wide along the bay and its tributaries in which activities are significantly restricted. Vegetation "effective in retarding runoff, preventing erosion, and filtering nonpoint source pollution" is to be retained if present or established if it does not exist. The act further requires that local governments use their land-use authority to, for example, subject proposed development within the 100-ft buffer to a water-quality impact assessment and to limit development to facilities that are "water-dependent" or that constitute redevelopment of existing facilities. The width requirements are lessened for agricultural lands enrolled in a government-funded BMP program or a soil and water-quality conservation plan. Special provisions apply to forestry within Streamside Management Zones—areas 50–200 feet from a stream.

In 1984, Maryland established the Chesapeake Bay Critical Area Program. The "critical area" consists of all land within 1,000 feet of the mean high-water line of tidal waters or the landward edge of tidal wetlands or tributary streams. The program's goals are to minimize adverse impacts on water quality that result from pollutants; conserve fish, wildlife, and plant habitat in the critical area; and establish land-use policies for the Chesapeake Bay Critical Area that accommodate growth. Similar to Virginia, Maryland has established criteria to minimize the adverse effects of human activities on water quality and natural habitats and to foster consistent, uniform, and more sensitive development activity within the critical area. These criteria involve classifying land as "intensely developed areas," "limited development areas," or "resource conservation areas" and regulating activities accordingly. In addition, local jurisdictions are required to designate habitat protection areas, which include the naturally vegetated 100-ft buffer along waterways; nontidal wetlands; the habitats of threatened and endangered species and species in need of conservation; significant plant and wildlife habitat; and anadromous fish-spawning areas. Even agricultural lands within the critical area are required to control nutrient runoff by establishing a 25-ft vegetated filter strip along tidal waters, wetlands, or tributary streams or by using equivalent BMPs.

Local Land-Use Regulation. Far more common than statewide or special area regulatory programs are local government regulations for land use adjacent to water. State statutes authorizing local governments to regulate land use within their jurisdiction often include language specifically authorizing protection of environmental values. Such regulations usually establish setbacks along streams or around lakes within which certain types of land uses—most commonly, the building of homes or other structures—are discouraged or prohibited. For example, to protect its drinking water supply, New York City has entered into a Watershed Memorandum of Agreement with local governments in the upstate

TABLE 4-2 Riparian Management Approaches on Private Forest Lands in States and Territories

State	Approach	Regulatory Guidelines	Voluntary Guidelines	Management Area Width	Harvest Practices
Alabama	BMP	None	Primary	35–50 ft	Not limited
Alaska	Regulations	Primary		100 ft	Partial
Arizona	BMP	None	Primary		Not limited
Arkansas	BMP	Secondary	Primary	35–150 ft	Not limited
California	Regulations	Primary	Secondary	100 ft	Partial
Colorado	BMP, Reg.	Secondary	Primary		Not limited
Connecticut	WQ Std	None	Primary		Not limited
Delaware	BMP	None	Primary		Not limited
District of Columbia	None	None	None		Not limited
Florida	BMP	Secondary	Primary	35–200 ft	Not limited
Georgia	BMP	Secondary	Primary	15 ft	Not limited
Hawaii	WQ Std	None	None	35–160 ft	Not limited
Idaho	Regulations	Primary	Secondary	30–75 ft	Partial
Illinois	BMP	None	Primary		Not limited
Indiana	BMP	None	Primary	75–200 ft	Not limited
Iowa	WQ Std	None	None	50–150 ft	Not limited
Kansas	None	None	None		Not limited
Kentucky	BMP	Secondary	Primary		Not limited
Louisiana	WQ Std	None	None	35–100 ft	Not limited
Maine	BMP	None	Primary		Not limited
Maryland	BMP	Secondary	Primary	75–250 ft	Not limited
Massachusetts	BMP	None	Primary	50–450 ft	Partial
Michigan	BMP	Secondary	Primary		Not limited
Minnesota	Regulations	Primary	Secondary	50–250 ft	Partial
Mississippi	BMP	None	Primary	30–60 ft	Not limited
Missouri	None	None	None		Not limited
Montana	Regulations	Primary	Secondary		Partial
North Carolina	BMP	Secondary	Primary		Not limited
North Dakota	BMP	None	Primary		Not limited
Nebraska	BMP	None	Primary	50–200	Not limited

Agency[a]	Legislative Authorization
Alabama Forestry Commission	None
Alaska DNR	Alaska Forest Resources and Practices Act
Arizona DEQ	Arizona Nonpoint Source Pollution Program
Arkansas Forestry Commission	None
California Dept. of Forestry	Z'Berg-Nejedly Forest Practice Act of 1973
Colorado State Forest Service	Forest Management Definitions; Statute 39-1-102
Connecticut DEQ	Water Quality Standards
Dept. of Agriculture/ Forest Service	Watershed Protection Program
Dept. of Trees and Lands, District Government	None
Florida Division of Forestry	None
Georgia Forestry Commission	Best Management Practices for Forestry
Dept. of Health, Clean Water Branch	Administrative Rules; Water Quality Standards
Bureau of Forest Assistance	Idaho Forest Practices Act
DNR, Division of Forestry	Conservation and Water Resources Statutes
DNR, Division of Forestry	Water Quality and Wetland Statutes
DNR, Division of Forestry	Water Quality Standards
NA	None
Kentucky Division of Forestry	Kentucky Forest Conservation Act; Kentucky Agriculture Water Quality Act
Louisiana Dept. of Agriculture and Forestry	Water Quality Standards
Maine Forest Service	None
DNR, Forest Service	Maryland Best Management Practices for Forest Harvests; Maryland Forest Service Tree Laws; Forest Conservation Act, Maryland Seed Tree Law; Maryland Reforestation Law; Maryland Tree Expert Law; Roadside Tree Law
NA	None
DNR, Forest Management Division	Best Management Practices for Water Quality; Michigan Public Act of 1994; DEQ Regulations
DNR, Division of Forestry	Minnesota Sustainable Forest Resources Act of 1998; Forest Management Guidelines
Mississippi Forestry Commission	None
Dept. of Conservation	None
Dept. of Natural Resources & Conservation, Forestry Division	Montana Forest Practices Statutes; Montana Streamside Management Zone Law
North Carolina Division of Forest Resources	Sedimentation Pollution Control Act; Forest Practices Guidelines and Best Management Practices
North Dakota Forest Service	Forest Resource Management Program
Dept. of Forestry, Fish, & Wildlife	None

continues

TABLE 4-2 Continued

State	Approach	Regulatory Guidelines	Voluntary Guidelines	Management Area Width	Harvest Practices
Nevada	BMP	Primary	Secondary		Partial
New Hampshire	WQ Std	Secondary	Primary	50–150 ft	Not limited
New Jersey	WQ Std	None	None		Not limited
New Mexico	BMP	Secondary	Primary		Not limited
North Carolina	BMP	None	Primary	20–200 ft	Partial
New York	None	None	None		Not limited
Ohio	BMP	None	Primary		Not limited
Oklahoma	BMP	None	Primary		Not limited
Oregon	Regulations	Primary	Secondary	20–100 ft	Partial
Pennsylvania	BMP	None	Primary	35–250 ft	
Puerto Rico	None	None	None		Not limited
Rhode Island	None	None	None		Not limited
South Carolina	BMP	None	Primary	40–160 ft	Not limited
South Dakota	BMP	Secondary	Primary		Not limited
Tennessee	BMP	None	Primary		Not limited
Texas	BMP	None	Primary	50 ft	Not limited
Utah	BMP	None	Primary	35–100 ft	Not limited
Vermont	None	None	None		Not limited
Virginia	BMP	None	Primary	35–50 ft	Not limited
Washington	Regulations	Primary	Secondary	50–200 ft	Partial
West Virginia	BMP	None	Primary	50–100 ft	
Wisconsin	WQ Std	None	Primary	35–100 ft	Not limited
Wyoming	BMP	None	Primary		Not limited

[a]DNR: State Department of Natural Resources; DEQ: State Department of Environmental Quality.

NA: Information not available.

Agency[a]	Legislative Authorization
Dept. of Conservation & Natural Resources, Division of Forestry	Nevada Administrative Code for Forest Practices; Nevada Administrative Code for Forest Practices and Division Reforestation
Division of Forests and Lands	New Hampshire Rivers Management and Protection Program; Water Quality Standards
State Forestry Service	Water Quality Standards
New Mexico Energy, Minerals, and Natural Resources Dept.	None
NA	NA
Division of Lands and Forests	None
DNR, Division of Forestry	Ohio Sustainable Forestry Initiative
Oklahoma Forestry Services	Extension Circulars
Oregon Dept. of Forestry	Oregon Forest Practices Act of 1972
Bureau of Forestry	None
Forest Services Bureau	None
Division of Forest Environment	None
Forestry Commission	State Endangered Species Act, Best Management Practices Guidelines
Division of Resource Conservation and Forestry	Cooperative Forestry Assistance Act of 1978
Division of Forestry	Tennessee Forest Practices Guidelines
Texas Forest Service	Texas Water Code; State Water Plan
DNR, Division of Forestry, Fire, and State Lands	Utah Code/Water Quality Act; State Water Plan
Dept. of Forest, Parks, and Recreation	Water Quality Issues
Dept. of Forestry	None
Dept. of Natural Resources	Forest Practices Act of 1976
Forestry Division	NA
DNR, Division of Forestry	Nonpoint Source Abatement Program, Water Quality Standards
Office of State Lands and Investments, Forestry Division	Wyoming Rules and Regulations Database

Catskill, Delaware, and Croton watersheds that includes a variety of setback regulations for activities considered potentially harmful to water quality. Activities such as the placement of septic systems, siting of wastewater treatment plants, pesticide application, and construction of new impervious surfaces are prohibited on lands directly adjacent to reservoirs, streams, and wetlands in the watershed, although agricultural activities are exempt (NRC, 2000).

Incentive-Based Approaches

A growing number of inducements are available to encourage private landowners to protect riparian areas. These inducements take the form of direct payments to landowners not to develop riparian lands, payments to encourage use of environmentally compatible practices, payments or tax benefits for placing a conservation easement on the property, funding for restoration or demonstration projects, stewardship education and technical assistance, and outright purchase of the lands. To be effective, incentives generally must at least equal the value of other use options available to the landowner.

Payments to the Landowner from Public Programs

Typically, there have been no economic incentives for private landowners to protect the ecological functions of riparian areas. However, an increasing number of public programs are offering some form of payment in return for such protection. Since 1985, Congress has been authorizing Farm Bill programs that provide for retirement of croplands for environmental benefits in return for annual payments. The largest of these Farm Bill programs is the Conservation Reserve Program (CRP) under which highly erodible or environmentally sensitive cropland may be retired for a period of years in return for annual rental payments and cost-share assistance for converting and maintaining the land. About 36 million acres of farmland are enrolled in CRP (which has a legal limit of 36.4 million acres). Annual payments amount to approximately $1.8 billion, an average of approximately $50 per acre. Since its inception, the program has increasingly emphasized the importance of water quality and wildlife habitat benefits as priority objectives for land retirement. Lands providing filter strips and riparian buffers adjacent to waterbodies have been given special attention, including a 10 percent incentive payment to landowners to enroll.

The Conservation Reserve Enhancement Program (CREP) brings federal funding to state-initiated and supported programs focused on converting agricultural lands for conservation benefits. Maryland established the first CREP program to create buffer strips within 150 feet of a stream. The impact of the CRP and CREP programs on protection of riparian areas is considered in detail later in this chapter.

The Wetlands Reserve Program (WRP) provides payments to agricultural landowners who enter into conservation easements under which croplands are turned back into wetlands for at least 30 years. In addition, a restoration cost-share agreement provides up to 75 percent of the cost of wetland restoration. If the landowner agrees to a permanent easement, the U.S. Department of Agriculture (USDA) will pay 100 percent of the costs of restoring the wetland. As of July 2000, about 915,000 acres were enrolled in this program.

Some of the incentives-based programs discussed above require landowners to protect highly erodible lands, wetlands, and riparian areas as a condition for participating in federal farm programs that subsidize agricultural production. This "conservation compliance" approach has resulted in the protection of many environmentally sensitive areas and particularly riparian areas, as farming was not profitable in many areas of the country without participation in the subsidy programs. Conservation compliance was based on the conviction that, because the agricultural community was receiving tens of billions of dollars per year in subsidies, it was reasonable for the agricultural community to protect water quality. Unfortunately, conservation compliance ended in 1995 with the passage of the Freedom to Farm Act, which in theory terminated most agricultural commodity subsidy programs along with the incentive to participate in conservation programs. There may be an opportunity to reintroduce conservation compliance in the near future because agriculture has not been very profitable since the end of crop subsidy programs, and there is some support for reintroducing a formal federal price support system.

Cost-Share Programs

Cost-sharing programs are used alone or in combination with land retirement as a means of generating conservation benefits. In return for making improvements or utilizing management practices deemed environmentally beneficial, the landowner receives back some share of the associated costs. An example is the Environmental Quality Incentive Program, established in the 1996 Farm Bill. The landowner submits a conservation plan, prepared with assistance from the Natural Resources Conservation Service (NRCS), proposing practices to address environmental concerns in priority areas. Up to 75 percent of the costs may be reimbursed during the period of the contract. In some instances, special incentive payments are available to pay 100 percent of the associated costs. Under the Wildlife Habitat Incentives Program, cost-sharing assistance is available to landowners to develop wildlife habitat. Partners for Fish and Wildlife, administered by the U.S. Fish and Wildlife Service and state wildlife agencies, also provides cost-share funding to landowners interested in protecting or restoring wetland habitat on their lands. Still another source of federal cost-share funding is the Rivers, Trails, and Conservation Assistance Program, administered by the National Park Service (NPS). Cooperative projects directed toward river and trail enhancement are eligible for funding and technical assistance.

In an effort to encourage riparian buffers in agricultural areas, the USDA initiated the National Conservation Buffer Initiative (NCBI) in 1997. Designed as a private–public partnership, this effort has established a goal of 2 million miles of buffer by 2002. The initiative includes corporate support through a subsidiary group—the National Conservation Buffer Council. To date the primary emphases of NCBI have been marketing and education; no additional funds have been made available through the initiative to support buffer implementation. As of March 1, 2001, NCBI reported that over 1 million miles of buffers[2] had been established through many buffer programs (e.g., CRP, CREP, WRP). However, there is little landowner knowledge of NCBI (NRCS, 1999), and there is no way to determine whether the efforts of NCBI have had any appreciable effect on the rate or extent of buffer installation.

Conservation Easements

Apart from public programs, nearly every state allows for conservation easements. The use of such easements for environmental protection of private lands has expanded exponentially in recent years (NRC, 1993). An easement is a legally enforceable agreement between a landowner and another party to maintain private lands for specified conservation purposes in perpetuity (Land Trust Alliance, 1996). Potential incentives to the landowner are the ability to limit future uses of land (e.g., to keep the land in agricultural use or as permanently protected open space), receipt of the fair market value of the easement (generally the development value of the property), and various tax incentives (described below). A nonprofit land trust, a conservation organization such as The Nature Conservancy, or a government agency typically holds the easement and is responsible for ensuring its implementation. The land remains in private ownership. No organization acquires conservation easements specifically for protection of riparian areas. Yet, because in the West riparian areas provide essential habitat for a large number of plant and animal species, conservation efforts in this region often emphasize protection and restoration of riparian lands.

North Carolina enlisted the use of conservation easements as a means of voluntarily moving swine operations out of the 100-year floodplain following the massive flooding caused by Hurricane Floyd in 1999. Landowners were invited

---

[2]This figure includes riparian buffers, filter strips, grassed waterways, shelterbelts, windbreaks, living snow fences, contour grass strips, cross-wind trap strips, shallow water areas for wildlife, field borders, alley cropping, herbaceous wind barriers, and vegetative barriers. These buffers are not equivalent to fully functioning riparian areas as described in Chapter 2. In addition, some of the buffers are "in field" rather than in the riparian area.

to submit a bid to the North Carolina Department of Environment and Natural Resources for purchase of a perpetual easement that would prohibit use of the land as a feedlot or for associated animal waste operations, prohibit non-agricultural development, require a soil and water conservation plan, and require implementation and maintenance of a minimum 35-ft forested riparian buffer along adjacent perennial or intermittent streams.

Tax Incentives

A growing number of tax incentives encourage landowners to protect the environmental values of their property. An important incentive for landowners to voluntarily donate a conservation easement on their property to a qualified organization is that federal income tax law allows a deduction over a 6-year period of the fair market value of the easement from the landowner's adjusted gross income (not to exceed 30 percent of the total income in any year) (Small, 1995). Many states offer a similar incentive. In addition, federal estate taxes may be reduced because the easement is deemed to have limited the value of the inherited property. Finally, some states authorize reduced local property tax assessments on properties covered by conservation easements.

A number of states have tax incentives directed specifically at, or applicable to, riparian lands. For example, Virginia authorizes local governments to exempt "riparian buffers" from property taxation (Va. Stat. § 58.1-3666). Under this statute, a riparian buffer is defined as "an area of trees, shrubs or other vegetation, subject to a perpetual easement permitting inundation by water, that is (i) at least 35 ft in width, (ii) adjacent to a body of water, and (iii) managed to maintain the integrity of stream channels and shorelines and reduce the effects of upland sources of pollution by trapping, filtering, and converting sediments, nutrients, and other chemicals." Idaho offers income tax credits to landowners equal to half of the costs of fencing riparian areas for managing livestock grazing or for controlling erosion (Id. Stat. § 63-3024B).

Restoration/Protection/Demonstration Programs

A number of federal and state programs provide funding for projects that can restore impaired or lost ecological functions of riparian areas. Funding made available under Section 319 of the Clean Water Act is intended to encourage demonstration projects addressing nonpoint source pollution. In many cases, these projects focus on improvements in riparian areas. For example, the Lake Champlain Basin Agricultural Watersheds Project utilized grazing management, livestock exclusion, and streambank protection as tools to reduce concentrations and loadings of phosphorus, nitrogen, suspended solids, and bacteria in two treatment watersheds. Riparian fencing was used to exclude cattle access, and bioengineering techniques were used to restore streambanks. After two years, significant reductions in all parameters were documented. In North Carolina's Neuse River Basin, a 319 project is evaluating the effectiveness of vegetative riparian buffers and controlled drainage in order to determine the preferred best management practice for reducing nutrients.

Since 1986, Congress has been increasing the role of the Corps in river channel and wetlands restoration. The 1986 Water Resources Development Act declared that all environmental improvements included by the Corps within their projects are to be considered economically justifiable. The act specifically authorized fish and wildlife mitigation measures, and it provided authority for the Corps to undertake restoration activities as needed to offset the environmental problems created by previous flood-control and navigation projects. A dramatic example of this new role is the half-billion-dollar Kissimmee River Restoration Project, which is reestablishing wetland conditions in the river's historic floodplain (see Chapter 5).

BOX 4-3
Riparian Restoration at Bear Creek, Iowa

Management agencies often struggle with designing effective riparian programs because of funding constraints within individual agency programs and the limited nature of agency expertise. This suggests that development of interdisciplinary research teams (including soil scientists, geologists, ecologists, economists, fisheries and wildlife scientists, and others) and partnerships among local, state, and federal agencies, non-governmental organizations (NGOs), and private industry could expedite protection of riparian areas. The Bear Creek, Iowa, riparian demonstration project along with the development of the Riparian Management System (RiMS) exemplifies both of these characteristics and has become a role model for subsequent research, demonstration, and management.

Initiated simply as an effort to accomplish research in an agricultural setting, the project has grown from working with a single landowner on a modest section of Bear Creek to its present state of nearly eight miles of riparian restoration. It is run by a partnership of diverse research and funding entities. The RiMS project has also been adopted as the guiding principle for buffer development by the nonprofit group Trees Forever, which has a five-year goal of 100 riparian demonstration or project sites in Iowa and which has just recently expanded into an Illinois buffer initiative of comparable magnitude.

The Bear Creek watershed is a 3,100-acre (7,661-ha) agricultural area located just northeast of Ames, IA. Narrow (2–4 miles or 3–6 km) and long (22 miles or 35 km), the watershed is 87 percent row crops, primarily in private ownership, and has a highly modified hydrology. In general, the watershed is typical of many Midwestern water-

States also have established programs to support restoration and protection efforts, including ones benefiting riparian areas. Arizona's Water Protection Fund provides funding for measures to enhance and restore rivers and streams and associated riparian habitats. California has established the Riparian Habitat Conservation Program, which uses funding from federal grants, private donations, and other sources to preserve and enhance riparian habitat. As part of a statewide program for developing water basin plans for wetlands and riparian area restoration, North Carolina established a Riparian Buffer Restoration Fund for restoration, enhancement, or creation of riparian buffers. Oregon has an extensive program promoting watershed management in which management and protection of riparian areas are the principal goals, with funding made available from the Oregon Watershed Enhancement Board. As described in Box 4-3, restoration and demonstration projects in riparian areas often require multiple sources of funding and interdisciplinary participation to achieve success.

---

sheds from Ohio to the eastern Dakotas. Presettlement conditions in the Bear Creek watershed consisted of rolling prairie, marshes, and very limited forest, with stream flows being intermittent and seasonal. Defined stream channels were difficult to find in many areas of the watershed based on 1847 land survey notes. As early as 1902, major changes to the watershed were brought about by drainage and land conversion. Significant channelization occurred into the 1970s, resulting in vastly altered hydrology and in changes in the biota, reflecting degraded habitat, increased turbidity, and warmer water temperatures (Schultz et al., 1995; Isenhart et al., 1997).

In 1990, initial headway was made by developing a partnership with a local conservation-minded landowner willing to allow experimentation with riparian management systems on his property. Ironically, many of the funding programs such as continuous CRP that are currently available to landowners were not available to the owners of the original study property. Hence, riparian management has expanded into upstream areas supported by funding that was not available to the people who took the initial risk. Nonetheless, working in the production agriculture setting forced the development of a riparian management system that is economically viable and practical to implement.

Also vital to the success of the Bear Creek project has been the cooperative nature of the research partners. Partners involved in the research component of Bear Creek include scientists from Iowa State University, the Leopold Center for Sustainable Agriculture, National Soil Tilth Laboratory, EPA, NRCS, and others. Current research and demonstrations are being conducted on riparian lands owned by ten farmer cooperators. In addition, 11 professional consultants from various government agencies and NGOs are cooperating with the team. The Bear Creek project demonstrates the importance of a collegial atmosphere among participants, obtaining funding from a wide variety of sources, and patience and perseverance.

Summary

Table 4-3 summarizes the many regulatory and nonregulatory approaches used by the states to address protection of privately owned riparian areas. The extreme variation in approaches is remarkable. A significant limitation of many of the approaches is that their success is measured by the number of practices implemented and rarely by actual environmental improvements. For example, the National Conservation Buffer Initiative, the goal of which is to install 2 million miles of new riparian and upland buffers by 2002, is being touted as a means of improving water quality. Rather than requiring the collection of water-quality data to determine if water quality is actually improving, program success is measured by counting the "number of miles of buffers installed." The program has no systematic and scientific means of targeting efforts to the riparian areas that would most protect water quality.

Such indirect metrics of success are typical of state and federal conservation programs, and they are partially justifiable given the ease with which such measurements can be made. Indeed, actually measuring improvements in water quality and habitat may take decades, given the lag time between implementation of restoration activities and riparian system response. To convince policy makers and the public that progress is being made—even if there is no measurable improvement in water or habitat quality—these indirect measures are often used. Because of these uncertain metrics, and because many restoration programs are relatively new, it is difficult to know whether the federal, state, and local programs described above have been or will be effective in restoring structure and functioning to riparian areas on privately owned land.

PROTECTION OF FEDERAL LANDS

Nearly 40 percent of the land area of the United States is in public ownership, primarily federal (Figure 4-2). In 1911, the U.S. Supreme Court in *Light v. United States* stated: "All of the [federal] public lands of the nation are held in trust for the people of the whole country." Congress can legislate to protect the public lands or any of the components thereof (e.g., wildlife, water, vegetation, or other resources) and to provide for the use, management, or disposal of the lands or their contained resources. States retain broad police powers over federal lands, subject to the federal government's power to preempt any state law that conflicts with a federal law or regulation. For example, states retain general authority to set hunting and fishing seasons and limits. Although only Congress may establish federal land policy, administrative agencies have extremely broad latitude to determine and conduct the day-to-day management of federal lands and resources (*United States v. Grimaud*, 1911).

All laws governing federal lands potentially affect riparian areas on those lands. Very few federal statutes, however, provide expressly for riparian area

TABLE 4-3 Types of State Laws and Policies for Protecting Privately Owned Riparian Areas in Representative States[a]

State	Zoning or "Buffer" Regulations[b]	Permits or Preconstruction Approval[c]	Tax Incentives	Other Incentives to Private Owners[d]	Conservation Easements[e]	Timber Harvest Standards	Incentives ($ and/or Tech. Asst.) to Cities[f]	Land Acquisition Funds	Watershed Planning (nonregulatory)	General Policy Statement
Alaska						X				
Arizona	X									X
California				X			X	X		X
Florida								X		
Kansas				X	X					
Idaho						X				
Maine									X	
Maryland				X			X			
Massachusetts	X			X			X	X		
Michigan		X		X				X	X	
Montana						X				
Nebraska				X						
N. Mexico								X		
N. Carolina	X	X		X				X	X	X
Oregon	X	X	X	X		X	X	X	X	X
Virginia										
W. Virginia										X
Wisconsin	X			X						

[a]This table is merely illustrative of the range of approaches states take to recognizing and protecting riparian values. It does not purport to catalogue all state laws and policies.
[b]Zoning regulations and buffer requirements may be difficult to distinguish. Each connotes regulation of development within a specified distance or a mapped district, and may include performance standards. Excluded from this category are timber harvest standards or similar "streamside management zone" regulations, which are referenced in the "Timber Harvest Standards" column of the table.
[c]Refers to something less comprehensive than zoning regulations or buffer requirements, e.g., Michigan's requirement of a permit and soil erosion plan for developments (other than logging or agriculture) within 500 feet of a water body.
[d]Includes economic incentives, education, and technical assistance.
[e]Note: Nearly all states provide by law for conservation easements. This column refers to those states that provide for state acquisition of an easement for such purposes. Several states authorize the expenditure of funds to acquire lands or waters, or interests therein, in riparian areas. These laws are referenced in the "Land Acquisition Funds" column of the table.
[f]Includes planning and restoration assistance, Greenbelt programs, training, etc.

FIGURE 4-2 Percentage of land in public domain by state.

management or protection. (An exception is the National Wild and Scenic Rivers Act.) Thus, it is not surprising that federal agencies are not required to coordinate their riparian management activities. Conversely, no law prohibits them from doing so.

Most agencies have adopted rules or policies governing activities in, or potentially affecting, riparian areas. But vague or nonexistent legislative mandates and the lack of any coordination requirement have resulted in inconsistent riparian management—as exemplified in the Greater Yellowstone area (Harting and Glick, 1994). Moreover, federal agency responsibilities with respect to riparian areas vary greatly. Agencies may have regulatory authority but no land management authority (e.g., EPA). They may manage some lands but have only regulatory authority over other lands (e.g., FWS and the Corps). They may possess full regulatory and land management authority over lands entrusted to them by Congress (e.g., BLM, USFS, and NPS), or they may exercise service, technical, or advisory functions but possess no regulatory or land management powers (e.g., NRCS). Several federal regulatory or management regimes may apply to a given tract of land, and more than one federal agency may be responsible for implementing or overseeing those laws. For example, several agencies exercise authority under a panoply of statutes on the national forests. The USFS has primary responsibility for managing all surface resources, but BLM is responsible for managing and conveying interests in minerals. FWS and/or National Marine Fisheries Service have technical, regulatory authority with respect to federally listed threatened or endangered species (under the ESA), and the Corps is responsible (under the CWA) for regulating dredge-and-fill activities in wetlands and streams. To further complicate matters, certain federal subsidies and policies, such as those related to flood prevention and control and agricultural practices, actually encourage the destruction of riparian habitat (Kusler, 1985).

Each of the land management agencies has some mandate to manage public lands with the national interest in mind. But "national interest" has no immutable, absolute meaning; indeed, its interpretation varies by land system. Moreover, such directives are usually paired with provisions allowing or requiring multiple uses and permitting individuals to obtain property or other interests in federal resources (such as timber contracts, mining claims, oil and gas leases, national park concessions, and federal grazing permits). Thus, riparian ecological concerns are balanced against (or reconciled with) other agency objectives for land and resource management. But unlike on private property, no individual user of federal land resources can maintain that a public riparian area should be managed for his benefit to the disadvantage or exclusion of others.

Finally, the federal government is required to comply with applicable state environmental laws and regulations—governing, e.g., water and air pollution—which may impact use and management of riparian areas on federal lands. Land-use planning is the domain primarily of states, but the federal government has prescribed planning and management standards for each system of federal public lands. In addition, federal agencies are generally directed to "coordinate," where possible, their planning activities with those of local and tribal governments.

This section describes (in alphabetical order) the major federal land systems and the respective agencies, statutes, and programs that encompass riparian area/resource considerations. The discussion is not exhaustive, as a complete survey of regulations and policies was not feasible. Statutes of minor importance have been omitted. Moreover, several statutes that affect management of federal lands and may bear on riparian area protection (e.g., NEPA, ESA, CWA, SMCRA) were discussed in the prior section. This section ends with a brief examination of state-owned submerged lands.

Bureau of Land Management Lands

The Bureau of Land Management (BLM) manages approximately 267 million acres of land (including both surface and mineral estate), chiefly in the 11 western states and Alaska, along with an additional 300 million acres of subsurface mineral resources under lands owned or managed by others. Including Alaska, BLM lands contain 24 million acres of "riparian wetlands" and 178,000 miles of "riparian streams," 174,000 of which are fishable. Only about 1.2 million acres managed by BLM outside Alaska are riparian. More than 2,000 miles (25 miles in New Mexico with the rest in Oregon) of rivers on BLM lands are designated wild, scenic, or recreational. These designations encompass nearly 1 million acres of land (BLM, 2000a).

Management of BLM lands is governed by the Federal Land Policy and Management Act (FLPMA, 43 U.S.C. §§ 1701–1784). FLPMA prescribes sustained-yield management principles, which apply to an open-ended list of "renewable and nonrenewable resources," including fish and wildlife, recreation, timber, range, wilderness, minerals, watershed, and natural, scenic, scientific, and historical values. The act further directs BLM, in managing the public lands, "to take any action necessary to prevent unnecessary and undue degradation of the lands."

Although riparian habitat is not expressly mentioned in FLPMA, many of its provisions implicitly authorize protection of riparian areas, and several BLM rules address riparian management and protection. Perhaps of greatest significance is FLPMA's requirement that the BLM "give priority to the designation and protection of areas of critical environmental concern," or ACECs. ACECs are defined to include areas "where special management attention is required . . . to protect and prevent irreparable damage to important . . . scenic values, fish and wildlife resources or other natural systems or processes." As of the end of 1999, BLM had designated nearly 13.1 million acres of ACECs, 5.9 million of which are in Alaska (BLM, 2000b). Riparian values are frequently a motivating factor for designating ACECs, especially in the Southwest. In a San Miguel River (Colorado) ACEC, for example, BLM proposed to "close relic [sic] riparian communities to all Bureau authorized actions," leaving the rest of the ACEC "open only to those Bureau authorized actions with an overriding public need that would not cause long-term visual impacts or damage riparian systems" (BLM, 1992). In 1986 the Alaska BLM proposed five ACECs in the Central Yukon Planning Area, one of which was established to "protect crucial riparian habitat associated with known peregrine falcon nesting areas." Four other ACECs with fish protection objectives included 300 feet along each side of the designated river (BLM, 1986).

Although ACECs often protect stream-related resources, riparian resources or values are often not expressly referenced as such in the designation notices. For example, in 1984 the Oregon BLM Office designated 35 special management areas, including ACECs, in part to protect riparian habitat or values. The descriptions of the areas used terms such as river canyons containing relict stands of old-growth trees, a river corridor with important scenic, fisheries, wildlife and botanic values, and a lake and bog ecosystem (BLM, 1984).

FLPMA requires BLM to prepare resource management plans (RMPs), which describe generally how an area will be managed to provide resources and services demanded by the public and commodity groups, while protecting the lands. In preparing and revising plans, the BLM is to use an interdisciplinary approach, incorporate current resources inventory data, weigh long-term benefits to the public against short-term benefits, provide for compliance with applicable environmental laws, and coordinate with state, local, and tribal land-use plans. Although FLPMA does not specify that RMPs consider riparian area values, at least since 1990, BLM has pursued riparian management in planning documents and on-the-ground activities. In a 1991 initiative, BLM established national goals for wetland-riparian resources on public lands, including the objective of achieving proper functioning condition (PFC) for 75 percent of riparian areas by 1997 (BLM, 1991a). Furthermore, in 1994 the agency committed to several ecosystem management principles, including reconnecting isolated parts of the landscape: "Rivers will be managed in association with floodplains, and management activities in upland habitats will be considered for their effects on riparian areas" (BLM, 1994).

Since issuing its riparian initiative, BLM has been restoring 23.7 million acres of riparian wetlands. In 1993 the agency revised 180 site-specific management plans, surveyed nearly 2,000 miles of streams, constructed 567 riparian habitat improvement projects, acquired nearly 37,000 acres of riparian habitat, and implemented management plans on 145 riparian acres through partnerships with state and private cooperators. The agency's Fish and Wildlife 2000 initiative also focuses attention on riparian restoration in its range administration program. Finally, BLM, NRCS, and USFS participate in a National Riparian Service Team, which, along with its private and local government partners, is pursuing restoration efforts across the country on rivers such as the Mississippi, Sacramento, and San Pedro Rivers.

In 1994, BLM promulgated significant new rules governing livestock grazing. The rules were preceded by an environmental impact statement, in which the BLM and USFS concluded that "watershed and water quality would improve to their maximum potential" if livestock were removed entirely from federal lands (BLM and USFS, 1994). Instead of eliminating livestock grazing, however, the BLM required its managers to take appropriate action to ensure that:

(a) watersheds are in, or are making significant progress toward, properly functioning physical condition, including their upland, riparian-wetland, and aquatic components; soil and plant conditions support infiltration, soil moisture storage, and the release of water that are in balance with climate and landform and maintain or improve water quality, water quantity, and timing and duration of flow.

b) ecological processes, including the hydrologic cycle, nutrient cycle, and energy flow, are maintained, or there is significant progress toward their attainment, in order to support healthy biotic populations and communities.

(c) water quality complies with state water quality standards and achieves, or is making significant progress toward achieving, established BLM management objectives such as meeting wildlife needs.

(d) habitats are, or are making significant progress toward being, restored or maintained for federal threatened or endangered species, federal proposed species, Category 1 and 2 federal candidate species and other special status species.

These "fundamentals of rangeland health" are followed by standards and guidelines for grazing administration. The regulations prescribe several minimum requirements, including "maintaining, improving, or restoring riparian-wetland functions, including energy dissipation, sediment capture, groundwater recharge, and stream bank stability" (43 C.F.R. § 4180.2(e)(3)). "Fallback" requirements apply where state or regional standards and guidelines had not been developed by late 1997. (Standards were not completed for all western states until late 2000.) A key fallback standard requires that riparian-wetland areas be in properly functioning condition.

Riparian area protection has also been proposed or accomplished on BLM lands through land exchanges and by withdrawing river corridors from mining. In addition, oil and gas operators on both BLM lands and national forests are required by regulation to protect riparian areas, floodplains, and wetlands.

National Forests

The U.S. Forest Service (USFS) is responsible for managing approximately 140 million acres of national forests and 51 million acres of national grasslands. The bulk of USFS lands are managed for multiple uses—outdoor recreation, grazing, timber, watershed, and wildlife and fish. Uses of some national forest areas, however, are more restricted. Approximately 49.9 million acres (26 percent) of USFS lands were managed primarily for "conservation" as of 1994 (GAO, 1996). This included 34.6 million acres within the National Wilderness Preservation System, 128,000 miles of rivers and streams, 239,000 acres of national rivers (as of 1991) and 618,000 acres of designated wild and scenic rivers (as of 1994), 2 million acres of lakes and reservoirs, and 14 million acres of wetlands and riparian areas (half of which are in the West). National forests provide more than half of all steelhead and salmon spawning habitat—15,000 miles of anadromous fish habitat in the Columbia River Basin alone (Feldman, 1995).

Under the 1897 Organic Administration Act, national forests are established to "furnish a continuous supply of timber" and "secure favorable conditions of water flows," the latter of which is highly relevant to riparian area protection. This act authorized the USFS to issue regulations to meet these objectives and to prevent the destruction of the national forests. The National Forest Management Act of 1976 (NFMA) governs administration of the national forests. Most NFMA provisions are procedural, relating to the management plans that the USFS is required to prepare for each forest. Planning must be interdisciplinary, incorporate NEPA procedures, and reflect multiple-use, sustained-yield principles. The NFMA requires that the agency (1) determine which lands are physically and economically suitable for logging and (2) maintain "diversity of plant and animal species," including the "diversity of tree species similar to that existing in the region." In a number of places, the NFMA calls for consideration and protection of watercourses and watersheds. One NFMA implementing regulation requires management of fish and wildlife habitat so as to "maintain viable populations of existing native and desired non-native vertebrate species" (36 C.F.R. § 219.19). Given the dependence of a high proportion of the West's fauna on riparian habitats, this mandate necessitates protecting riparian areas.

Despite the absence of explicit references to riparian areas in USFS authorizing legislation, riparian protection is an important objective on national forests and grasslands. The agency's 1990 Strategic Plan, which focused on enhancing recreation and fish/wildlife resources among other things, identified the effect of management actions on riparian areas as one of the most important issues facing the agency (Mohai and Jakes, 1996). As of 1993, the agency had made riparian wetlands management a priority and was increasing use of watershed analysis and assessment, modifying management practices, and undertaking an aggressive restoration program. Moreover, riparian ecosystem research is an important component of USFS biodiversity research efforts (Keystone Center, 1991).

National Forest Plans

Though not in its guiding legislation, riparian protection is explicitly mentioned in USFS regulations, which prescribe minimum specific management requirements to be met in accomplishing goals and objectives for the National Forest System (36 C.F.R. § 219.27). Specifically, forest plans are directed to give "special attention" to

> land and vegetation for approximately 100 feet from the edges of all perennial streams, lakes, and other bodies of water. This area shall correspond to at least the recognizable area dominated by the riparian vegetation. No management practices causing detrimental changes in water temperature or chemical composition, blockages of watercourses, or deposits of sediment shall be permitted within these areas which seriously and adversely affect water conditions or fish habitat.

Management prescriptions also must "protect streams, streambanks, shorelines, lakes, wetlands, and other bodies of water," "maintain diversity of plant and animal communities," and "provide for adequate fish and wildlife habitat to maintain viable populations of existing native vertebrate species." Vegetative manipulations, including silvicultural practices, must ensure conservation of soil and water. For the Toiyabe National Forest in Nevada, the forest plan states that "in the event of conflicts between resource uses, the protection of riparian areas would be given 'preferential consideration'" (*Nevada Land Action Association v. U.S. Forest Service*, 1993). In 1997, the secretary of agriculture appointed a Committee of Scientists (COS) to provide technical and scientific advice on land and resource planning on the national forests and grasslands. Interestingly, the first component of a six-part strategy for conserving and restoring watersheds for purposes of meeting ecological sustainability goals was to "provide conditions for the viability of native riparian and aquatic species" (COS, 1999).

Forest plans delineate categories of management areas, one of which typically emphasizes riparian management. Thus, some national forests establish riparian habitat conservation areas along perennial and intermittent streams to which specific management constraints apply. In the Idaho Panhandle National Forests, for instance, timber harvest is not allowed in these areas (USFS, 1998). The USFS's most stringent riparian area protections are found in the Northwest Forest Plan, which governs management of both BLM lands and national forests and prescribes requirements for land uses and management activities within ri-

---

BOX 4-4
Northwest Forest Plan

The Northwest Forest Plan for federal lands within the range of the Northern spotted owl represents the largest scale application of riparian conservation and restoration within the framework of a landscape-management plan. This plan for federal lands uses site-potential tree heights as a functional basis for delineating boundaries for riparian reserves. Site-potential tree height, which varies by species and region, represents the height of the dominant overstory trees in late succession. Riparian reserves are two site-potential tree heights wide (approximately 300–450 ft) on each side of perennial streams and one site-potential tree height wide on intermittent and ephemeral streams. All floodplains are protected, and no timber harvest occurs within them. Default management criteria call for no timber harvest within riparian reserves unless silviculturally required to attain desired ecological conditions along stream corridors. Thinning of young stands is permitted only to accelerate recovery of riparian forests and development of more natural patterns of forest structure. These riparian reserves account for 2.6 million acres (approximately 11 percent) of the land base under the Northwest Forest Plan.

Part of the plan calls for the identification and protection of key watersheds or strongholds to help maintain landscape and aquatic integrity and to provide refugia. Key watersheds are basins where objectives for aquatic resources, wildlife, and ecological functions are the primary management criteria. These watersheds include areas for forest harvest, but timber production is secondary to watershed function. The primary key watersheds account for 8.1 million acres of the 24.4 million acres of USFS and BLM lands within the Northwest Forest Plan.

The Northwest Forest Plan applies to federal forests within the range of the Northern spotted owl in western OR, western WA, and northern CA. Federal agencies in the western United States have developed several other regional conservation strategies that protect and restore riparian systems and watersheds. PACFISH applies to federal forests outside the range of the Northern spotted owl but within the range of anadromous salmonids in eastern OR, eastern WA, ID, and parts of CA. INFISH is a management plan for USFS lands outside the ranges of both the Northern spotted owl and anadromous salmonids but for drainages with native fish in eastern OR, eastern WA,

parian reserves (see Box 4-4). Riparian protection is also mandated by 1990 legislation specific to the Tongass National Forest in Alaska. Commercial timber harvesting is prohibited (unless the timber had already been sold) within a buffer not less than 100 feet wide on each side of certain streams.

Riparian restoration is an implicit objective of the use of "range betterment" funds, which are collected from western national forest grazing permittees. Regulations governing the Hells Canyon National Recreation Area in Idaho, managed in part by the USFS, require grazing permits to "provide for terms and conditions which protect and conserve riparian areas." Similarly, USFS rules governing grazing fees in the eastern United States authorize fee credits for range improve-

---

ID, western MT, and parts of CA. Like the Northwest Forest Plan, these strategies call for combinations of riparian reserves and key watersheds to provide ecological strongholds and landscape connectivity. These strongholds retain the habitats of anadromous and resident fish stocks and support aquatic and terrestrial biodiversity at a landscape scale. This emphasis on identification of strongholds and their conservation is a unique aspect of all these programs—one that is rarely found in other riparian management approaches. Similar approaches have been recommended in conceptual aquatic conservation strategies (Moyle and Sato, 1991; Doppelt et al., 1993; Bradbury et al., 1995). Riparian practices in the Northwest Forest Plan are compared to the other regional federal riparian policies in the table below.

One of the potential barriers to implementing these progressive strategies is legal challenge based on monitoring requirements. The Northwest Forest Plan requires that more than 70 species be monitored for before any ground-disrupting activities occur. Legal challenges have been based on not conducting a full survey, even for small activities over several square meters. Legal challenges based on requirements designed for larger landscape elements but applied to far smaller spatial extents undermine the intent and successful application of the regional strategy embodied by the Northwest Forest Plan.

Riparian Management Criteria Under Various Federal Plans

	Width (in Site-Potential Tree Heights) on One Side of Bank-full Channel		
Waterbody Type	NW Forest Plan	PACFISH	INFISH
Fish-bearing streams	2 tree height	2 tree height	2 tree height
Perennial streams w/o fish	2 tree height	1 tree height	1 tree height
Ephemeral/intermittent streams	1 tree height	1 tree height	1 tree height
Lakes	2 tree height	1 tree height	1 tree height
Wetlands	1 tree height	1 tree height	1 tree height

SOURCES: USFS and BLM (1994a,b); USFS (1995).

ments that meet specified requirements, including enhancement or protection of riparian values. Finally, oil and gas leasing operations are prohibited in national forest riparian areas and wetlands.

National Parks

The National Park Service (NPS) manages a nationwide system consisting of 376 units and about 80 million acres, two-thirds of which are located in Alaska. The system consists of designated areas of "land and water," including national parks, monuments, seashores, trails, and recreation areas. System units are highly diverse, ranging from the 2.2 million-acre Yellowstone National Park to urban parks such as the Washington Monument. The agency is responsible for protecting the natural, historical, and cultural resources of these areas, while promoting recreation opportunities.

The NPS was created by the National Park Service Organic Act of 1916, which states that the fundamental purposes of national parks and monuments are to conserve their scenery, natural and historic objects, and wildlife and to provide for their enjoyment so as to leave them unimpaired for future generations. The act also states that regulation of park system units must be consistent with these purposes. Although many parks contain rivers and streams with associated riparian areas that clearly contain the above-mentioned resources (e.g., scenery, natural objects, and wildlife), the act contains no express reference to riparian areas. The mandate to conserve natural objects and wildlife thus serves as the foundation of any NPS responsibilities with respect to riparian areas. The act further authorizes the secretary of the interior to issue regulations deemed necessary for use and management of the parks, in particular concerning activities on or relating to waterbodies located within the national park system. A general management plan is required for each park system unit, although the act makes only very general prescriptions about plan contents.

In at least one area, the NPS has developed policies that may bear on riparian area management. The agency's Floodplain Management Guideline (NPS, 1993) was developed to comply with federal directives concerning floodplain management. The guideline applies to all NPS actions that "have the potential for adversely impacting the regulatory floodplain or its occupants, or which are subject to potential harm by being located in floodplains." Guideline objectives include defining the regulatory floodplain, defining or assessing hazards, and providing guidance for managing activities that modify or occupy floodplains or impact their values. Once the regulatory floodplain is determined, the NPS develops information concerning flood conditions and hazards, then it designs actions to manage flood conditions (e.g., by selecting alternative sites, using mitigation, warning users, and developing contingency and evacuation plans). The NPS also has a separate wetland management policy.

At least one court case suggests that NPS attention to riparian management may have been deficient. A federal appellate court in California rejected the NPS's determination that highway reconstruction-related impacts to the Merced River and its riparian corridor were acceptable under the National Wild and Scenic Rivers Act (*Sierra Club v. Babbitt*, 1999).

National Wildlife Refuges

The U.S. Fish and Wildlife Service (FWS) oversees the National Wildlife Refuge System, administers fish and wildlife research and habitat conservation activities, manages migratory species hunting and conservation activities, and implements and enforces the Endangered Species Act (ESA). The agency manages about 93 million acres of land, including 530 national wildlife refuges, 111 research and field stations, and 75 wilderness areas.

The National Wildlife Refuge System is a complex of federal lands designated in 1966 in conjunction with the passage of the first ESA and managed principally for wildlife conservation purposes. President Teddy Roosevelt established the earliest individual refuges in the first decade of the twentieth century. Refuges have been reserved from multiple-use (or public domain) lands or acquired with funds from various sources, including the sale of duck stamps. They range in size from a few acres (e.g., Pelican Island, Florida) to more than 1 million acres (e.g., Alaska's Arctic National Wildlife Refuge). Wetland habitats (marine, estuaries, rivers, lakes, and marshes) comprise almost 37 percent of the refuge system (Keystone Center, 1991).

Management of the refuge system is guided primarily by the National Wildlife Refuge System Administration Act of 1966, as amended by the National Wildlife Refuge System Improvement Acts of 1997 and 1998. These statutes do not refer expressly to riparian resources. However, the agency clearly has management authority over riparian areas on refuge lands, and it has ample authority (if not an obligation) to manage them to maintain and restore their animals, plants, and habitats, to "ensure their biological integrity, diversity and environmental health," and to monitor their status. According to the acts, the mission of the refuge system is to "administer a national network of lands and waters for the conservation, management, and where appropriate, restoration of the fish, wildlife, and plant resources and their habitats . . . for the benefit of present and future generations." The law directs that each refuge be managed to fulfill this mission, as well as the specific purposes for which that refuge was established. It further directs the interior secretary to maintain adequate water quantity and quality to fulfill the mission of the National Wildlife Refuge System and the purposes of each refuge and to acquire water rights needed for refuge purposes. To carry out such objectives, a comprehensive conservation plan is required for each refuge, although to date few plans have been prepared. Plans are supposed to identify the distribution and abundance of fish, wildlife, and plants and their related habitats; significant problems affecting those species and habitats; and opportunities for compatible wildlife-dependent recreational uses.

The following examples reveal FWS authority to manage riparian areas on refuge lands. In Hart Mountain National Wildlife Refuge, Oregon, refuge managers had determined that eroded stream channels and deficiency of riparian vegetation along a majority of streams, among other resource-related problems, were preventing the Refuge's goals from being achieved. They proposed a comprehensive management plan that included discontinuing livestock grazing for 15 years and allowing passive restoration of riparian areas (except in limited areas where prescribed burnings, willow plantings, and check dams would be used) (FWS, 1994). Although FWS cautioned that improved soil productivity and native plant community restoration might not occur for 100 years, according to the Oregon Natural Desert Association, significant improvements in riparian vegetation and streambank conditions were evident within a few years after removal of livestock. Unfortunately, funding for an ecological study of streamside vegetation and bird populations was terminated by the FWS after four years, and in fact FWS has removed livestock exclosures that had been in place for decades (Durbin, 1997a,b).

Riparian habitat concerns also prompted the FWS to phase out livestock grazing in the Cabeza Prieta National Wildlife Refuge in Arizona. Livestock grazing was deemed to conflict with the refuge's primary purpose—assisting in the recovery of desert bighorn sheep—as well as with providing crucial habitat for one of the last remaining herds of the endangered Sonoran pronghorn antelope. The agency noted the importance of "dry riparian habitats" to refuge wildlife, including pronghorn antelope and bighorn sheep. Riparian and tidal restoration were also identified as major issues to be addressed in Washington's Nisqually National Wildlife Refuge's comprehensive conservation plan in order to ensure sanctuary for migratory birds. Finally, in Alaska's Arctic National Wildlife Refuge, the agency forbids movement of any equipment associated with geological and geophysical exploration of the coastal plain through riparian willow stands, except with prior expressed approval (50 C.F.R. § 37.31(b)(3)).

The FWS recently acquired new responsibilities that may result in greater riparian area protection on lands managed by federal agencies. A presidential executive order issued January 2001 directs all federal agencies whose actions may affect migratory birds to work with FWS to develop an agreement to conserve these birds. The order also establishes a Council for the Conservation of Migratory Birds to serve as a clearinghouse for information. It directs agencies to ensure that their NEPA analyses consider potential effects on migratory birds, and it requires agencies to control the introduction and spread of nonnative animals and plants that might harm migratory birds. Many shorebirds, waterfowl, and neotropical migrants that depend heavily on riparian areas and wetlands stand to benefit if the new order is vigorously implemented.

Wild and Scenic Rivers

The national Wild and Scenic Rivers Act (WSRA) of 1968 sets forth congressional policy that certain rivers, which, "with their immediate environments, possess outstandingly remarkable scenic, recreational, geologic, fish and wildlife, historic, cultural, or other similar values, shall be preserved in free-flowing condition and . . . protected for the benefit and enjoyment of present and future generations." As of 1994, the system exceeded 1 million acres (see http://www.nps.gov/rivers/index.html for an update). River designations usually extend one-quarter mile from the normal high-water line on each side of the river, for an average of 320 acres of land per mile of river. Federal agencies are authorized to acquire nonfederal lands within the boundaries of any designated river segment, subject to certain area limits. The act necessarily, though only implicitly, governs management of designated rivers' riparian areas.

The WSRA designated the initial components of the system and established the criteria and procedures by which additional river segments could be added. Three designations were created—wild, scenic, and recreational. Wild rivers are free of impoundments and generally inaccessible except by trail, with watersheds or shorelines primitive and waters unpolluted. Scenic rivers differ from wild rivers only in that they may be accessible in places by roads. Recreational rivers are readily accessible by road or railroad and may have some development along their shorelines, or they may have undergone some impoundment or diversion in the past.

USDA and the Department of the Interior are charged with studying rivers identified by Congress and recommending whether they should be included in the system. These agencies are further directed to consider potential additions to the system in the course of their regular land and resource planning. Rivers may be designated by Congress, or they may be included in the system if designated by state legislation after review and approval of the interior secretary, in consultation with the heads of other federal departments.

A key provision that affects riparian areas is the restriction on water projects in or affecting designated river segments. The act prohibits construction (and federal assistance for construction) of any dam, water conduit, or other water project "that would have a direct and adverse effect on the values for which the river was established." In addition, the Federal Energy Regulatory Commission is forbidden to license any water project on or directly affecting any of the rivers identified by Congress for possible inclusion in the system.

Another significant provision of the act is the withdrawal of all public lands within designated components "from entry, sale, or other disposition under the public land laws." Subject to valid existing rights, lands within one-quarter mile of designated rivers are thereafter unavailable for mining. Miners may continue to develop existing, valid claims (subject to regulation) after designation of a river, but they may not patent their claims unless they had already applied for a patent and met all the patenting requirements. Prospecting can continue, also subject to regulations imposed by the land manager.

Components of the WSRA system are to be managed to protect the values for which they were designated, with primary emphasis on a river's "esthetic, scenic, historic, archeological, and scientific features." Land management agencies prepare management plans for rivers under their jurisdiction and manage them using their general statutory authorities (e.g., the National Park Service Organic Act for the NPS). The act specifies that federal agencies should pay particular attention to timber harvesting, road construction, and similar activities that might be contrary to the purposes of the WSRA. Managing agencies are further directed to cooperate with EPA for the purpose of combating water pollution.

Recent court cases have examined some of these planning and management requirements. First, failure to prepare a comprehensive management plan violates the WSRA (*National Wildlife Federation v. Cosgriffe*, 1998; *Sierra Club v. Babbitt*, 1999). Such plans must contain management prescriptions not just for the area within the river's designated boundaries, but also for other lands managed by the agency if activities conducted there might impact designated river segments. Impacts to riparian areas have been at issue in several cases. An Oregon court held that the BLM had violated substantive requirements of the WSRA by failing to provide, in a river management plan, protection against the effects of livestock grazing in riparian areas (*Oregon Natural Desert Association v. Green*, 1997). Further, the agency should have examined, in an environmental impact statement, the effects of both grazing and planned road and parking improvements. Another court concluded that the "WSRA gives the BLM authority to eliminate cattle grazing, or any other commercial use, if doing so is consistent with the mandate to protect and enhance the river values" (*Oregon Natural Desert Association v. Singleton*, 1999).

Submerged Lands

"Submerged lands" is a shorthand reference for the state-owned beds and banks of navigable waters. Title to these submerged lands passed to states upon their admission to the Union (*Pollard's Lessee v. Hagan*, 1845). Federal law determines navigability; a body of water is navigable if it was usable by customary modes of travel, in its natural and ordinary condition, at the time of statehood. "Natural and ordinary" means without artificial improvement through activities such as dredging or impoundment. At least one court has held that present transportation methods may be considered (*Alaska v. Ahtna*, 1989).

These submerged lands are subject to a public trust, first enunciated by the U.S. Supreme Court in 1892. In *Illinois Central Railroad Co. v. Illinois*, the Court ruled that the Illinois state legislature could not transfer title of the Chicago waterfront and part of the submerged lands in the Chicago harbor because those lands, which the state had acquired at statehood, were held in trust for the benefit of Illinois citizens. The Court held that a state could permissibly convey a temporary interest in, or even sell, parcels of submerged lands if such conveyances were in the public interest. But a state may never transfer "the whole property." The Illinois Central Court distinguished the nature of the state's title to lands beneath navigable waters from the title that the United States holds in the public lands that are open to preemption and sale.

Identifying what waters were navigable at the time of statehood, and hence which lands are held in trust by the states, has become an issue in determining title to minerals in submerged lands, in establishing management responsibilities on rivers within national parks or other federal lands, and in evaluating the legality of disposal of state lands. For example, in *Alaska v. United States* (1997), the U.S. Supreme Court concluded that the United States had expressly reserved title to submerged lands in the National Petroleum Reserve and what is now the Arctic National Wildlife Refuge in Alaska. Thus, Alaska did not acquire title to these lands at statehood, and the United States had authority to offer these lands for mineral leasing. All 38 states that have considered the issue have concluded that the state holds lands beneath navigable waters in trust for the people. In *Arizona Center for Law in the Public Interest v. Hassell* (1991), an Arizona appeals court invalidated a state law that relinquished the State's interest in riverbed lands, holding that it violated the public trust doctrine and the gift clause of the Arizona Constitution. Although it did not adjudicate the navigability of any Arizona streams, the court did find substantial evidence that "portions of Arizona rivers and streams other than the Colorado" met the federal test at statehood in 1912. Ironically, the invalidated Arizona legislation included a requirement that record owners of lands under the Colorado, Gila, Salt, and Verde Rivers pay a quitclaim fee to the state of $25 per acre, with the revenue to be used to acquire riparian lands for public benefit.

It seems plain that the existence of a public trust in submerged lands has implications for riparian area management and protection. Each state must develop its own jurisprudence for administering these lands. Although the trust is plainly enforceable, practical obstacles include the difficulty of identifying rivers that were navigable at statehood, determining which lands should be retained and which lands (or interests therein) could be relinquished, and how public interests can best be protected.

Summary

The use and management of public lands and resources are governed by both federal and state laws. The specific federal laws that apply depend on which system (e.g., national forest, BLM land, wild and scenic river) the land is included within and what resources are at issue. Each federal land system is impressed with some broad, national interest criterion. Each managing agency affords some consideration to riparian areas and resources, whether by regulation or in an internal manual or policy handbook. Few specific provisions for riparian areas have been established in congressional legislation by or in executive orders, but agencies have considerable latitude to decide how and to what extent their planning and management activities will account for these areas. One result is that individual districts or units within agencies may vary in their interpretation and implementation of riparian measures established administratively. Thus, while different and additional constraints apply to management of federal riparian lands compared to privately owned riparian lands, the constraints on federal lands are not uniform from system to system, nor uniformly interpreted and applied within systems, and for the most part have been established principally by administrative action, not by legislation, and thus are subject to administrative change.

PROTECTION OF WATER RESOURCES

Protection of riparian areas often is premised on their importance in protecting waterbodies. However, the reverse can also be true—protecting stream flows and lake levels can help preserve the structure and functioning of adjacent riparian areas. This section explores how laws relating to instream flows and uses of water can benefit riparian areas.

Current Water Law

Riparian Doctrine

The legal system regards water as a public resource. Allocation and use of water are governed by state law. In those states following the riparian doctrine (generally those located east of the ninety-eighth meridian), the owner of land adjacent to a waterbody (the riparian landowner) is regarded as holding a legal right to make use of the water. Riparian water rights are a product of English common law, generally adopted by the eastern states. The right to use water is simply an extension of the ownership of land adjacent to that water. It is an acknowledgment of access to waterbodies that riparian land ownership provides.

The nature of the legal interest in adjacent water is somewhat ill defined. The riparian landowner has the ability to enjoy the benefits of using water—a so-called usufructory right—but does not own water. The purposes for which a riparian owner may use water are open-ended, constrained only by the limitation that the use may not unreasonably interfere with the equivalent right of all other riparian owners to their enjoyment of the resource. In short, it is a correlative system in which the interests of all riparian landowners in use of the shared water resource are to be balanced. Generally this means that riparian uses must be "reasonable." Ultimately, in a common law system, the arbiter of what is reasonable is a judge.

This system works reasonably well in places with an abundance of water and when rivers and lakes are valued by humans primarily for their navigation and power-generation (instream) benefits. As demands have grown for out-of-stream uses of water (such as for drinking water, industrial cooling water, and irrigation), many eastern states have developed a permit system. Persons desiring to use river or lake water must apply to a state agency, specifying the purpose(s) of use and the quantity of water. Permits authorize the use for a specified term, at the conclusion of which the user must request a renewal.

Increasing out-of-stream uses have prompted some states to begin establishing base flow levels for rivers and lakes regarded as necessary to protect existing fish populations, water quality, or other considerations. Permits for out-of-stream diversions are constrained such that they do not reduce flows or water levels below the established minimum. For example, Florida law directs the regional water management districts to establish a minimum flow for all surface watercourses and a minimum water level for all aquifers. These minimum flows and levels are to be established in the amount "at which further withdrawals would be significantly harmful to the water resources [or ecology] of the area" (Florida Statutes § 373.042(1)).

Prior Appropriation Law

Settlement of the American West in the 1800s depended on active control and use of the limited water resources available in streams and rivers. Because the United States owned nearly all the land in the West at that time, riparian land ownership made little sense as the basis for determining legal rights of individuals to use water. Early gold miners vied for control of streams just as they vied for control of land containing mineral deposits. A common custom emerged—the first to take control of and actively develop and use the resource held a protectable legal right to the land and water against all other claimants. For water, this custom of "first-in-time, first-in-right" came to be called the prior appropriation doctrine.

Under the prior appropriation doctrine, the legal basis of an individual's right to use water is physical possession of water and its application to a beneficial use. In virtually all western states, a would-be user applies to a state agency for permission to use some quantity of water from a particular source for a particular purpose. Once the water has actually been put to beneficial use, the permit holder's interest ripens into a permanent property interest in the right to make continued beneficial use of the water. The core of the right is the priority date. The more senior the right, the more likely it is that the right holder can divert and use their full entitlement of water. Unlike riparian rights, appropriative rights are not correlative—the senior appropriator is entitled to use all of the water authorized by their right ahead of more junior appropriators from the same source of water.

A prior appropriation water right traditionally required the physical control of water with a dam or a diversion structure. Only through this physical appropriation of water could a legal right be established. Moreover, beneficial use has been viewed in utilitarian and economic rather than ecological terms, as demonstrated by *Empire Water & Power Co. v. Cascade Town Co.*, 205 Fed. 123 (8th Cir. 1913). The Cascade Town Company owned a tourist resort located at the base of a waterfall on Cascade Creek, near Colorado Springs. The company sought to prevent a hydroelectric project from diverting the creek upstream of the resort, partly on the basis that the diversion would harm the "exceptionally luxuriant growth of trees, shrubbery, and flowers ... produced by the flow of Cascade Creek through the cañon and the mist and spray from its falls." The court noted the substantial investment made by the resort and its dependence on this natural setting but found that "the laws of Colorado are designed to prevent waste of a most valuable but limited natural resource [The Cascade Town company] cannot hold to all the water for scant vegetation which lines the banks but must make the most efficient use by applying it to lands."

Incorporating Natural Systems Needs into Water Law

Existing water law does not readily address the water-related needs of riparian areas. This is most apparent in the preference for economical uses of water common to prior appropriation states, as discussed above. The riparian doctrine was meant to limit uses to those that would leave the stream "undiminished in quantity and unimpaired in quality," and indeed the reasonable use standard has been applied to prevent users from withdrawing substantially all water from a stream during a drought (*Collens v. New Canaan Water Company*, 234 A.2d 825, 1967). Nonetheless, courts have chosen not to protect the right of riparian landowners to continue to benefit from natural inundation of their bottomlands in preference to the expressed need for upstream storage of water that would eliminate such flows (*Herminghaus v. Southern Cal. Edison Co.*, 252 P. 607, 1927).

From a water rights perspective, the only clear strategy for assuring that riparian areas receive sufficient water is to divert and deliver water directly to such areas. In theory, such irrigation would attempt to mimic the seasonal rewatering of naturally occurring overbank flows or groundwater recharge. In a few instances, western states have granted water rights for the purpose of "irrigating" wetlands or have determined that water rights for irrigation purposes may also apply to providing water for wetlands. For example, policy guidance recently issued in Colorado supports the use of an irrigation water right to provide water for wetlands (Stenzel, 2000). The guidance recognizes wetland irrigation as a beneficial use "as long as the water is diverted from a stream and applied for the purpose of the growth, irrigation, and maintenance of wetland plants." Riparian vegetation presumably could be irrigated in this same manner.

As appreciation of the instream benefits of water has grown, including water's essential role in supporting the ecological functioning of river corridors, water law has broadened. Riparian law states are developing permit systems, many of which account for environmental considerations, and instream purposes are increasingly being recognized as legitimate uses for river water. As shown in Table 4-4, some states are allowing federal agencies to acquire instream water rights, which could be important to the protection of riparian areas, as federal agencies tend to have different environmental objectives than those reflected in state law.

Instream Flow and Groundwater Protection Programs

Virtually all prior-appropriation states now have programs that authorize the use of water for "instream" purposes (Gillilan and Brown, 1997). Approaches vary considerably, but the core of these programs is to authorize legal protection for maintaining designated flows of water between points along a stream channel or maintaining designated lake levels. Such flows must not have already been appropriated for out-of-stream use. In practice, protecting the minimum flow of water necessary to sustain fish populations has been the primary purpose of these programs to date. Generally only a state agency may hold an instream flow water right.

State instream flow programs typically set aside flows on the basis that they are the minimum needed for protection of an existing fishery—e.g., flows sufficient to allow fish passage through riffles in a stream channel. Generally such claims are made to a single minimum level of flow year-round rather than to flows that change throughout the year to mimic the natural hydrograph. That is, such flows prevent a stream from being totally dewatered, but they do little to benefit riparian areas that depend on variable flows to regenerate and maintain native vegetation or to ensure the occurrence of channel processes necessary to maintain instream habitats (e.g., sediment transport and pool formation).

Nevertheless, these programs have the potential to protect a river's existing hydrograph or selected portions of the hydrograph. An instream flow claim could be made for all unappropriated or otherwise unclaimed water in a specified reach of a river or stream. In this case, no additional water development would be possible. Alternatively, additional water development and use could be limited to periods of time and portions of the hydrograph regarded as less critical to system functioning. Perhaps only certain designated peak flows could be protected along with acceptable minimum flows that mimic the natural hydrograph.

A few state instream programs have allowed such claims. The Nature Conservancy, for example, has been successful in obtaining certified instream water

TABLE 4-4 Ability of Federal Agencies to Protect In-Place Values and Uses of Water Within Western State Water Law Systems

State	In-Place Beneficial Uses Recognized Under State Law
Alaska	Protection of fish and wildlife habitat, migration, and propagation; recreation and parks purposes, navigation and transportation purposes; sanitary and water quality purposes, Alaska Stat. §46.15.145 (1992)
Arizona	Recreation and wildlife including fish, Ariz. Rev. Stat. Ann. §45.151.A (1987)
California	Preserving or enhancing wetlands habitat, fish and wildlife resources, or recreation in or on the water, Cal. Water Code §1707 (West Supp. 1993)
Colorado	To preserve the natural environment to a reasonable degree, Colo. Rev. Stat. §37-92-102 (1990)
Idaho	Protection of fish and wildlife habitat, aquatic life, recreation, aesthetic beauty, transportation and navigation values and water quality, Idaho Code §41-1501 (1990)
Kansas	Water quality, fish, wildlife, aquatic life, recreation, general aesthetics, domestic uses, and protection of existing water rights, Kan. Stat. Ann. §85a-928i (1989)
Montana	Fish and wildlife, recreational uses, and maintenance of water quality, Mont. Code Ann. §85-2-316 (1993), Mont. Admin. R. §36-16.102(3)

EXISTING LEGAL STRATEGIES FOR RIPARIAN AREA PROTECTION 271

Special Restrictions on In-Place Water Protection	Can Federal Agencies Hold In-Place Appropriative Rights?[a]	Provision for Federal Agency Involvement in State In-Place Water Protection Program
—Reservations only —Reviewable every 10 years for continuing need	Yes, reservation is regarded as an appropriation	Instream flow statute lists the federal government as a party allowed to apply for instream reservation
—Special administrative review procedure	Yes (two such permits issued to date)	There is no state program for protecting instream flows
—Applies only to changes of use of existing water rights	Only by changing the use of existing water rights	Only through participation as protestant in state water rights proceedings
—"Minimum" streamflow —Restricted to Colo. Water Conservation Board (CWCB) —Must be a natural environment that can be preserved	Yes, but only if a diversion of water is involved	CWCB does "request recommendations" from the Departments of Agriculture and Interior Recommendations must be made "with specificity and in writing"
—The minimum amount required to protect beneficial uses, which is capable of being maintained —Approved by the legislature	Not settled	Federal agency may request Idaho Water Resources board to consider a minimum flow
—Minimum desirable streamflow —Approved by the legislature	No	No
—The amount must be necessary for purpose and cannot exceed 50 percent of average annual flow on gaged stream —Reservation only —Reviewed at least once every 10 yrs; may be modified in 5 yrs	Yes, but only if a diversion of water is involved	The U.S. or any agency thereof may apply to reserve a minimum stream flow

continues

TABLE 4-4 Continued

State	In-Place Beneficial Uses Recognized Under State Law
Nebraska	Recreation, fish and wildlife, Neb. Rev. Stat. §46-2,108 (1988)
Nevada	Any recreational purpose, Nev. Rev. Stat. Ann. §533.030(2); wetland protection, Nev. Rev. Stat. §502.322.
New Mexico	No statutory in-place protection
North Dakota	No statutory in-place protection
Oregon	Conservation, maintenance and enhancement of fish and wildlife habitat, Or. Rev. Stat. §537.336 (1988)
South Dakota	No statutory in-place protection
Utah	Propagation of fish, public recreation, the reasonable preservation or enhancement of the natural stream environment, Utah Code Ann. §73-3-3 (1993 Supp.)
Washington	Protecting fish, game, birds, or other wildlife resources, or recreation or aesthetic values of said public waters whenever it appears to be in the public interest, Wash. Rev. Code. Ann. §90.22.010 (West 1992)
Wyoming	To establish or maintain new or existing fisheries, Wy. Stat. Ann. §41-3-1001 (1993 Supp.)

[a]Generally, federal agencies cannot hold an instream water right issued under state law. This column indicates if they can hold such rights. In some cases, the instream water right is only given if a diversion structure exists on the river. Normally, such structures would remove water from the river, but in the case of instream rights, they are simply placeholders for the right. For example, in

Special Restrictions on In-Place Water Protection	Can Federal Agencies Hold In-Place Appropriative Rights?[a]	Provision for Federal Agency Involvement in State In-Place Water Protection Program
—Minimum amount necessary —Available only to Game and Parks Commission or a natural resources district	No	Only as a party to state allocation decision process
	Yes (at least 2 such permits granted)	There is no state program for protecting instream flows
	Perhaps, if a diversion is involved	There is no state program for protecting instream flows
	Perhaps, if a diversion is involved	There is no state program for protecting instream flows
—Minimum perennial streamflows —Restricted to Dept. of Fish and Wildlife, Dept. of Environmental Quality, and State Parks and Recreation Dept.	No	There is no state program for protecting instream flows
	Yes, but only if a diversion of water is involved	There is no state program for protecting instream flows
—Limited to transfer of existing rights only —Restricted to Dept. of Wildlife Resources and State Parks and Recreation Dept.	Yes, but only if a diversion of water is involved	
—Restricted to Dept. of Ecology, but Dept. of Fisheries and Dept. of Wildlife may request consideration	Perhaps, if a diversion is involved	
—Minimum flow necessary —Need identified by Game and Fish Commission, application made by Water Development Commission	Yes, but only if a diversion of water is involved	In Clarks Fork River via special congressional legislation

Colorado, structures used to control the flow for kayaking purposes have qualified as diversions for the purposes of appropriating rights, even though no flow is removed from the river.

SOURCE: Reprinted, with permission, from MacDonnell and Rice (1993). © 1993 by Natural Resources Law Center.

rights in Arizona for streams running through its Hassayampa River Preserve and Muleshoe Ranch properties. These rights protect a significant portion of the hydrograph on a year-round basis. BLM has filed an application for an instream right on the San Pedro River in Arizona, claiming a flow of 18,200 cubic feet per second for 24 hours. The claim is based on an analysis indicating that this is the flow rate needed to inundate the banks and benches adjacent to the channel and to support riparian vegetation. The application states:

> The natural morphological processes that have shaped the San Pedro and other river systems are largely driven by periodic high flows that flush and redistribute sediment and rock. These high flows inundate the floodplain spreading sediment and seed, scour, and shape vegetation on banks and floodplain. The long-term improvement and maintenance of the San Pedro River environment is [sic] absolutely dependent on the recurrence of sufficiently high peak flows. The concept of a riparian conservation area is meaningless if artificial structures eliminate the natural variability that once created and now maintains the fluvial biome. (BLM, 1991b)

Since Colorado began appropriating water for instream uses in 1973, such appropriations have been established on more than 8,000 miles of Colorado streams and on 486 lakes. In 1996, the Colorado Water Conservation Board appropriated all of the unappropriated water in Hanging Lake and Dead Horse Creek—a heavily visited watershed within the White River National Forest containing scenic waterfalls and unusual flora and fauna. The Hanging Lake appropriation is the first to appropriate all the remaining water.

Related to the protection of instream flows are laws that restrict groundwater pumping. Although riparian vegetation is as dependent on access to underlying groundwater as it is to surface water flow, state laws do not generally provide a means of protecting this water supply from the adverse effects of groundwater pumping. Legal rights to pump groundwater are associated with ownership of the overlying land in most states. Public supervision, if any, generally concerns only conflicts between different groundwater users or, in a few states, groundwater and surface water users. Currently, there are no limitations placed on groundwater development with the intent to protect riparian areas.

One approach that has been used to lessen the impact of groundwater pumping on the San Pedro is the purchase of lands being pumped and the retiring of wells. A more comprehensive legal approach would be for states to limit groundwater pumping in and adjacent to riparian areas as necessary, something that is being considered by the state of Arizona. The Arizona Groundwater Management Act in its present form provides little direct protection of riparian areas, although its goal of aquifer stability could indirectly be protective. To date, no legal changes have been made to the law that would expressly recognize riparian groundwater needs.

Public Trust Doctrine

The concept of public trust has been applied beyond protection of submerged lands to include protection of navigation, fishing, public recreational uses of water, and water-dependent environmental functions. One prominent example of the public trust doctrine being used to protect water resources and, indirectly, riparian areas involved California's Mono Lake. Los Angeles has held state-granted rights to divert virtually the entire flow of water from several tributaries to Mono Lake. By the 1970s, exercise of these rights had dramatically lowered the water level of the lake. In *National Audubon Society v. Superior Court* (1983), the California Supreme Court found that the public trust doctrine imposes a duty on the state to "protect the people's common heritage of streams, lakes, marshlands and tidelands." The ruling allowed the state to reconsider past allocations of water if necessary to protect the public trust. In subsequent proceedings, the State Water Resources Control Board developed a plan to restore Mono Lake and the flow regimes of several tributaries to the lake—thus reducing the city's diversion rights.

No other state has invoked the public trust to alter existing water rights, but several have applied the doctrine in considering decisions regarding new uses of water. For example, the Idaho Supreme Court found that public trust concerns such as assuring minimum stream flows, encouraging conservation, and protecting aesthetic or environmental qualities of particular areas should be taken into consideration when evaluating applications for water rights (*Shokal v. Dunn*, 1985). The North Dakota Supreme Court held that the public trust doctrine must be a consideration in state planning regarding allocation of water (*United Plainsmen v. North Dakota Water Conservation Commission*, 1976).

With the exception perhaps of several tributaries to the Mono Basin, the public trust doctrine has not yet been extended to riparian water needs, which may reflect a limited understanding of the essential role of water in supporting and maintaining riparian functions. As a common law concept, the doctrine could conceivably evolve to include this public value of water. Alternatively, states might choose to statutorily recognize the need to consider this important function of water in their water planning and allocation procedures.

Federal Reserved Water Rights

The reserved rights doctrine emerged in the context of water needs of Indian tribes on their reservations. The U.S. Supreme Court held that when the United States reserved land for a tribe, it also implicitly reserved an amount of water necessary to meet the purposes for which the land had been set aside. The legal basis of the right was determined to be independent of, and superior to, subsequent state authority to allocate water resources (*Winters v. United States*, 207 U.S. 564, 1908). This doctrine, which was extended in 1963 to other federal reservations of land such as national parks and forests, could be used to protect both water resources and their associated riparian areas on all federal reservations.

Several national parks have been accorded reserved water rights that essentially preclude any out-of-stream development. For example, streams within Glacier and Yellowstone National Parks are dedicated to instream flow uses (Amman et al., 1995). In 1993, a Colorado water court awarded the United States all unappropriated flows on the east side of Rocky Mountain National Park; another water court decreed the same result for the west side of the park in 2000 (Silk et al., 2000). In a 1996 negotiated settlement for Zion National Park, Utah agreed to a federal reserved right to all stream flows except for a designated amount committed to future depletion by users upstream of the park (USA et al., 1996). Similarly, negotiated agreements in Montana for several national park units and for wild and scenic river areas established fixed levels of upstream consumptive use, with the remainder of the water being dedicated to instream flow.

To date, no reserved right specifically for protection of the riparian functions of a federal reservation has been legally recognized. The United States has asserted reserved right claims to water for environmental purposes with limited success—primarily because the U.S. Supreme Court has determined that the water claimed must be necessary to achieve the primary purpose(s) for which the reservation was expressly created. Thus, the Supreme Court upheld the need for water to protect the desert pupfish in a national monument specifically set aside for this purpose (*Cappaert v. United States*, 1976). But it denied an instream flow right for the Rio Mimbres in the Gila National Forest on the basis that the primary purpose for which national forests were established was not environmental protection (*United States v. New Mexico*, 1978). This latter decision is odd given that the two primary purposes in the 1897 Organic Act are "securing favorable conditions of water flows and furnishing a continuous supply of timber." In the future, it may be possible for the USFS to convince a court that "favorable conditions of water flows," and hence downstream yields of water, depend on streams and riparian areas that are in good functioning condition. This could enable the successful assertion of federal reserved rights for protection of riparian areas.

CRITICAL EVALUATION OF THE POTENTIALLY MOST INFLUENTIAL PROGRAMS

Although a diverse array of programs may be used to promote, protect, or implement riparian management, two relatively new programs may have substantial and unique impacts nationwide because of their scope and scale. The first is the Conservation Reserve Enhancement Program (CREP), which is a voluntary, incentive-based program authorized by the 1996 Farm Bill. CREP is essentially a variation on the Conservation Reserve Program (CRP), but with a state partnership component that allows for a variety of additional options, including more flexible width requirements, permanent easements, and additional cost sharing. The second is the Total Maximum Daily Load (TMDL) program originating in Section 303 of the Clean Water Act. A TMDL defines the maximum amount of pollution that a waterbody can receive from various sources and still meet water-quality standards. As of 1998, over 20,000 waters (including over 300,000 miles of rivers and 5 million acres of lakes) are in violation of water-quality standards and require a TMDL calculation. The CREP and TMDL programs impact riparian systems through different administrative mechanisms and funding sources.

Conservation Reserve Enhancement Program

The Conservation Reserve Program (CRP), established by the 1985 Food Security Act, is a voluntary program that offers annual rental payments, incentive payments for certain activities, and cost-share assistance to establish approved cover on eligible cropland. The program allows lands meeting certain characteristics, such as highly erodible or cropped wetlands, to be removed from agricultural production for 10–15 years. Continuous CRP is a form of the program that allows highly vulnerable lands, such as riparian areas, to be enrolled without the same administrative requirements as CRP.

The potential for CRP to restore and protect riparian areas has been greatly enhanced by the development of the Conservation Reserve Enhancement Program (CREP), which is an extension of CRP. Once landowners are enrolled in CRP (a federal program), they can take advantage of additional services offered by state CREP programs. A primary goal of CREP is to create an opportunity where the resources of a state government and the federal Commodity Credit Corporation can be targeted in a coordinated manner to address specific conservation and environmental objectives of that state and the nation, providing greater flexibility to address regional problems. For example, a state and the NRCS might decide that a certain percentage of state CREP money must be used to restore riparian areas in a specific watershed. CREP is meant to improve water quality, erosion control, and wildlife habitat in specific geographic areas that have been adversely impacted by agricultural activities, with emphasis on nonpoint source pollution control. Conservation of species listed as threatened or endangered or identified as candidates for listing are included under these objectives. The federal–state partnership is a unique aspect of CREP.

CREP broadens the federal payment options for landowners enrolled in CRP programs (annual rental payment, maintenance payment, and a cost share of 50 percent for the installation of conservation practices). For example, many of the states that currently have an established CREP will supplement the cost-share payments and pay for extensions on the easement contract beyond the regular 10–15 years. There is no cap on expenditures in CREP from the federal side, although acreage limitations have been imposed (e.g., Illinois is limited to 132,000 acres). Nonetheless, funding can pose a barrier because of the cost-share requirement of 20 percent, which can be difficult for states to meet.

Eligible Lands

Because each state CREP agreement is unique, the definition of eligible lands varies considerably. In general, lands eligible for CREP must meet some of the same criteria that CRP land must meet. That is, the land must have been owned or operated by the applicant for the previous 12 months, must have been planted in crops two of the last five years, and must be physically and legally capable of being planted in a normal manner. There are numerous exceptions granted to individual states to meet specific needs. For example, the New York City CREP is directed towards protection of the city's water supply watersheds. Each state agreement designates specific watersheds, counties, or other areas that are targeted. Some also include specific definitions of eligible land such as 100-year floodplains (Illinois CREP), highly erodible lands (many states), and sinkholes (Kentucky CREP). In addition to the floodplain demarcation requirement, Illinois also allows lands anywhere within the watershed that are farmed wetlands, prior converted wetlands, or wetlands farmed under natural conditions. Finally, additional acres have been designated for permanent easements in Illinois. These lands are defined as non-cropped acres or land in another CRP signup that meet criteria based on riparian definitions, erodibility, or adjacency to the primary land enrolled in CREP. A significant advantage of CREP programs is that conservation practices can go beyond CRP guidelines regarding buffer width and structure. For example, if a CREP program defines eligible lands as those within the 100-year floodplain boundary, such a designation may, depending on the region, include lands well beyond the 234-ft maximum buffer width allowed under CRP. As a consequence, CREP can encompass and potentially protect larger riparian areas than CRP.

Current Status

As shown in Table 4-5, CREP is currently established in 21 states, and nine more are in the process of preparing applications. Continuous CRP and CREP account for 418,000 miles of riparian land, while general CRP has enrolled over 333,000 miles. There is a tremendous variety in the types of management practices utilized by the state CREP programs, with the most common being filter strips, riparian buffers, wetland restoration, and restoration of prairie and tallgrass prairie/oak savanna ecosystems for rare and declining wildlife habitat. However, a variety of other practices such as tree planting and providing shallow water areas for wildlife are frequently utilized in riparian areas.

The goals of the existing CREP programs include sediment, nutrient (nitrogen and phosphorus), pesticide, and pathogen loading reductions, fish and wildlife habitat enhancement, streambank stabilization, and hydrologic restoration. For example, in North Dakota the program goal is to develop 20-acre blocks of habitat (cover-locks) for upland game. The New York CREP is designed to protect the water supply of New York City through riparian buffers and dairy cattle management. The Maryland CREP has established the goals of reducing annual agricultural-based pollution by 5,750 tons of nitrogen, 550 tons of phosphorus, and 200,000 tons of sediment while increasing wildlife and habitat. A fundamental concept of CREP is to allow states to identify their most pressing resource issues and to develop unique strategies to address these issues.

All current CREP programs offer incentive payments, but they vary considerably. Typical payments include supplemental cost sharing (such as the incentive payment of 130 percent of the county rental rate in Delaware), payments for the remainder of the costs for establishing the conservation practice (e.g., New York, North Carolina), and additional lump-sum payments for certain practices such as installing filter strips (Ohio) or planting trees (North Carolina). The first CREP program was initiated in Maryland in 1997; there are now 21 CREP states with 16,492 contracts for over 265,511 acres and a projected lifetime federal cost of over $540 million (for current information, see http://www.fsa.usda.gov/dafp/cepd/crep.htm). Additional state payments for cost sharing and extensions of easements will add to the total cost. In addition, many existing programs (e.g., Minnesota and Illinois) are expanding beyond their initial eligible land areas. Thus, final program cost and enrolled acreage will substantially increase as the program matures.

Program Evaluation

Although related programs such as CRP have been evaluated in numerous studies (e.g., Allen, 1996), the effectiveness of CREP has not been given much attention. There are several reasons for this. First, each state's program is unique, making documentation difficult. Other than the number of miles or the acreage enrolled, there is no single program-wide characteristic that permits CREP to be evaluated for environmental impact on a consistent national basis. Second, CREP rules state that each program should include a comprehensive monitoring program (with an identified funding source) that will include specific objectives, measurement descriptions, and a process for program refinement based on monitoring results. Unfortunately, most states support very little monitoring specific to their program, making it difficult to judge the effectiveness of the program. (See NRC, 2000, for a description of the monitoring necessary for determining agricultural program performance.) Most states do have some baseline monitoring in place that will allow superficial assessment of the program, but such monitoring is often not designed to track specific program inputs. It is probable

TABLE 4-5 Current Listing of Conservation Reserve Enhancement Programs and Their Intended Goals and General Eligibility Designations

State	Major Goals
Arkansas	Reduce sediment loading up to 10,000 tons/year; increase wildlife populations; establish 200 miles of riparian forest buffers
California	Enhance wildlife habitat (possible increase of 27,000 ducklings and 20,000 pheasants); reduce soil erosion; improve surface and groundwater quality; improve air quality
Delaware	Reduce nutrient and sediment loading, improve temperature and dissolved oxygen; increase wildlife habitat and create wildlife corridors
Illinois	Reduce input of sediment, nitrogen, phosphorus; enhance wildlife, fish, mussels, threatened and endangered species
Iowa	Reduce nitrogen loading to streams by 300–600 tons/year; reduce sediment delivery to Lake Panorama by 80,000 tons/year; reduce or maintain soil erosion at or below 2–5 tons/acre; enhance wildlife habitat; increase recreation
Kentucky	Reduce by 10 percent the sediment, pesticide, and nutrient loads entering the Green River and Mammoth Cave system; protect wildlife habitat, restore riparian habitat, and restore subterranean ecosystem by targeting 1,000 high priority sinkholes
Maryland	Reduce nutrient loadings into the tributary streams of the Chesapeake Bay
Michigan	Reduce sediment inflow by 784,000 metric tons over 20 years, nitrogen by 1.6 million pounds, and phosphorus by 0.8 million pounds; protect water supplies; protect 5,000 linear miles of streams from sedimentation; improve wildlife habitat
Minnesota	Reduce sediment and nutrient loading into the Minnesota River
Missouri	Reduce by 50 percent the pesticides in 58 drinking water supplies; reduce sediment inflow by 50 percent; reduce soil erosion rate to less than 5 tons/acre; help ag. Producer meet nutrient goals; improve wildlife habitat
New York City	Reduce soil erosion by 36,000 tons/year; reduce levels of nutrients and pathogens; enhance wildlife habitat

Eligible Areas and Initial Acres	Eligible Practices and Riparian Specific Conditions[a]	State Contract Extensions
4,700 acres in the Bayou Meto of central Arkansas	CP22	Unknown
12,000 acres in the North Central Valley	CP1, CP2, CP4D, CP9, CP10, CP12, CP21, CP22, CP23	Yes, but not defined
6,000 acres in the watersheds of Chesapeake Bay, Delaware Bay, and the Inland Bays basin area (goal is approximately 1,200 miles of buffers)	CP3A, CP4D, CP21, CP22, CP23	None
132,000 acres in 100-year floodplain and adjacent highly erodible lands of the Illinois River Basin	CP2, CP3, CP3A, CP4D, CP12, CP21, CP22, CP23, CP25	15- and 35-yr extensions or permanent
9000 acres in 37 counties in north-central Iowa	CP7, CP21, CP22, CP23	Yes
100,000 acres in the Green River watershed	CP1, CP2, CP3A, CP4D, CP8, CP10, CP11, CP21, CP22, CP23	Extensions and permanent easements
70,000 acres of riparian buffers, 10,000 acres of wetlands, and 20,000 acres of highly erodible lands in Chesapeake Bay watershed	CP4, CP5, CP8, CP16, CP17, CP18, CP21, CP22, and wellhead protection areas	Permanent easements
80,000 acres in the Macatawa, River Raisin, and Saginaw Bay watersheds	CP1, CP2, CP5, CP21, CP22, CP23	Voluntary easement
100,000 initial acres (190,000 final) in Minnesota River floodplain, tributaries, cropland filter strips, and wetlands.	CP9, CP21, CP22, CP23, CP25	20 years or permanent
50,000 acres along streams in the watersheds of 83 water supply reservoirs	CP1, CP2, CP3A, CP4D, CP15A, CP21, CP22, CP23	Not specified
5,000 acres of riparian areas or highly erodible croplands	CP1, CP2, CP3, CP4, CP21, CP22, CP23	None

TABLE 4-5 Continued

State	Major Goals
North Carolina	Reduce the excessive nutrient and sediment loading from agricultural runoff (15 percent nitrogen reduction); improve anadromous fish habitat; enhance wildlife habitat
North Dakota	Create wildlife habitat; improve water quality; and reduce soil erosion
Ohio	Reduce sediment entering Lake Erie by 2,325,000 metric tons over 20 years; reduce nutrients and pesticides entering Lake Erie and tributaries; protect 5,000 miles of streams from sedimentation; improve wildlife habitat
Oregon	Restore salmon habitat through restoration of riparian forests; reduce sediment and nutrient input; stabilize streambanks; restore water temp. to natural ambient conditions; restore natural hydraulic and geomorphic conditions
Pennsylvania	Reduce nutrients and sediment delivery to the Potomac and Susquehanna Rivers and the Chesapeake Bay
Vermont	Reduce phosphorus loading into Lake Champlain by 48.3 tons/year; enhance wildlife and aquatic habitat
Virginia	Reduce nutrient input into the Chesapeake Bay and non-Bay watersheds; modify hydrology through wetland restoration; enhance wildlife habitat
Wisconsin	Reduce sediment loading by 335,000 tons/year, phosphorus loading by 610,000 pounds, and nitrogen by 305,000 pounds; establish 3,700 miles of riparian buffers and 15,000 acres of grassland habitat.

[a]CP 1 Planting of introduced grasses
CP 2 Planting of native grasses
CP 3 Tree planting (3A is hardwood trees)
CP 4 Restoration of wildlife habitat via prairie ecosystem restoration and tallgrass prairie/oak savanna ecosystem restoration
CP 4D Restoration of permanent wildlife habitat
CP 5 Field windbreaks
CP 8 Grassed waterways
CP 9 Shallow water areas for wildlife
CP 10 Established grass
CP 11 Established trees

Eligible Areas and Initial Acres	Eligible Practices and Riparian Specific Conditions[a]	State Contract Extensions
100,000 acres in the Albermarle-Pamlico Estuarine system (85,000 acres riparian; 15,000 acres wetland)	CP3A, CP21, CP22, CP23, CP25	15-year and permanent
1,000 20-acre "cover locks" in six watersheds. Cover locks consist of 5 acres of trees, 10 acres of herbaceous cover, and 5 acres of winter food. Each cover lock will have 140 acres of associated conservation easements	CP4D, CP12, CP16	Easement length not specified
Western Lake Erie Watershed	CP3A, CP4, CP5, CP21, CP22, CP23	None
100,000 acres along designated salmonid streams and 5,000 acres of wetlands	CP21, CP22, CP23	None
100,000 acres	CP1, CP2, CP3A, CP4D, CP8, CP9, CP12, CP15A, CP21, CP22, CP23	None
Lake Champlain watershed of Vermont	CP8, CP21, CP22, CP23	
30,500 acres of riparian lands and 4,500 acres of wetlands	CP21, CP22, CP23	10- and 15-yr contracts; up to 8,000 ac of permanent easements
100,000 acres in 51 counties	CP1, CP2, CP8, CP21, CP22, CP23, CP25	Permanent

CP 12   Wildlife food plots
CP 15A   Contour grass strips
CP 16   Shelter belts
CP 17   Snow fences
CP 18   Salt tolerant vegetation
CP 21   Filter strips
CP 22   Riparian buffers
CP 23   Wetland restoration
CP 25   Restoration of rare and declining wildlife habitat

that CREP, like other conservation programs, will have to be evaluated indirectly. One current exception is the Illinois CREP, which has established a paired watershed design that will allow some degree of differentiation between areas with intense practice implementation and areas with a lesser degree of land-use change (see Box 4-5).

The length of easement contracts among state CREP programs varies greatly. The basis of the program is a federal contract for 10 to 15 years. As part of the state agreement most states add on voluntary easement extensions ranging from 15 years to permanent, and generally landowners are compensated at rates linked to the length of easement extension. Maryland, for example, has established a target of obtaining permanent easements on 25 percent of its contracts. In Illinois, three easement extensions are offered—15-year, 35-year, and permanent. Thus, the opportunity for long-term modification of the landscape are certainly available through CREP, but landowners do not always take advantage of these options even though they may include payments that can equal or exceed the fair market value of the land.

The program is limited by its reliance on voluntary landowner participation. Thus, although states target lands whose restoration will provide the greatest impact (e.g., riparian or highly erodible lands in specific watersheds), owners of severely degraded land may choose to not participate because of personal preference, insufficient financial incentives, lack of information, or various other reasons. (Where there have been concerted efforts to contact individual landowners to inform them of the available funding, participation increases sharply, as experienced in the Maquoketa River watershed in Iowa.)

Within the eligible areas defined for each CREP, there is generally no prioritization of areas such that critical parcels of land are targeted for enrollment. This is particularly troublesome when the chosen watershed is large, such that certain sections are experiencing greater degradation than others. In addition, the effectiveness of some restoration practices is dependent on stream size. For example, Chapter 2 discusses the disproportionate pollutant removal abilities of first- and second-order streams compared to larger-order streams. In none of the CREP programs reviewed was prioritization by stream size apparent, although there appear to be no restrictions to doing so. Fortunately, in some CREP programs where nongovernmental organizations such as The Nature Conservancy are involved, staff has been directed to identify priority areas to help attain their conservation objectives.

Finally, the federal cap on acreage enrolled in CRP is 36.4 million acres. Current contracts account for nearly 33.7 million acres, and state CREP programs are authorized for a substantial amount of the remaining acreage. Each state CREP program was initially limited to no more than 100,000 acres. Though most state CREP programs have not reached their limits, at least one has (IL) and it is expected that others will attain this limit as their marketing programs and landowner awareness increases. With other states in the process of developing CREP agreements, there is no doubt that increasing the federal acreage cap on CRP will be an important aspect of ensuring program success.

Total Maximum Daily Load Program

Section 303(d) of the CWA requires states to identify waters that are not attaining ambient water-quality standards (i.e., waters that are impaired). States must then establish a priority ranking for such waters, taking into account the severity of the impairment and the uses to be made of such waters. For impaired waters, the states must establish total maximum daily loads (TMDLs) for pollutants necessary to meet water-quality standards and develop an implementation plan that will allow the TMDL to not be exceeded. The CWA further requires that once water-quality standards are attained, they must be maintained.

The term TMDL has essentially two meanings (EPA, 1991):

- The TMDL process is used for implementing state water-quality standards—i.e., it is a planning process that will lead to the goal of meeting the water-quality standards.
- The TMDL is a numerical quantity determining the present and near future maximum load of pollutants (from point and nonpoint sources as well as from background sources) to receiving waterbodies that will not violate the state water-quality standards with an adequate margin of safety. The permissible load is then allocated by the state agency among point and nonpoint sources.

In 1998, the national list of impaired waters—the 303(d) list—included 21,845 waters with 41,318 associated impairments that will require TMDLs (EPA, 2000a). Additional waters are being added annually. The distribution of these impairments is summarized in Figure 4-3 and Table 4-6. Every state has impaired waters and is affected by the TMDL program.

More than 200 distinct types of water impairments have been identified through the 303(d) reporting process. By combining similar impairments (e.g., combining fecal coliform, bacteria, and *E. coli* as "pathogens"), EPA has reclassified the impairments into approximately 50 categories. The top 15 pollution categories, which encompass 91 percent of the total impairments, are shown in Table 4-7, which indicates that sediment, pathogens, and nutrients are the major pollutants with regard to water-quality degradation nationwide. To help address this concern, EPA recently released protocols for developing sediment, nutrient, and pathogen TMDLs (EPA, 1999a,b,c).

The TMDL program does not explicitly require the protection of riparian areas. However, implementation of the TMDL program will have a substantial impact because most of the TMDL implementation plans that have been, are being, or will be developed call for restoration of riparian areas as one of the required management measures for achieving reductions in nonpoint source pol-

BOX 4-5
Conservation Reserve Enhancement Program in the Illinois River Basin

The Illinois CREP is a restoration program for the stream and river floodplains in the Illinois River Basin (see map). As of September 2001, Illinois has enrolled over 88,000 acres in CREP, making it the most successful in the country on the basis of acreage.

The Illinois River Basin covers about 44 percent of the state, resulting in a landscape that is both intensely urban (e.g., Chicago) and agricultural. The Illinois River transports about 14 million tons of sediment, much of which is deposited into the Illinois River valley and the Mississippi River. The resultant sedimentation has led to severe habitat loss and commercial navigation problems, and water quality continues to need attention. The Illinois CREP has four goals: (1) reduce silt and sediment entering the main stem of the Illinois River by 20 percent, (2) reduce the amount of phosphorus and nitrogen in the Illinois River by 10 percent, (3) increase by 15 percent the population of waterfowl, shorebirds, non-game grassland birds, and state and federally listed threatened and endangered species such as bald eagles, egrets, and herons, and (4) increase by 10 percent the native fish and mussel stocks in the lower reaches of the Illinois River.

In the Illinois CREP, the current program authority calls for enrolling up to 132,000 acres for a minimum of 15 years. To do this, landowners enter into 15-year federal CRP contracts and then may extend their enrollments through a supplemental state contract for an additional 15-year, 35-year, or permanent conservation easement. Of the 88,426 enrolled by September 2001, the state options included 58,287 acres and of this, 91.5 percent (53,319 acres) was placed in permanent easements. In addition to receiving annual payments, a federal CREP contract entitles a landowner to cost-share assistance for 50 percent of the establishment costs associated with the selected conservation practice. Landowners who enter into an additional state CREP contract receive a lump-sum incentive payment during the first year of the enrollment. They are also eligible to receive up to another 50 percent cost share for establishment costs. Thus, up to 100 percent of these costs are paid for by state and federal agencies. Enrollments in CREP have been on a continuous basis since the program's inception in May 1998.

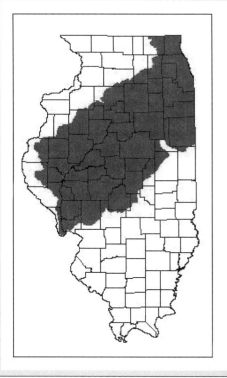

Eligible Lands

Although the current program calls for enrolling 132,0000 acres, the final program goals call for 232,000 acres to be enrolled. The eligible riparian areas must be either within the 100-year floodplain of the stream or river or be a farmed wetland, prior converted wetland, or a wetland farmed under natural conditions. Also eligible are highly erodible lands, defined as those lands with a weighted erodibility index of greater than or equal to 12, and which also must be adjacent to a stream corridor, have the riparian areas in a conservation practice, or have become an uneconomic remnant (i.e., land not profitable to farm).

Landowner Payments

Like CRP, the CREP uses soil rental rates to determine payments to landowners but provides an additional incentive of 30 percent above these rates for riparian lands and a 20 percent bonus for highly erodible lands. In addition, Illinois will make lump-sum payments for extensions beyond the initial 15-year federal contract. As an example, for land with a soil rental rate of $130/acre, the federal CREP will pay annually $169/acre ($130 plus a 30 percent bonus). In addition, the landowner will receive 50 percent of the cost of establishing trees and a $5/acre maintenance fee. If the landowner chooses to utilize one of the state extensions, such as a permanent easement, the state will make a one-time lump-sum payment equal to the rental rate times 15 years times 30 percent. For example, if the soil rental rate is $130/acre, the landowner will receive a payment of ($130 x 15 x 0.30) = $585/acre. In addition, Illinois will pay the remaining 50 percent of the practice cost share. Thus, total payments to the landowner summed over the 15 years will include $2,535/acre from the federal government (15 years at $169/acre, not including practice cost share and maintenance payments and other incentives such as the current Signup Incentive Payment (SIP) and Practice Incentive Payment (PIP)) and $585/acre from the state (also not including practice cost share).

Practices Available

Conservation practices funded by Illinois CREP are designed to reduce soil erosion, improve water quality, and create or enhance wildlife habitat. Emphasis is placed on the use of native vegetation with the choice dependent upon historic vegetation, soil types, water levels, and slope. Eligible practices include wetland restoration, riparian forest buffers, filter strips, establishing permanent native grasses, tree planting, permanent wildlife habitat development, and wildlife food plots. Riparian buffers (CP22) and wetlands (CP23) are the most common practices.

Monitoring

To determine if the Illinois CREP will have the impacts projected, the state has installed or supplemented monitoring on the main river and several tributaries, and it has supported modeling that predicts both erosion reduction and the economic value of the program. Because the Illinois CREP area is so expansive, the assessment has also focused on two smaller watersheds within the eligible area where intensive stream gaging, habitat monitoring, and land-use mapping are being conducted. These assessments are based upon a paired watershed design and incorporate site-specific studies to assess certain practices.

288

FIGURE 4-3 Distribution of impaired waters in the United States requiring TMDLs. SOURCE: EPA (2000b). Note: States with high rates of impairment do not necessarily have poorer water quality than states with lower impairment rates. Impaired waters are a function of water-quality standards that vary from state to state. Because of these state-by-state differences, waters that would be classified in one state as impaired might not be classified as impaired in other states. Thus, states that have the most comprehensive and strictest water-quality standards often have the most impairments.

TABLE 4-6 Number of Impaired Waters in the United States

Percentage of Impaired Water Miles within Watersheds[a]	Number of Watersheds[b]	Percentage of Total U. S. Watersheds
No Waters Listed	550	24.3
<5%	670	29.7
5%–10%	360	15.9
10%–25%	480	21.3
>25%	199	8.8
Total	2,259	100.0

[a]This is the percentage of total water miles within a given watershed that are impaired. "No waters listed" may imply that no waters are impaired or that the watershed has not been monitored.
[b]8-digit Hydrologic Unit Code (HUC) watersheds

SOURCE: EPA (2000a).

lutant loadings. For example, water-quality impairments caused by nonpoint source pollutants released in riparian areas or in adjacent uplands may be best remedied by enhancing the pollutant removal and assimilation functions of existing riparian areas or by creating new riparian areas. Indeed, riparian buffer zones have been used and promoted as management measures to address all the impair-

TABLE 4-7 Top 15 Categories of Impairment Requiring TMDLs (from 1998)

Cause of Impairment	Number of Impaired Waterbodies
Sediments	6,133
Pathogens	5,281
Nutrients	4,773
Metals	3,984
Dissolved Oxygen	3,758
Other Habitat Alterations	2,106
Temperature	1,884
pH	1,798
Impaired Biologic Community	1,440
Pesticides	1,432
Flow Alterations	1,099
Mercury	1,088
Organics	1,069
Noxious Aquatic Plants	831
Ammonia	752

NOTE: "Waterbodies" refers to individual river segments, lakes, and reservoirs. A single waterbody can have multiple impairments. Because most waters are not assessed, there is no estimate of the number of unimpaired waters in the United States. SOURCE: EPA (2000a).

ments listed in Table 4-7, with the possible exception of mercury and other point source-dominated impairments.

The potential impact of the TMDL program on riparian area protection was demonstrated in the *Pronsolino v. Marcus* case (1999), in which the TMDL implementation plan in question required substantial reductions (60 percent) in sediment loading. The Garcia River TMDL identified logging operations as a significant cause of excessive sediment, and thus limits were placed on allowable sediment losses from forestry operations. When the Pronsolinos filed for a permit to harvest timber, the California Department of Forestry required (as a result of the Garcia River TMDL implementation plan) that they reduce sediment losses by reducing harvesting within 100 feet of streams, refraining from construction or using skid trails on slopes greater than 40 percent within 200 feet of streams, and not removing trees from certain unstable areas, which have the potential to deliver sediment to a watercourse. Thus, the sediment load reductions required protection of existing riparian areas.

Several dozen TMDLs for sediment, nutrient, temperature, and fecal coliform impairments have been developed since 1999, many of which require riparian area protection and restoration as part of their proposed implementation plans (e.g., EPA Region IX, 1999; Indiana Department of Environmental Management, 2000). For fecal coliform-impaired waters in Virginia, every implementation plan for watersheds where cattle grazing is significant and in which cattle have access to streams has required an 80 percent to 100 percent reduction in cattle access to streams and riparian areas (Virginia Department of Environmental Quality, 1999; EPA, 2000c; Virginia Tech, 2000a,b,c,d). Exclusion of cattle from streams was specifically mentioned in each case. Although these TMDLs do not have explicit plans for riparian area restoration, the required exclusion of cattle from streams and riparian areas will result in significant improvement in riparian area functioning. However, how much functioning will be restored is uncertain because the TMDL plans do not specify how much fencing will be required or the widths of protected riparian areas. It is possible that fences will be installed immediately adjacent to streams, which is unlikely to promote functioning riparian areas. Most cost-share programs such as CRP that help landowners with fencing and off-stream water system costs require a riparian area with a minimum average width of 35–100 ft. Consequently, much of the fencing installed for TMDL implementation will involve the restoration of riparian areas 35–100 ft wide.

TMDL implementation plans developed for other impairments such as nutrients, benthic impairment, *Cryptosporidium*, and pesticides are also likely to require some level of cattle exclusion from streams. In addition, some of these plans may recommend restoration of riparian areas as a means of reducing nonpoint source pollutant loadings to streams from upland areas. For example, the fecal coliform TMDL for Pleasant Run in Virginia calls for a 25 percent reduction in nonpoint source loadings from pasture and cropland (Virginia Tech,

2000c). If cattle are excluded from the streams by fencing, as required by the TMDL, it is possible that the nonpoint source loading reduction goal will be met as a consequence of the riparian areas becoming reestablished. This assumes that the fences are located so that the resulting protected widths are adequate to achieve desired pollutant reductions. Width and vegetative composition need to be based on site-specific topographic and hydrologic conditions and the pollutant reductions required for each site (see Chapter 5).

The TMDL program also may lead to protection and restoration of riparian areas in those parts of the country where summertime stream temperature is an important water-quality issue. Some of the earliest stream temperature research in forested stream systems was undertaken in the late 1960s in Oregon by Brown (1969), and a significant body of knowledge has been acquired (e.g., Beschta et al., 1987) and stream temperature models (e.g., Boyd, 1996) have been developed for understanding and predicting the effects of vegetation removal on stream temperature. Temperature TMDLs are required for those waters where the instream water temperatures deviate from the state temperature standard (which in Oregon is a numeric standard based on seven-day maximum temperatures). The exercise involves identifying potential sources that contribute to increased water temperatures in conjunction with modeling efforts to evaluate the extent to which temperature improvements can be attained through improved riparian management. For example, along reaches normally occupied by a riparian forest, site potential vegetation (e.g., assumed to be late seral conifers) is utilized in a stream temperature model to indicate the potential improvements in temperature that might be realized if revegetation were to occur. Results of these analyses (e.g., Boyd et al., 1998) can be used to formulate TMDLs on a basin-by-basin basis.

The TMDL program is currently the nation's most comprehensive attempt to restore and improve water quality (NRC, 2001). Though not a primary stated goal of the program, TMDL implementation should protect many functioning riparian areas and restore thousands of miles of degraded riparian areas along the streams and shorelines of the United States. TMDL plans for the restoration of waterbodies impacted by livestock will likely involve streamside fencing and the reestablishment of riparian vegetation. For forested stream systems, the use of riparian reserves or stream buffers of unharvested trees will become increasingly common. In addition, TMDL implementation plans for waterbodies with impairments caused at least in part by nonpoint source pollutants from cropland and pasture will likely recommend the protection of existing riparian areas that are in relatively good condition and the restoration of those that have been degraded.

CONCLUSIONS AND RECOMMENDATIONS

As reflected in the foregoing materials, a variety of laws offer mechanisms to help protect some riparian areas or aspects of riparian areas. Few of these laws, however, reflect awareness of riparian areas as landscapes supporting multiple important functions and warranting special management and protection as unique physical and natural systems in their own right. Rather, protection of riparian areas is an indirect consequence of other objectives, such as water-quality protection or habitat management.

Protecting riparian areas in private ownership is especially challenging. Willingness of states and local governments to regulate land use in riparian areas for general ecological benefits varies widely. Striking examples of state and local programs that provide significant protection of riparian areas are relatively few in number. Interest seems to be growing in conservation easements and other incentives to induce landowners to hold riparian areas as buffers, natural areas, or open space, as well as in the purchase of riparian lands for greenways or wildlife areas. Enactment of laws such as the Conservation and Reinvestment Act (considered by Congress in 2000) could make available several billion dollars annually for purchases of such areas.

Many states have been willing to regulate or manage timber harvesting on private lands in riparian areas. They have not, however, been nearly as willing to restrict other agricultural activities, except in some areas with demonstrated water-quality problems. Instead, the preference has been to induce change in farming practices through incentives provided by programs such as the Conservation Reserve Program.

Riparian areas on federal lands are seldom managed as natural systems, though they may receive management attention or protection when they support resources of concern (such as wildlife or fisheries) and are threatened by certain land uses (such as livestock grazing or mining). Federal statutes contain very little guidance for land managers who face conflicts between riparian area protection and permissible land uses. Only if a federal agency proposes an activity in or affecting a riparian area that would jeopardize threatened or endangered species or violate water-quality requirements is the protection of riparian values clearly required.

Although the BLM is taking an increasingly active role in developing policies regarding the management and protection of riparian areas and in coordinating efforts to assess the condition of riparian areas, it has no clear mandate to do so. BLM's relatively new "fundamentals of rangeland health" regulations authorize the inclusion in livestock grazing permits of conditions to protect riparian areas, but these conditions are neither consistently included nor enforced. As a result of legal challenges, however, both the BLM and USFS have reduced or eliminated livestock grazing along some streams or have initiated consultation with the FWS regarding the impacts of grazing on threatened and endangered species. In the Pacific Northwest, the BLM and USFS have implemented significant riparian protections under the Northwest Forest Plan. In addition, the USFS has taken a number of forest- or stream-specific actions to protect riparian areas, but there is no clear agency policy guiding these actions. Neither the NPS nor the

FWS has a policy regarding protection of riparian areas, even though this would seem consistent with the predominantly protection-oriented missions of these two agencies.

State water laws have been concerned almost solely with the allocation of water for human uses and not for ecological needs. New programs directed at the protection of unutilized instream flows have the potential to address the water-related needs of at least some riparian areas.

In sum, existing legal and management protection of the ecological functions and values of riparian areas is inadequate. Even on federal lands, uses of riparian areas are not singled out for special consideration by statute or regulation. Uses of riparian areas on private lands are addressed, if at all, as a matter of local land-use regulation or through a mix of incentive programs. Several suggestions for strengthening and improving the legal framework governing protection of riparian areas are offered below. In the absence of making such legal and regulatory changes, it is unlikely that the degradation of riparian areas documented in Chapter 3 will be halted or even slowed.

Management guidelines and regulations differ drastically among forest, range, agricultural, residential, and urban lands on private lands. No state has a general land-use law or framework to coordinate management of the landscape for multiple uses (e.g., forest harvesting, grazing, agriculture, mining, urban development). Fragmentation of policy has contributed to vastly different levels of protection for, and degradation of, riparian areas across individual watersheds and regions. This phenomenon will only increase with increased population growth and continued economic development.

States should consider designating riparian buffer zones adjacent to waterbodies within which certain activities would be excluded and others would be managed. The broad importance of protecting riparian areas for water quality and fish and wildlife benefits calls for state-level programs of land-use regulation to accomplish this objective. A statewide program such as the Massachusetts Riverfront Protection Act treats all riparian landowners equally in providing these important public benefits. At the very least, states should consider establishing such buffers for sensitive areas (as has been done for the Chesapeake Bay). In the absence of a statewide program, local governments should be encouraged to develop riparian buffer zones.

Increased federal and state funding should be directed toward encouraging private riparian landowners to restore and protect riparian areas. At the federal level, this means increased funding for riparian buffers under Farm Bill programs, for wildlife habitat under Partners for Fish and Wildlife, and for watershed restoration under the Clean Water Action Plan and other federal agency initiatives. Both federal and state funding should be made available to land trusts, soil and water conservation districts, watershed groups, and others working with private riparian landowners to protect and improve their riparian areas.

Few, if any, federal statutes refer expressly to riparian area values and as a consequence generally do not require or ensure protection of riparian areas. Even the National Wild and Scenic Rivers Act refers only to certain riparian values or resources; it does not consider riparian areas as natural systems, nor does it require integrated river corridor management. Moreover, statutes governing federal land management do not direct agencies to give priority to riparian area protection when conflicts among permissible land uses arise. This absence of a national riparian mandate stands in stark contrast to the existence of a federal wetlands law.

Federal land management agencies should promulgate regulations requiring that the values and functioning of riparian areas under their jurisdiction be restored and protected. This goal is consistent with the ecological benefits of riparian areas and the overarching principle that public lands are to be managed in the national interest. Such regulations should account for the full spectrum of riparian values and services—habitat-related, hydrological, water quality, aesthetic, recreational. At a minimum, agencies should assess the condition of riparian areas, develop and implement restoration plans where necessary, exclude incompatible uses, and manage all other uses to ensure their compatibility with riparian area protection. Clear, enforceable regulations are necessary because existing rules and policies are inconsistent, vague, and/or only advisory. Alternatively, agencies could protect riparian areas using existing rules for special management areas, such as BLM's ACECs.

Ideally, Congress should enact legislation that recognizes the myriad values of riparian areas and directs federal land management and regulatory agencies to give priority to protecting those values. A mandate from Congress would establish riparian protection as a federal priority and ensure consistency both within and among agency regulatory and land management programs. Absent federal legislation, a presidential executive order could promote consistent riparian management by federal agencies.

Federal agencies should coordinate riparian management activities to improve efficiency and help ensure that protection of riparian values and functioning does not vary across jurisdictional boundaries. Many streams and other waterbodies, especially in the West, are located on private, state, and/or federal lands. Frequently, streams traverse lands managed by different federal agencies. If intact riparian areas are to be restored and maintained, management and protection of riparian values and functioning cannot be left to the vagaries of political systems, but will require coordination of management policies and prescriptions.

States should administer the public trusts in water and state-owned submerged lands to protect the public interests in properly functioning and ecologically healthy riparian areas. States are obligated to protect public interests in recreation, fisheries, water yield, and other values and services of state waters, a responsibility that cannot be carried out without regard to riparian functioning. Each state has the authority to decide how it will administer these trusts, subject to judicial review according to standards established by the respective state and by the U.S. Supreme Court.

Instream flow laws can help protect riparian areas if river and stream flows are managed to mimic the natural hydrograph. Water allocation has historically favored human claims to water over using it for environmental needs. Recently, the needs of natural systems have been addressed in some cases by preserving minimum stream flows. Because riparian functioning is dependent on the full range of variation in the hydrologic regime, the reintroduction or maintenance of such flow regimes (in addition to minimum stream flow) is essential for restoring and sustaining healthy riparian systems.

Implementation of the CREP and TMDL programs has the potential to protect existing and restore degraded riparian areas nationwide. State involvement in CREP is increasing exponentially, with the potential for taking millions of miles of riparian land out of agricultural production. Many TMDL plans developed to date call for the restoration of riparian areas to reduce nonpoint source pollutant loadings and to restore streamside shading.

REFERENCES

Allen, A. W. 1996. Northern Prairie Science Center conservation reserve program bibliography. Jamestown, ND: Northern Prairie Wildlife Research Center.

Amman, D., B. Cosens, and J. Specking. 1995. Negotiation of the Montana–National Park Service Compact. Rivers 5(1):35–45.

Beschta, R. L., R. E. Bilby, G. W. Brown, L. B. Holtby, and T. D. Hofstra. 1987. Stream temperature and aquatic fish habitat: fisheries and forestry interactions. Pp. 191–232. In: Streamside management: fisheries and forestry interactions. E. O. Salo and T. W. Cundy (eds.). Contribution No. 57. Seattle, WA: University of Washington, Institute of Forest Resources. 471 pp.

Boyd, M. S. 1996. Heat source: stream, river and open channel temperature prediction. M.S. Thesis, Oregon State University, Corvallis. 148 pp.

Boyd, M., B. Kasper, and A. Hamel. 1998 (draft). Tillamook basin temperature; total maximum daily load (TMDL). Portland, OR: Oregon Department of Environmental Quality. 42 pp. + appendices.

Bradbury, W., W. Nehlsen, T. E. Nickelson, K. Moore, R. M. Hughes, D. Heller, J. Nicholas, D. L. Bottom, W. E. Weaver, and R. L. Beschta. 1995. Handbook for prioritizing watershed restoration to aid recovery of native salmon. Portland, OR: Pacific Rivers Council.

Brown, G. W. 1969. Predicting temperatures of small streams. Water Resources Research 5:68–75.
Bureau of Land Management (BLM). 1984. Federal Register 49:162.
BLM. 1986. Federal Register 51:12,747.
BLM. 1991a. Riparian-Wetland Initiative for the 1990's. BLM/WO/GI-91-001+4340. Washington, DC: BLM.
BLM. 1991b. Application to Appropriate Public Water, Attachment A.
BLM. 1992. Federal Register 57: 53,924.
BLM. 1994. Ecosystem Management in the BLM: From Concept to Commitment. BLM/SC/GI-94-005+1736. Washington, DC: BLM.
BLM. 2000a. Public Land Statistics 1999. Volume 184, BLM/BC/ST-00/001+1165. Washington, DC: BLM.
BLM. 2000b. Final Notice of Issuance and Modification of Nationwide Permits: Correction. Fed. Reg. 65:14,255.
Committee of Scientists (COS). 1999. Sustaining the peoples' lands: recommendations for stewardship of the national forests and grasslands into the next century. Washington, DC: U.S. Department of Agriculture. 193 pp.
Doppelt, B., M. Scurlock, C. Frissell, and J. Karr. 1993. Entering the watershed: a new approach to save America's river ecosystems. Washington, DC: Island Press.
Durbin, K. 1997a. Restoring a refuge: cows depart, but can antelope recover? High Country News 29 (Nov. 24, 1997).
Durbin, K. 1997b. Selling science to the agencies: an ecologist's story. High Country News 29 (Nov. 24, 1997).
Environmental Protection Agency (EPA). 1991. Guidance for water quality-based decisions: the TMDL process. Washington, DC: EPA Assessment and Watershed Protection Division.
EPA Region IX. 1999. South Fork Eel River total maximum daily loads for sediment and temperature. San Francisco, CA: EPA Region IX Water Division. 62 pp.
EPA. 1993. Guidance specifying management measures for sources of nonpoint pollution in coastal waters. 840-B-92-002. Washington, DC: EPA Office of Water.
EPA. 1999a. Protocol for developing sediment TMDLs. First Edition. EPA 841-B-99-004. Washington, DC: EPA Office of Water.
EPA. 1999b. Protocol for developing nutrient TMDLs. First Edition. EPA 841-B-99-007. Washington, DC: EPA Office of Water.
EPA. 1999c. Protocol for developing pathogen TMDLs. First Edition. EPA 841-B-00-002. Washington, DC: EPA Office of Water.
EPA. 2000a. 1998 Section 303(d) list fact sheet: national picture of impaired waters. Washington, DC: EPA Office of Water.
EPA. 2000b. Atlas of America's polluted waters. EPA 840-B-00-002. Washington, DC: EPA Office of Water.
EPA. 2000c. Fecal coliform TMDL modeling report: Cottonwood Creek, Idaho County, Idaho. Washington, DC: EPA Office of Water. 139 pp.
Federal Interagency Floodplain Management Task Force. 1992. Floodplain management in the United States: an assessment report—volume 2: full report. Washington, DC: Federal Emergency Management Agency.
Federal Energy Regulatory Commission (FERC). 1992. Office of Hydropower Licensing, Paper No. DPR-4. Washington, DC: FERC.
Feldman, M. D. 1995. National forest management under the Endangered Species Act. Natural Resources and Environment 9(Winter 1995):32–37.
General Accounting Office (GAO). 1996. Land ownership: information on the acreage, management, and use of federal and other lands. GAO/RCED-96-40. Washington, DC: GAO.
Gillilan, D., and T. Brown. 1997. Instream flow protection: seeking a balance in western water uses. Washington, DC: Island Press.

Harting, A., and D. Glick. 1994. Sustaining Greater Yellowstone: a blueprint for the future. Greater Yellowstone Coalition.

Houck, O. A. 2000. The Clean Water Act TMDL program: law, policy, and implementation. Washington, DC: Environmental Law Institute.

Indiana Department of Environmental Management. 2000. Dissolved oxygen and ammonia TMDL development for Kokomo Creek, Indiana (draft). Indianapolis: Indiana DEM. 49 pp.

Interagency Floodplain Management Review Committee. 1994. Sharing the challenge: floodplain management into the 21st century. Washington, DC: Administration Floodplain Management Task Force.

Isenhart, T. M., R. C. Schultz, and J. P. Colletti. 1997. Watershed restoration and agricultural practices in the Midwest: Bear Creek of Iowa. Pp. 318–334 In: Watershed restoration: principles and practices. J. E. Williams, C. A. Wood, and M. P Dombeck (eds.). Bethesda, MD: American Fisheries Society.

Keystone Center. 1991. Final consensus report of the keystone policy dialogue on biological diversity on federal lands.

Kimball, K. D. 1997. Using hydroelectric relicensing in watershed restoration: Deerfield River watershed of Vermont and Massachusetts. Pp. 179–97 In: Watershed restoration: principles and practices. J. E. Williams, C. A. Wood, and M. P. Dombeck (eds.). Bethesda, MD: American Fisheries Society.

Kusler, J. A. 1985. A call for action: protection of riparian habitat in the arid and semi-arid west. In: Riparian ecosystems and their management: reconciling conflicting uses. G. Bingham, E. H. Clark, L. V. Haygood, and M. Leslie (eds.). April 16–18, 1985. Tucson, AZ: USFS.

Land Trust Alliance. 1996. The conservation easement handbook. Washington, DC: Land Trust Alliance.

MacDonnell, L. J., and T. A. Rice. 1993. The federal role in in-place water protection. In: Instream flow protection in the West, Revised Edition. Boulder, CO: Natural Resources Law Center.

Mohai, P., and P. Jakes. 1996. Forest Service in the 1990s: is it headed in the right direction? Journal of Forestry 94:31–37.

Moyle, P. B., and G. M. Sato. 1991. On the design of preserves to protect native fishes. Pp. 155–173 In: Battle against extinction: native fish management in the American West. W. L. Minckley and J. E. Deacon (eds.). Tucson, AZ: University of Arizona Press.

National Park Service (NPS). 1993. Floodplain management guideline. Washington, DC: Department of the Interior.

National Research Council (NRC). 1993. Setting priorities for land conservation. Washington, DC: National Academy Press.

NRC. 1995. Wetlands: characteristics and boundaries. Washington, DC: National Academy Press.

NRC. 2000. Watershed management for potable water supply: assessing the New York City strategy. Washington, DC: National Academy Press.

NRC. 2001. Assessing the TMDL approach to water quality management. Washington, DC: National Academy Press.

North Country Resource Conservation and Development Area Inc. 1995. A guide to developing and re-developing shoreland property in New Hampshire. Meredith, N.H.

Schultz, R. C., J. P. Colletti, T. M. Isenhart, W. W. Simpkins, C. W. Mize, and M. L. Thompson. 1995. Design and placement of a multi-species riparian buffer strip system. Agroforestry Systems 29:201–226.

Silk, N., J. MacDonald, and R. Wigington. 2000. Turning instream flow water rights upside-down. Rivers 7(4):298–313.

Small, S. J. 1995. The federal tax law of conservation easements. Second Supplement. Washington, DC: Island Press.

Smith, D. S., and P. C. Hellmund, eds. 1993. Ecology of greenways. Minneapolis, MN: University of Minnesota Press. 222 pp.

Stenzel, R. L. 2000. Policies concerning wetland vegetation. Colorado Stream Lines 14(1). Denver, CO: Colorado Division of Water Resources.

United States of America, the State of Utah, the Washington County Water Conservancy District, and the Kane County Water Conservancy District. 1996. Zion National Park water rights settlement agreement.

U.S. Department of Agriculture Forest Service (USFS) and Department of Interior Bureau of Land Management (BLM). 1994a. Northwest Forest Plan. Washington, DC: USFS.

U.S. Department of Agriculture Forest Service (USFS) and Department of Interior Bureau of Land Management (BLM). 1994b. PACFISH. Washington, DC: USFS.

U.S. Department of Agriculture Forest Service (USFS). 1995. Inland native fish strategy (INFISH). Washington, DC: USFS.

U.S. Department of Agriculture Forest Service (USFS). 1998. Blue mountains biodiversity project v. Pence. Federal Register 63:63,831. Washington, DC: USFS.

U.S. Department of Agriculture Natural Resources Conservation Service (NRCS). 1999. The National Conservation Buffer Initiative: a qualitative evaluation. Washington, DC: USDA NRCS.

U.S. Department of the Interior Bureau of Land Management (BLM) and U.S. Department of Agriculture Forest Service (USFS). 1994. Rangeland reform '94: draft environmental impact statement executive summary. Washington, DC: US DOI and USDA.

U.S. Fish and Wildlife Service (FWS). 1994. Federal Register 59:30,035.

Virginia Department of Environmental Quality. 1999. Fecal coliform TMDL development for Muddy Creek, Virginia. Richmond. 111 pp.

Virginia Tech Biological Systems Engineering and Biology Departments. 2000a. Fecal coliform TMDL for Dry River, Rockingham County, Virginia. Final Report prepared by Virginia Tech Departments of Biological Systems Engineering and Biology and submitted to Virginia Departments of Environmental Quality and Conservation and Recreation, Richmond. 122 pp.

Virginia Tech Biological Systems Engineering and Biology Departments. 2000b. Fecal coliform TMDL for Mill Creek, Rockingham County, Virginia. Final Report prepared by Virginia Tech Departments of Biological Systems Engineering and Biology and submitted to Virginia Departments of Environmental Quality and Conservation and Recreation, Richmond. 113 pp.

Virginia Tech Biological Systems Engineering and Biology Departments. 2000c. Fecal coliform TMDL for Pleasant Run, Rockingham County, Virginia. Final Report prepared by Virginia Tech Departments of Biological Systems Engineering and Biology and submitted to Virginia Departments of Environmental Quality and Conservation and Recreation, Richmond. 108 pp.

Virginia Tech Biological Systems Engineering Department. 2000d. Fecal coliform TMDL for Sheep Creek, Elk Creek, Machine Creek, Little Otter River, and Big Otter River in Bedford and Campbell Counties, Virginia. Final Report prepared by Virginia Tech Department of Biological Systems Engineering and submitted to Virginia Departments of Environmental Quality and Conservation and Recreation, Richmond. 302 pp.

5

Management of Riparian Areas

The condition of the nation's riparian areas represents the outcome of decades of local and basinwide land use, often with little understanding of how various practices might impact these valuable and productive systems. With an increasing body of scientific knowledge regarding riparian areas—their ecological processes and functions, their diversity at local and landscape levels, and their productivity and utility for a variety of human uses—the nation is now in position to protect, improve, and restore many of its riparian systems. This chapter outlines approaches for improving the ecological functioning of riparian areas—an opportunity for landowners, irrigation districts, watershed councils, professional societies, government at local, state, and federal levels and their associated regulatory agencies, and the public at large. According to Verry et al. (2000), "The acid test of our understanding is not whether we can take ecosystems apart on paper...but whether we can put them together in practice and make them work."

The restoration of riparian areas and their associated aquatic ecosystems has become a topic of intense scientific interest. For example, the experimental flood of the Colorado River in the southwestern United States in the spring of 1996 focused worldwide attention on alternative methods for managing and restoring river and riparian ecosystems (Collier et al., 1997). Reinstating flooding and overbank flows on a river where flow regulation has been in place for decades is now seen as a potential means for partially restoring fluvial geomorphology and riverine habitats for threatened and endangered species in this human-impacted landscape. Similarly, the initiation of restoration efforts on the channelized and flow-regulated Kissimmee River in south Florida is a major undertaking designed to restore 70 km of river channel and 11,000 ha of wetland over the next

15 years (Cummins and Dahm, 1995; Dahm et al., 1995; Toth et al., 1998). The goal of this long-term project is to reestablish 104 km^2 of river-floodplain ecosystem and return a more normal hydrograph to the river. These ambitious and expensive projects represent historic initiatives in ecosystem restoration; however, they are a small part of the challenges that remain in restoring rivers and riparian areas throughout the United States.

Because degradation of riparian areas varies in areal extent, severity, and proximity to streams and other waterbodies, attempts at restoring these areas will entail more than simply understanding the workings of a narrow strip of land along a stream, river, or other body of water. Upslope and upriver land uses must necessarily be considered. Understanding the watershed context is often essential in undertaking restoration efforts that are targeted at improving streamside areas (Kershner, 1997). Unfortunately, although watersheds as geographic areas are "optimal organizing units" for dealing with the management of water and related resources such as riparian areas (NRC, 1999), the natural boundaries of watersheds (and their riparian systems) rarely coincide with legal and political boundaries. City, county, state, and federal jurisdictions provide a mélange of authorities across the landscape. Thus, comprehensive watershed approaches to riparian restoration, by necessity, will need to involve numerous landowners, a cross section of political and institutional representations, and coalitions of special interest groups.

GOALS OF MANAGEMENT

Strategies and practices that reflect a spectrum of goals will likely be needed for maintaining and improving the ecological functions of existing riparian areas and for improving their sustainability and productivity for future generations. This section identifies several broad management approaches that have different objectives and expected outcomes.

Protection

Protection (also referred to as preservation or maintenance) of intact riparian areas is of great importance, both environmentally and economically. It is distinct from restoration, which addresses degraded systems. Intact riparian areas represent valuable reference sites for understanding the goals and the efficacy of various restoration approaches and other management efforts. In some cases they are important sources of genetic material for the reintroduction of native biota into areas in need of restoration. For these reasons and others, riparian areas in a natural state warrant a high level of protection (NRC, 1992, 1995; Kauffman et al., 1997).

As a management strategy, riparian protection may entail more than simply preventing human-induced alterations. For example, actions such as prescribed fire, management of exotic species invasions, and large herbivore management may be necessary to maintain natural characteristics and functions and to sustain them over time. Because degraded riparian areas are so prevalent in many portions of the nation, protecting any that remain relatively uninfluenced by human perturbations should be a high priority. Measures to protect intact areas are often relatively easy to implement, have a high likelihood of being successful, and are less expensive than the restoration of degraded systems (NRC, 1992; Cairns, 1993).

Restoration

Definitions of the verb *restore* commonly include to reestablish, to put back into existence or use, to bring back into the former or original state, to renew, to repair into nearly the original form, and to bring back into a healthy state. These definitions point to the reestablishment of former conditions, processes, and functions (i.e., making healthy again). Although seemingly simple in concept, the restoration of degraded riparian areas is often a scientific and social challenge. In some instances, the natural or pristine conditions of a particular riparian area may no longer exist or may not be known with certainty. In others, multiple causes of degradation may have occurred over long periods of time—hence, cause-and-effect relationships that define existing conditions may not be well known or easy to decipher at either local or landscape scales.

Restoration may refer both to the process of repairing degraded riparian areas and to the desired end goal of such actions, although the term is sometimes used to refer only to the latter. Thus, for example, NRC (1992) defined restoration of aquatic ecosystems as representing the "re-establishment of pre-disturbance aquatic functions and related physical, chemical, and biological characteristics." It further indicated that "restoration is different from habitat creation, reclamation, and rehabilitation—it is a holistic process not achieved through the isolated manipulation of individual elements." This definition has the stated goal of regaining predisturbance characteristics, which this report categorizes specifically as *ecological restoration*. Thus, a working definition of ecological restoration for riparian areas, based upon the above as well as upon definitions within Jackson et al. (1995), Kauffman et al. (1997), and Williams et al. (1997) might be:

> The reestablishment of predisturbance riparian functions and related physical, chemical, and biological linkages between aquatic and terrestrial ecosystems; it is the repairing of human alterations to the diversity and dynamics of indigenous ecosystems. A fundamental goal of riparian restoration is to facilitate self-sustaining occurrences of natural processes and linkages among the terrestrial, riparian, and aquatic ecosystems.

Ecological restoration of riparian areas results in the reestablishment of functional linkages between organisms and their environment, even though these systems may be continually responding to the natural dynamics of various environmental conditions.

Across the nation, there are many riparian areas where ecological restoration is possible. For example, riparian areas in forests and rangeland areas throughout the western United States represent likely candidates for ecological restoration if the adverse effects of historical or ongoing land uses can be significantly reduced, controlled, or eliminated. Success is more likely where fundamental disturbance regimes continue to occur relatively unhindered by human influence. Ecological restoration of riparian areas that border low-order streams or other small bodies of water is also possible where human impacts have involved relatively benign land uses. Tributary junctions of streams and rivers represent additional landscape locations where disturbance regimes often remain in a relatively natural state. In such situations, it may be possible to recover nearly the full array of riparian composition, structure, and functions that existed before significant human alterations or impacts occurred.

Although ecological restoration may be an achievable and desired goal for some areas, it obviously cannot be attained everywhere. For example, permanent or irreversible changes in hydrologic disturbance regimes (e.g., via dams, transbasin diversions, irrigation projects, extensive landscape modification), natural processes (e.g., global climate change, accelerated erosion), channel and floodplain morphology (e.g., channel incision, rip-rap, levees), and other impacts (e.g., extirpation of species, biotic invasions) may preclude our ability to precisely or completely re-create the composition, structure, and functions that previously existed. Riparian areas adjacent to large rivers may represent a greater challenge than those associated with smaller streams and rivers because of the greater number of factors affecting flow regimes at these larger scales (Gore and Shields, 1998). Nevertheless, even in such situations, there are often numerous opportunities to effect significant ecological improvement of riparian areas and to restore, at least in part, many of the functions they formerly performed.

Based on the above considerations and others, this report classifies as *restoration* those efforts that lead to the recovery of some of the previously existing riparian composition, structure, and functions. As shown in Figure 5-1, restoration represents a reversal in the decline of ecosystem health and movement of a degraded system toward its historical conditions and functions. Although the predisturbance composition, structure, and functions of the riparian area (i.e., ecological restoration) may not be the final outcome of a restoration effort, the primary intent of such efforts is nevertheless to shift a riparian area in that direction.

This chapter considers many of the scientific and social challenges to be faced in restoring riparian areas that have been significantly altered or degraded by human activities. Distinguishing between natural disturbances and the effects of human-induced modifications to riparian areas is an important aspect of restoration. Understanding the values of society is equally important, as they will

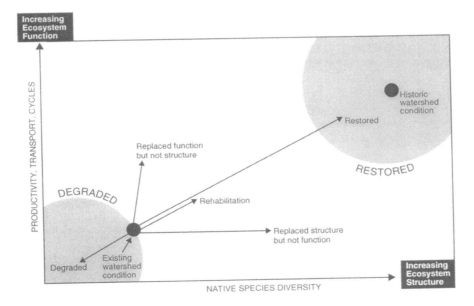

FIGURE 5-1 Restoration is dependent on ecosystem structure and function. A primary goal of restoration is to redirect the trajectory of a degraded area, in relation to both its structure and function. Restoration refers broadly the moving towards the upper right corner. Ecological restoration is represented by the historic watershed condition. SOURCE: Reprinted, with permission, from Williams et al. (1997). © 1997 by American Fisheries Society.

likely need to change and adapt over time if restoration efforts are to proceed. Because riparian areas represent an entire suite of organisms, physical features, processes, and functions, a species-only or single-process approach will likely fail to achieve a significant degree of restoration. For example, the reintroduction of an extirpated plant species into a degraded riparian area is likely to fail if the underlying causes of extirpation have not been addressed. Focusing on those human influences that affect multiple ecological processes is more likely to attain greater restoration of riparian habitat and species of interest.

Alternatives to Protection and Ecological Restoration

Across the United States, a large number of aquatic and riparian projects are implemented each year, many of them having "restoration" as one of their expressed goals. Although ecological restoration may be a nominally important objective of some projects, many others are simply altering aquatic and riparian systems with little emphasis on understanding or attempting to benefit long-term ecological processes or functions (Goodwin et al., 1997); improvements in ecological functions are typically not specified nor necessarily expected. Terms such as *creation, reclamation, rehabilitation, replacement, mitigation, enhancement,* and *naturalization* have been coined to describe the wide variety of land management approaches (NRC, 1992, 1996). These approaches typically emphasize altering ecosystem components to serve a particular human purpose, but generally are not intended to restore the full suite of ecological functions that would normally be associated with a particular riparian area (Kauffman et al., 1997). Although these terms have different meanings to various disciplines (legal, political, and scientific), the appropriate characterization of riparian management options and goals is more than a matter of semantics. It is important to properly distinguish between a wide range of management approaches so that interested parties have realistic expectations regarding their potential outcomes.

Creation

Creation is the establishment of a new riparian system on a site where one did not previously exist; it is generally associated with the establishment of a "new" reach of stream. For example, the repositioning of a section of stream or river channel will inherently cause the "creation" of a new riparian area that may or may not be ecologically similar to the section of channel lost by such a repositioning. Often the newly created channel will be less sinuous than the original one and less likely to be hydrologically connected to former floodplains. In other instances, channels may have been unintentionally developed or created as a result of long-term land-use practices. As discussed in Chapter 3, conversion of native forests and grasslands to agricultural crops throughout much of the Midwest was commonly accompanied by altering field drainage patterns (e.g., tiling and ditching), such that new channels eventually developed. An extended network of intermittent and ephemeral streams has become established in many agricultural areas where they did not previously exist; many of these streams could support riparian plant communities.

Reclamation

Reclamation has traditionally been defined as the process of adapting natural resources to serve utilitarian human purposes (NRC, 1992). Historically, it often involved the conversion of wetlands and riparian areas to agricultural, industrial, or urban uses. More recently, however, reclamation has been defined as a process resulting in a stable, self-sustaining ecosystem that may or may not include some exotic species. The structure and functions of reclaimed sites may be similar, although not identical, to those of the original land (Jackson et al., 1995).

Rehabilitation

Rehabilitation implies rebuilding or making part of a riparian area useful again after natural or anthropogenic disturbances. For example, the mechanical excavation and reconfiguring of an eroding bank could represent rehabilitation. Although the resulting bank configuration might assist in retarding subsequent erosion, its configuration and other properties might be quite unlike that of a natural channel. Restoration of predisturbance processes and functions is neither required nor implied in the definition of rehabilitation; rehabilitation efforts typically do not focus on reproducing conditions characteristic of functionally intact riparian systems or on meeting regional ecological goals.

Mitigation

Mitigation is an attempt to alleviate some or all of the detrimental effects or environmental damage that arise from human actions. Mitigation is commonly used with regard to wetlands—e.g., the creation of a new wetland is often proposed as mitigation for natural wetlands that are to be impacted by dredging, filling, or other human alterations. However, constructed wetlands seldom display the full complement of structural and functional attributes of the native wetlands they replace (Quammen, 1986; Kusler and Kentula, 1990; NRC, 2001). Mitigation with regard to riparian areas focuses on minimizing potential detrimental impacts from a particular human action. For example, where levees may be needed along a river to protect human developments, mitigation might require the levees to be set back some distance from the channel edge to retain some riparian functions of the streamside vegetation and to maintain hydrologic connectivity of the near-channel floodplains and side channels. Where rip-rap is to be employed along a streambank, mitigation might require that measures be taken to ensure that riparian plants can become established and survive along the structure. In forested systems, large wood could be placed in channels in an attempt to mitigate the effects of prior harvesting practices that removed all trees along streams.

Replacement

Replacement represents the substitution of a native species or ecosystem feature with an alternative species (e.g., exotic species) or foreign object. An example would be the replacement of native conifers or deciduous trees with non-native species. Sometimes the replacement can be structural; for example, rip-rap may be used where floodplain or meadow streambanks have begun to erode because land uses have removed streamside vegetation or reduced the ability of the remaining vegetation to retard fluvial erosion. Replacement approaches are generally narrow in scope and seldom successful in promoting a wide range of ecological goals.

Engineered approaches that reconstruct or greatly modify a particular stream and its riparian system to meet specific human ideas regarding what they should look like or how they should function are also considered replacement. Such approaches are often employed in urban areas where significant alterations to a stream and riparian area have occurred and where the hydrologic regime has been significantly altered (e.g., where increased amounts of impervious surface contribute more surface runoff and higher pollutant loads to a stream). Although these designed systems may provide many benefits (e.g., stabilized channel morphology, permanent streamside vegetation), they seldom have the features of more natural streams and thus do not provide the full range of functions associated with natural systems.

Enhancement

Enhancement represents an attempt to accentuate or improve a *specific component* of riparian areas. Thus, enhancements may come at the expense of other components and may create conditions that are uncharacteristic of a natural riparian system. A widespread example is the employment of structures of various types and sizes in channels and on streambanks (e.g., log weirs, gabions, large rocks) to enhance fisheries habitat (e.g., Wesche, 1985; Hunter, 1991; Seehorn, 1992). These structures can alter streambank structure, sediment transport dynamics, and hydrologic connectivity with riparian vegetation, often resulting in disruption of riparian–stream linkages. Similarly, when spoils, rocks, or boulders are removed from streams and added to streambanks and floodplains to enhance local channel stability, conditions may no longer be suitable for the natural establishment of riparian vegetation or for adjustments in channel morphology in response to streamflow and sediment transport. In-channel enhancement projects are unlikely to provide long-term or sustainable improvements for riparian/aquatic systems (Platts and Rinne, 1985; Elmore and Beschta, 1989; Beschta et al., 1994).

Naturalization

Naturalization, an alternative to ecological restoration, attempts to accommodate watershed-scale human influences in environmental designs of channels by establishing stable, self-sustaining geomorphologic systems with abundant and diverse ecological communities that are fundamentally different from those that existed before. The concept of naturalization was developed for specific application to agricultural streams that have been significantly modified, often by deepening and straightening previously existing channels (Rhoads and Herricks, 1996). Where headwater channel gradients are low, as in the Midwest, such channelized and modified streams have developed relatively stable configurations over many decades; the goal of naturalization would be to maintain that new stable configuration.

Naturalization assumes that the pristine stream network may not be the best restoration objective for stream management because (1) adequate information on the pristine state of streams is not available, (2) environmental conditions in most watersheds are far removed from the pristine state, and (3) restoration at the watershed scale is economically impractical. Because most streams in agricultural settings are not regulated by dams or lined with concrete, they retain some of their capability to morphologically adjust to changing flow and sediment regimes, implying there is some potential for these areas to support other riparian functions such as habitat provision. Management that might be used to achieve naturalization includes not only vegetated riparian buffers (discussed later), but also off-channel wetlands, side-slope reduction of streambanks, increased stream sinuosity, and other practices that provide improved ecological and water-quality benefits (Petersen et al., 1992).

The alternative approaches described above differ from protection and ecological restoration in their ultimate goals and consequently in the amount of ecological functioning that a degraded riparian area might eventually attain. While it is not the objective of this report to advocate ecological restoration as a goal for all degraded riparian areas, it is important to understand the trade-offs between restoring an area to full functioning vs. partial functioning. Much more important than the setting of a challenging goal (e.g., ecological restoration) is continual progress toward a more functional system. When conceptualized as a series of activities that improve both ecosystem structure and function, restoration can be monitored over time and at specific milestones. Box 5-1 illustrates the restoration of Bear Creek, Oregon—i.e., movement of this riparian area toward improved structure and functioning.

Passive Versus Active Approaches to Restoration

Once the necessary background information has been obtained for understanding the status, trends, and factors influencing a particular riparian area, perhaps the most critical step in undertaking restoration is to curtail those activities and land uses that are either causing degradation or preventing recovery. Such an approach is referred to as *passive restoration* (Kauffman et al., 1997). Removing human disturbances in degraded systems allows natural process to be the primary agents of recovery. Many riparian areas are capable of recovery following a reduction in or curtailment of human perturbations because the biota of these systems has evolved to reproduce and survive in an environment of frequent natural disturbances. In the absence of other types of management, natural disturbance regimes and ecosystem responses will dictate the speed of recovery for areas undergoing passive restoration (NRC, 1996).

BOX 5-1
Bear Creek, Oregon: A Restoration Case Study

Bear Creek provides a unique opportunity to observe the evolution of a riparian area over 21 years of changing management. It also demonstrates the resiliency of functioning riparian areas to management alternatives and high-flow events. Bear Creek is approximately 1,000 m (3500 ft) in elevation and located in the high desert of central Oregon. Although annual precipitation averages only 300 mm (12 in), the year-to-year variation in precipitation is quite high. Peak runoff from snowmelt typically occurs in mid to late February, and summer thunderstorms are common.

Livestock have grazed the Bear Creek area since the late 1800s; the permitted use in 1977 was 75 animal unit months (AUMs) from April until September. Surveys during 1977 revealed that the Bear Creek riparian area totaled 0.95 ha/km (3.8 ac/mile) of stream (representing an average riparian width of less than 5 m (16 ft) on each side of the stream) and was producing approximately 225 kg/ha (200 lbs/ac) of forage. That meant that if livestock consumed all the available forage and used 365 kg/AUM (800 lbs/AUM), 1.6 km (1 mile) of stream was required to support one cow for one month. As shown in Plate 5-1, by 1977 streambanks were actively eroding, the channel was deeply incised, and riparian vegetation was sparse. Flows were frequently intermittent, and runoff events contained high sediment loads.

The Bureau of Land Management (BLM) then changed the grazing rotation in the area such that in 1979 and 1980, the area was grazed for one week in September. From 1981 to 1984, none of the area was grazed. As shown in Plate 5-2, by May 1983, banks were stabilizing and the channel was narrowing and deepening. Sediment trapped by vegetation can be seen on the banks among newly emerging plants. Juniper trees in the floodplain seen in Plate 5-1 were cut down to see if this practice would affect willow reestablishment. (To date, willow reestablishment has been unsuccessful.) The large juniper indicated by the arrow was left, and it can be seen in the remaining photos.

By comparing Plates 5-1 and 5-2, it can be seen that over the six-year period of controlled grazing and livestock exclusion, riparian vegetation increased, the channel narrowed and deepened, and channel stability increased. Sediment, trapped by vegetation, can be seen on the banks in the reestablishing riparian area. These results were the result of natural recovery of the riparian area once livestock were excluded. Active restoration techniques, such as channel grading and planting, were not used.

During 1985, the pasture was divided into three pasture units, and controlled grazing was permitted from mid-February to mid-April. Vegetation was then allowed to grow to protect the stream system during the critical summer thunderstorm period and to provide livestock forage the following year. From 1983 to 1986, the channel continued to deepen and narrow, and nearly 460 mm (1.5 ft) of sediment was trapped on the floodplain because of increased riparian vegetation, which not only reduced channel scour but also reduced flow velocities and sediment transport capacity, as shown in Plate 5-3.

Plate 5-4, taken in June 1987, shows the effects of a large summer thunderstorm and resulting flood event on the riparian area. Compared to 1986 (Plate 5-3), it appears that much of the riparian vegetation has been inundated with sediment. The main channel widened some, but it is still narrower than it was in 1977 (Plate 5-1), and the channel

PLATE 5-1

PLATE 5-2

PLATE 5-3

PLATE 5-4

PLATE 5-5

PLATE 5-6

PLATE 5-7

PLATE 5-8

PLATE 5-9

PLATE 5-10

PLATE 5-11

and the stream banks appear stable. There are obvious sediment deposits on the streambanks.

By August 1987 (Plate 5-5), the riparian vegetation was recovering rapidly and was stabilizing sediment trapped during the flood event, although some bare areas were still present. By October 1998, 16 months after the June 1987 flood event, the riparian area appears to have fully revegetated (Plate 5-6). The floodplain now appears stable and has trapped over 600 mm (2 ft) of sediment since 1976.

By 1989, the increased productivity of the riparian area permitted grazing to increase to 354 AUMs, nearly five times the 1977 allotment of 75 AUMs. This reportedly reduced the livestock permittee's winter feeding costs by over $10,000 a year. Plate 5-7, taken in August 1994, shows the riparian area during a drought. Because of reduced channel flow, sedges and rushes seeking water occupy almost the entire channel. The formerly intermittent stream has become perennial because of increased infiltration and moisture storage in the reestablished riparian area.

By 1995, beavers had returned to the watershed, presumably attracted by the improved hydrologic regime and increasing riparian vegetation. This is another possible indication of improved riparian functioning, as beavers usually avoid streams in poor condition. The dam building activities of the beavers will further stabilize the stream and increase water storage within the stream system. Plate 5-8 shows a newly established beaver dam slightly downstream of previous photos.

By 1996, the riparian area had increased in size to 3 ha/stream km (12 ac/stream mile), and forage production had increased to 370 kg/ha (2,000 lbs/acre)—approximately a 10-fold increase since 1977. Sediment deposition in the riparian area raised the streambed by 0.75 m (2.5 ft), and channel storage increased eightfold to approximately 9,400 m^3/km (4,000,000 gal/mile) since 1977. Stream length (sinuosity) increased by 11 percent, and rainbow trout returned to the stream for the first time in decades.

In February 1996, the stream experienced another major flood caused by the rapid melting of the winter snow pack. As shown in Plate 5-9, the flood inundated a large portion of the floodplain. When the water receded, however, little damage was revealed, as shown in Plates 5-10 and 5-11, taken two and eight months later in April and October 1996, respectively. The established riparian vegetation was able to resist damage from this flood, protect the stream channel from scour, reduce flow velocities, and trap an additional 13 cm (5 in) of sediment in the floodplain.

The Bear Creek project demonstrated the potential of passive restoration in a riparian area long degraded by overgrazing. In this case, total exclusion of livestock from the riparian area occurred for several years, followed by controlled late winter–early spring grazing from February 15 to April 15 once most of the riparian vegetation was reestablished. Livestock were excluded from the riparian area at all other times of the year.

According to the BLM project manager, the timing and duration of grazing appeared to be more important than the number of livestock in maintaining the health of riparian vegetation once it had been reestablished. In addition, the most important factor in riparian area restoration was commitment by the operator to observe the livestock exclusion and the subsequent controlled grazing.

Photos and project description provided by Wayne Elmore.

Because passive restoration focuses on altering, reducing, or eliminating the primary causes or factors that have contributed to a degraded riparian system or have prevented recovery, its importance cannot be overemphasized. Passive restoration is the logical and necessary first step in any restoration program—and in many cases may be all that is required.

Although passive restoration is a relatively straightforward concept, it can sometimes be difficult to implement because doing so typically requires changing the types or extent of land or water uses within riparian areas or at other locations in a given watershed. Existing land uses that have occurred over many years or decades may be difficult to change. Often the most significant barriers to passive restoration are social, cultural, and political rather than scientific or biological. However, bypassing this step represents a major strategic flaw in any restoration program and may ensure the inevitability of project failure if ecological restoration is the goal.

In recent years, an increasing number of passive restoration activities have been implemented in portions of the American West. For example, most western states (and some eastern states) have implemented forest practices rules on industrial forestlands that identify riparian protection as an important management objective. Such rules often identify the dimensions of required no-harvest buffers and other practices (e.g., directional felling, limitations on ground skidders) designed to reduce and minimize forestry impacts to riparian areas. As discussed in Chapter 4, federal agencies [e.g., U.S. Forest Service (USFS), Bureau of Land Management (BLM)] have implemented a system of riparian reserves that often provide full no-harvest protection for areas one to two site-potential tree heights from a stream. In grazed areas, extended periods of non-use or exclosure fencing have begun to occur in some riparian areas on both federal and private lands. In the Mono Basin of northeastern California, the return of flows to Rush and Lee Vining Creeks and the removal of grazing along these streams by the city of Los Angeles, after a protracted legal battle, have resulted in a major recovery of riparian vegetation and functioning. (As of yet there has been little monitoring or research to document the ecological changes that are occurring as a result of these improved riparian management practices.)

After passive restoration is implemented, a riparian area may remain in an ecological condition that is significantly different from that of a comparable reference site, particularly if its inherent capacity to recover has been severely influenced or lost. To improve the likelihood of achieving restoration in such situations, active manipulations, herein referred to as *active restoration*, may be needed. Active restoration attempts to restore a degraded or dysfunctional riparian area by combining elements of natural recovery with management activities directed at accelerating the development of self-sustaining and ecologically healthy systems (NRC, 1996; Kauffman et al., 1997). This requires not only an understanding of the complex processes and linkages between the biotic and physical components of intact systems, but also of the range of active management practices that might be successfully implemented.

Factors that may prevent the return of a degraded riparian area to a more natural dynamic one via passive restoration include species extinction, exotic species invasion, significant structural modifications, and continued alteration of hydrologic flow regimes. Although some of these factors can be addressed via active restoration, others can be sufficiently severe in their magnitude, persistence, and spatial extent that full ecological restoration may not be technologically or economically feasible. Nonetheless, restoration can still achieve ecological improvement of a system so that specific ecosystem features (e.g., water quality), biotic species (e.g., endangered species), or channel morphology (e.g., reestablishment of historical channels) are improved.

Regardless of whether passive or active restoration is chosen, riparian and watershed activities that do not address the recovery of multiple ecosystem linkages and functions are likely to have only limited success, they may have no effect, or they may even exacerbate ecosystem degradation. Continued degradation following the implementation of restoration measures could occur because of an inadequate scientific basis for the established goals, institutional constraints such as insufficient funding or funding at an inappropriate time, or severe environmental conditions during the early phases of a restoration project (e.g., exceptionally large floods, drought, fire). Unfortunately, continued degradation not only suppresses the recovery of ecological functions, but it may further limit the capability of a riparian system to be restored.

The alteration and degradation of riparian areas at both local and regional scales generally reflect land uses occurring over extended periods. Similarly, the recovery of riparian areas after the cessation or removal of perturbations will require time. Time requirements for the recovery of a degraded riparian system are seldom mentioned in restoration projects. Some riparian functions can recover relatively rapidly, while others require long periods to achieve their full potential. Figure 5-2 illustrates the projected recovery rates, under passive restoration, for various components of riparian areas associated with salmonid habitats in the Interior Columbia River Basin. While the exact timing of individual features' recovery may vary by location, watershed, or region, the overall pattern is one of increasing functional interaction among riparian vegetation, channel morphology, and aquatic and riparian habitats over time.

Potential Conflicts Between Riparian Restoration and Other Management Goals

Given that human use of riparian areas has often been at the expense of maintaining their ecological processes and functions, attempts at improving and restoring riparian functions may encounter some degree of social, political, or

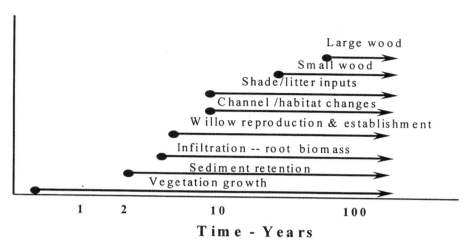

FIGURE 5-2 Projected recovery times of selected leading-edge and keystone components of salmonid habitats in the Interior Columbia River Basin following the cessation of activities causing degradation or preventing recovery (passive restoration). SOURCE: Reprinted, with permission, from Beschta and Kauffman (2000). © 2000 by American Water Resources Association.

institutional opposition. The following are some of the issues that may affect whether and how quickly restoration efforts move forward.

Short-term versus Long-term Restoration Goals

In nearly all restoration efforts, one is typically faced with trying to balance short- and long-term goals. For example, former riparian plant communities may have experienced a loss in species diversity and cover because of grazing or conversion to agricultural crops. In the first instance, many of the native plants may still be present, but their abundance and growth have been greatly curtailed. In the second, the original riparian vegetation may no longer exist except in small, isolated areas. Halting those land use practices causing degradation or preventing recovery, i.e., passive restoration, could be accomplished by removing grazing animals from the riparian areas in the first instance and by no longer cropping to the edge of the stream in the second. Vegetation in the previously grazed area would likely recover relatively quickly if most of the native plants were present. For the agricultural setting, however, it may be necessary to reintroduce native plants to help "jump start" native plant communities. Although both approaches would be directed at a reestablishment of native plant communities and their attendant physical, biological, and chemical processes, recovery in the cultivated agriculture setting would likely take longer and require some form of active restoration. In addition, where channel morphology has been altered via management practices, the recovery of channel dimensions and form is not likely to occur with the same rapidity as that of vegetation. Thus, practitioners involved in restoration efforts should realize that the short- and long-range goals of any restoration effort will be met sequentially if the restoration approach is ultimately successful.

Small-Scale versus Large-Scale Perspectives

Land managers typically view riparian issues at small scales—i.e., the size of their property or management unit slated for restoration. In doing so, they may fail to realize the extent to which their riparian areas have been altered over time, the role of off-site factors, or the impact that their management decisions can have on other areas. To help land managers better understand riparian issues, the condition of their land, and the potential for ecological improvement and restoration, a larger-scale landscape perspective is often required. Landscape assessments can be undertaken to provide multiple landowners with potential strategies for improving the ecological and social values of their specific riparian areas. It is also at these larger scales that scientific input can offer a crucial perspective regarding the magnitude of problems and the potential for improvement. Where monetary resources for restoration efforts are limited, having a large-scale perspective will allow for more effective allocation of resources to accomplish the greatest good. Watershed councils, state and federal agencies, and other groups are often of major assistance in developing basinwide perspectives that are useful to various landowners and land managers as they engage in riparian improvement and restoration across a given drainage basin.

Private Lands versus Public Resources

The vast majority of the nation's riparian lands are in private ownership. These lands provide a wide variety of economic and social benefits to landowners. However, many of the benefits derived from functional riparian areas also cross into the public domain. For example, intact riparian forests generally ensure high levels of stream shading and tend to reduce stream temperatures during summertime. Although the incremental impact to summertime stream temperatures is likely to be small if a single landowner were to harvest the riparian forest or convert it to another use (e.g., grazing, agriculture), the cumulative effect on water quality can be considerable if multiple landowners temporarily or permanently remove their riparian forests.

The regulation of instream water quality by states may affect landowners' management of their riparian areas (although the exemption of nonpoint sources of pollution from permit requirements means that many land uses are not directly regulated). In forestry, public concern about water quality has increasingly been codified into forest practice rules that, on a state-to-state basis, vary in the level of riparian protection and enforcement. Similarly, public concerns about fisheries and wildlife habitats may affect the management of riparian areas on private lands. The extent to which the public's need or desire to maintain and protect public resources (e.g., water quality, fish and wildlife habitat) outweighs landowners' rights to manage and alter riparian systems continues to be a hotly contested issue. In some instances, the core issue is the extent to which restrictions can be placed on traditional activities on private lands now understood to result in significant ecological damage.

Riparian Area versus Human Water Needs

One point of conflict in many western states regards the water needs of riparian vegetation versus those of humans. Because they represent sites of higher moisture availability than upslope terrestrial environments, riparian areas have relatively high evapotranspiration rates. In the arid and semiarid regions of the western United States, woody plant communities found along streams often have extensive root systems that allow them to extract water from the water table or from the capillary fringe immediately above (Brooks et al., 1991). Active removal and eradication of riparian plant communities in the name of water conservation has been a common practice in these areas, although actual water savings have rarely been quantified. In fact, the presumed water savings from the removal of riparian vegetation stands in stark contrast to reports that document the loss of perennial flow from rivers, streams, and springs when riparian vegetation is removed (e.g., Hendrickson and Minckley, 1984). Clearly, a simple projection of potential evapotranspiration gains associated with the removal of riparian plant communities is not adequate for evaluating the merits of such projects.

Reintroducing historical overbank flows at their customary timing and frequency of occurrence as a restoration strategy may sometimes lead to conflict with the water needs of human populations. Although research in both humid regions (Cummins and Dahm, 1995; Dahm et al., 1995) and arid regions (Lieurance et al., 1994; Molles et al., 1995, 1998) has shown beneficial responses of riparian areas to restored flooding, the water costs of such strategies have not been carefully documented. That is, how water availability for other purposes (e.g., hydropower generation, irrigation withdrawals, municipal or industrial use) changes when such approaches to restoration are used has seldom been quantified. Larger-scale use of restored flooding depends on assessing the amount of water that will be required and convincing multiple local, state, and federal agencies and various water users that the activity is sound management policy. Floods are not only one of the most common large natural disturbances that occur in the United States, but they are also one of the most costly in terms of property damage and loss of human life. A major challenge with such restoration efforts is to reestablish enough of the unregulated high-flow dynamics (magnitude, frequency, duration, and timing) so that characteristic riparian plant communities can be restored while not substantially increasing the risk to property or human life.

Eliminating exotic vegetation in riparian areas and reestablishing native plant communities is another approach to riparian restoration that can sometimes lead to conflicts between interest groups. For example, controversy marks recent proposals for biological control of exotic saltcedar in riparian areas of the western United States. In this case, one group of managers within the U.S. Fish and Wildlife Service pressed for release of insects to control invasive saltcedar populations (DeLoach, 2000). Meanwhile, a second group within the same agency worked to block the release of these biological control agents in order to protect nesting habitat for the southwestern willow flycatcher, a federally listed endangered species, which actively nests in pure stands of saltcedar in some areas (Leon, 2000).

* * *

Potential conflicts between contemporary land uses and the changes needed for improving the nation's riparian resources encompass the full range of land uses, water resource developments, and management policies. At least some of the current land and water use practices and management policies were initiated long ago when the nation was going through a period of expanding occupancy and settlement. Land development continues today under a policy climate that often encourages alteration of natural systems. Unfortunately, such policies were formulated and implemented during a time when impacts on riparian areas and their stream systems were not widely understood.

Because many of the options for improving riparian systems across watersheds encompass a wide range of individual and societal values, there is a great need to engage various stakeholders in broad-scale and collaborative restoration efforts. The potential success of collaborative efforts rests firmly on two foundations: credible scientific information and broadly inclusive participation where the full spectrum of perceptions, interpretations, claims, and contentions can be openly discussed, critiqued, and challenged. As a process for finding areas of agreement amongst all stakeholders (scientists, land managers, regulators, and the public), such deliberations need to ensure inclusiveness, openness, safety of expression, and respect for divergent views and positions. Although such an approach takes time and may not lead to full consensus, it is only through such a process that restoration can truly be a collaborative effort with substantive public support (Committee of Scientists, 1999).

Riparian Management as Part of Watershed Management

Because riparian areas are integral components of larger watersheds (drainage basins), management of riparian areas should attempt whenever possible to be incorporated into larger-scale watershed management plans. Watershed management refers to the managing of water resources (both surface water and groundwater) in a watershed or river basin context (rather than in a political or jurisdictional context) (NRC, 1999). Although instigated in the early part of the twentieth century, watershed management has found renewed support in the last 15 years for primarily water quality and ecological reasons (Adler, 1995). It is a holistic approach that addresses multiple sources of pollution within a watershed, such as urban and agricultural runoff, landscape modification, depleted or contaminated groundwater, and introduction of exotic species, to name just a few. As articulated by the U.S. Environmental Protection Agency (EPA), the watershed approach is a coordinating framework for environmental management that focuses public and private sectors on addressing the highest priority problems within hydrologically defined geographic areas (EPA, 1995). It targets those issues not adequately addressed by traditional point source programs—programs that for the most part have failed to protect watersheds from the cumulative impacts of multiple activities.

Although watershed management may vary in terms of specific objectives, priorities, elements, timing, and resources, it is based on the following principles, which necessarily should also characterize riparian area management:

• *Partnership.* All stakeholders affected by management decisions should be involved throughout watershed management and should shape key decisions. This ensures that environmental objectives are integrated with economic, social, and cultural goals. It also provides those who depend upon the natural resources within watersheds with information on planning and implementation activities.

• *Geographic Focus.* Activities should be specific to geographic areas, typically the areas that drain to surface waters or that recharge or overlay groundwater or a combination of both.

• *Science-Based Management.* Collectively, watershed stakeholders should employ high-quality scientific data, tools, and techniques in an iterative decision-making process including (1) assessment and characterization of the natural resources, (2) goal setting and identification of objectives based on the needs of the ecosystem and stakeholders, (3) prioritization of identified problems, (4) development of management options and action plans, (5) implementation of management options, and (6) effectiveness evaluation and plan revision (NRC, 2000).

Coordination of the many public and private interests implicated in watershed management is a major challenge. Institutional mechanisms for such coordination do not yet exist in most places, and where they have been developed, their effectiveness has been highly variable (Scurlock and Curtis, 2000). Fortunately, the involvement of stakeholders in watershed management has been aided by the emergence of local watershed groups—encouraged, in part, by EPA's emphasis on watershed approaches, but motivated also by rapidly developing ecosystem science and by frustration with the jurisdictional and programmatic piece-mealing of the landscape (Natural Resources Law Center, 1996). Although watershed groups form because of some specific, broadly shared concern, protection of riparian areas has not, by itself, been a primary or common focus. Rather, such things as improvement of water quality and protection of a fishery have been the motivating factors (Kenney et al., 2000). Yet it is precisely because of their important role in achieving many distinct objectives, such as healthy fish and wildlife habitat or floodplain management, that protection and restoration of riparian areas should be approached on a watershed scale, even though this may increase the complexity and timeline of the project.

Box 5-2 presents two examples of where riparian area management was incorporated into larger watershed management efforts. In the first case (the San Pedro Riparian National Conservation Area), it was recognized that restoration of the riparian area would not succeed without a more holistic understanding of the causes of degradation—most of which are outside the riparian area. In the second case (the Model Watershed Project in Idaho), activities in the riparian area were determined to be critical to achieving the overall goals of watershed management.

TOOLS FOR ASSESSING RIPARIAN AREAS

For decision-makers to be effective in managing riparian areas, they need information on the status and condition of these areas. The identification of riparian areas is a first step in accumulating information about their quantity and quality. Where they have been highly degraded, it may be difficult to identify riparian areas by remote sensing or even ground-based surveys. It is similarly difficult to identify wetlands that have been effectively drained. Yet their recognition is important precisely because former wetland areas are among the best opportunities for restoration. The same principle applies to riparian areas.

A wide variety of tools are available for assessing the condition of riparian areas. The assessment tool chosen will depend on the objectives of the program for which it is to be applied. For example, a program designed to identify priority areas for restoration might find useful a large-scale watershed assessment approach, such as the Hydrogeologic Equivalence or the Synoptic Approach discussed below. These approaches generally consider the existing condition of all riparian areas, the cumulative length of the various stream orders, and longitudinal connections that are necessary for migrations of fish and other organisms. For smaller-scale projects such as restoration of a reach of riparian corridor, knowledge is needed about what types of vegetation should be planted, the appropriate channel capacity for the stream, and the width of the riparian area necessary for carrying out various functions. In tracking the progress of individual restoration projects, detailed information on hydrology, seedling survival, and animal recruitment might become components of an assessment.

> **BOX 5-2**
> **Incorporating Riparian Areas into Watershed Management**
>
> **San Pedro Riparian National Conservation Area**
> Because of the unique quality of its riparian habitat and bird populations, Congress designated a 40-mile segment of the upper San Pedro River in Arizona as a National Conservation Area in 1988. Funding allowed the buy-out of irrigated farmlands within the area and retirement of the associated water rights. Despite this designation and the use of passive restoration, studies have documented a continued decline in stream flows and in the health of riparian vegetation.
> The San Pedro originates in Mexico and flows north into Arizona to its confluence with the Gila River. In most segments the San Pedro is a perennial stream, but surface flows sometimes disappear—especially in very dry years. Flood flows in the San Pedro are the product of large rainfall events, usually in late summer but sometimes in the winter. Base flow is the product of groundwater discharge—primarily from the more deeply underlying regional aquifer rather than from the shallow alluvial aquifer. Water use in the Mexico portion of the watershed is primarily for irrigation and includes significant groundwater pumping. Water use, also via groundwater pumping, in the southern Arizona segment is primarily for the needs of a military base (Fort Huachucha) and for urban growth in and around Sierra Vista (Commission for Environmental Cooperation, 1998).
> There is little doubt that riparian vegetation along the San Pedro will continue to decline unless ways can be found to limit additional groundwater depletions and, possibly, recharge the regional aquifer—actions that must be taken well outside the riparian

In general, complex restoration projects dealing with multiple impacts and a variety of riparian and wetland types might be better served with a unique assessment approach tailored to site-specific peculiarities. At the watershed scale, this could involve a large research and data gathering component, followed by modeling and validation, and it would include input from stakeholders in the region. An advantage of costly and involved large-scale assessments such as this is that the information is often transferable to similar physiographic regions.

This section focuses on standardized approaches that can be applied in a relatively short period (rapid assessment) and that do not require long-term training for practicing environmental professionals. Assessments fall into two basic categories: (1) functional assessments that estimate probabilities that a riparian function exists and (2) reference-based methods that estimate ecosystem condition. Rapid assessment methods for evaluating ecosystems have undergone dramatic development in the past three decades. Of particular relevance for riparian areas are methods that were developed for assessing wetlands, to which there has been considerable attention as a consequence of government regulatory programs. However, this section also evaluates methods developed specifically for riparian areas. It should be noted that there are various instream flow assessments

area of interest. No acceptable solutions have as yet emerged, but a watershed partnership of all the affected interests (Upper San Pedro Partnership) has formed with the goal of working out an agreement respecting water use. Significant progress has been made in water conservation practices at Fort Huachucha and in Sierra Vista. Although other actions still will be necessary to ensure the long-term health of the riparian system along the San Pedro, it is clear that no one entity can make this happen.

Model Watershed Project

The Model Watershed Project in Idaho started in 1992 to address local factors related to the decline of salmon and steelhead runs, particularly problems of fish habitat and passage related to irrigation water use. The project area encompasses the Lemhi River, the Pahsimeroi River, and the East Fork of the Salmon River in Idaho, which have a combined drainage area of approximately 2,735 square miles. The purpose of the project is to provide a basis of coordination and cooperation between local, private, state, tribal, and federal fish and land managers, land users, landowners, and other affected entities to protect and restore anadromous and resident fish habitat.

Watershed-wide, the land area is approximately 88 percent federal and 12 percent private. However, the stream corridor, which is most influential in providing salmon habitat, is 90 percent private and 10 percent federal. Beef cattle production is the dominant economic activity, and hay is the primary crop. As analysis of fish habitat conditions proceeded, it soon became evident that protection of riparian areas was essential. Because the analysis involved public/private partnerships, local landowners were willing to participate. Construction of fences creating a riparian buffer zone for grazing management has been a primary focus. In addition, projects have focused on streambank stabilization and riparian vegetation plantings, as well as instream structure work.

that target aquatic ecosystems. Although some of these methods take the condition of riparian areas in account and thus may be valuable for assessing riparian areas—such as the Stream Visual Assessment Protocol (USDA NRCS, 1998)—they are not the focus of this section.

General Characteristics of Assessment Tools

Need for Classification

Within a geographic region of interest, classifying riparian area types is one of the first steps in organizing information. The highest order of classification separates riparian areas by rivers, lakes, and estuarine/marine settings. Within each of those categories, further classification should recognize the amount of natural variation so that variation related to degraded conditions can be more easily identified. Most assessment methods discussed below use classification to identify what portion of the landscape is being evaluated and whether the riparian component needs to be further subdivided into more relatively homogeneous areas (although classification is not emphasized in the description of any particular procedure). Assessments done with little regard for initial classification are not likely to be successful.

Rather than adopt a classification system for use throughout the United States, regionally developed classifications may be more appropriate. Resource managers may already be familiar with them, thus reducing the time that otherwise would be used to create, learn, and adopt a new system. If no appropriate classification of riparian areas exists for the region, a preexisting classification from elsewhere might be adopted. It is important that the classification not rely exclusively on vegetation because vegetation is commonly modified by human activity. Rather, the underlying template for classification should be based on geomorphology and hydrology. Classifications should not be restricted to channel morphology, which is only one part (albeit a very important one) of riparian areas. Methods that rely on channel forms (e.g., Rosgen, 1995; Montgomery and Buffington, 1997) can be utilized where they are a critical feature of riparian areas (as described in Box 5-3). Other classification features that have been used in the past include stream order, stream slope, valley width, drainage basin size, and underlying lithology. Ideally, one would use a hierarchical approach that first places adjacent waterbody type (river, lake, estuarine/marine) at the highest level and vegetative cover (or lack of it) at the lowest level, that would have the flexibility of accommodating new information, and that would recognize fluvial and geomorphic forces as principal organizers of riparian systems (Brinson, 1993). Altered riparian areas should not be identified as core riparian classes, but rather recognized as departures from one of the established classes of riparian areas. This would be consistent with a restoration approach that uses relatively unaltered riparian areas as targets for restoration (NRC, 2001).

Reference Sites

Reference sites ideally represent relatively large and intact riparian systems that are self-sustaining and have not been markedly influenced by anthropogenic impacts. Their identification is crucial to the restoration of riparian areas for a number of reasons. First, sites with minimal alterations illustrate the natural interactions of hydrologic processes, geomorphic setting, and vegetation dynamics. Indeed, much of our understanding of ecology has been derived from studies of intact ecosystems. Second, to the extent that these sites can be placed in a successional sequence, they may be used effectively for insights into restoration goals. These sites provide excellent opportunities for locally "grounding" the available science, thus providing a knowledge base important for addressing restoration needs. Finally, reference sites can serve as demonstration areas where scientists, managers, regulators, and interested citizens can interact on a common footing when addressing restoration needs and priorities and the potential for successfully attaining restoration goals. Several of the assessment tools discussed in this section require identification of reference sites as a preliminary step.

Obtaining information on undisturbed riparian areas is not always easy or possible. However, through the use of Government Land Office records, historical vegetation surveys, soil maps or profiles that delineate floodplains, chronosequences of aerial photographs, geomorphic assessments of channel conditions and fluvial landforms, and other sources of information, the nature of intact riparian areas can often be revealed. Thompson (1961) provides an excellent example of this reconstruction for the riparian areas of the Sacramento Valley in California. Today these areas bear absolutely no resemblance to their former condition of marshes and streamside forests, the latter of which were sometimes several miles wide. Without some sense of the conditions and characteristics of the natural riparian systems prior to human intervention, it is not possible to evaluate project goals and determine the general direction of change that is needed for recovery.

Natural disturbances must be recognized as a fundamental property of riparian areas and must be accounted for by reference sites. The range of variation arising from natural causes such as climate, topography, and geomorphology can be assessed by considering a number of individual sites within riparian classes. This range should include the normally encountered differences in geomorphic and hydrologic conditions, as well as natural disturbance regimes that result in successional stages within a riparian class. A second source of variation important for effective restoration and other management programs is that arising from human-induced alterations. Thus, altered sites should become part of the reference system so that the full range of variation—including both natural variation and that associated with anthropogenic sources—is taken into account. It is particularly important to choose sites impacted by activities that will be common targets for restoration. This ensures that the reference system identifies what altered and degraded riparian areas might evolve toward if the appropriate restoration approaches are undertaken. If the degree or extent of riparian degradation relative to that of an appropriate reference site is severe, recovery is unlikely to be quickly or simply achieved.

Although there are no fixed protocols for setting up a reference system, one guideline is to choose a relatively homogeneous array of sites in terms of climate, stream order, species composition, and disturbance regimes. Where such sites are abundant, the task of capturing natural variation is relatively easy. Conversely, where unaltered sites are infrequent, small, or fragmented, one may have to rely on historical data and descriptions or on information gathered at similar sites outside the geographical region of interest. Even less-than-ideal reference sites can provide important information on species composition and structure of plant communities, frequency of inundation, and other characteristics.

Once a reference system is established, it becomes a valuable asset for restoration programs. Its value would be enhanced if reference sites were protected by acquiring covenants or conservation easements that ensure the perpetuation of natural disturbance regimes (Brinson and Rheinhardt, 1996). Restoration projects

BOX 5-3
Classification of Channel Types in the River Continuum

Methods for classification of natural channels across a wide array of landscapes may have utility in the assessment riparian areas because channel types influence processes within riparian areas such as sediment transport and deposition, flooding frequency, and flow dynamics. The simplest classification is the division of natural channels into braided, meandering, and straight channels by Leopold and Wolman (1957). Rosgen (1995) distinguished eight primary stream types in a classification that emphasized dimensional measurements such as number of channel threads, entrenchment ratio, width-depth ratio, and sinuosity (Figure 5-3). Montgomery and Buffington (1997) defined seven channel types based on similar criteria but were more forthright in recognizing the qualitative decision rules needed (Figure 5-4).

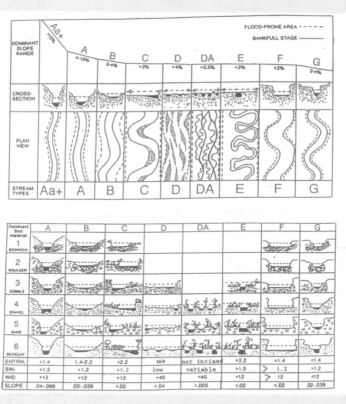

FIGURE 5-3 Rosgen Stream Classification. SOURCE: Reprinted, with permission, from Rosgen (1995). © 1995 by Wildland Hydrology.

MANAGEMENT OF RIPARIAN AREAS

Other classifications have attempted to move beyond static descriptions of channel form to the dominant processes in a channel that determine the likelihood of further channel changes. A process-based classification by Whiting and Bradley (1993) replaces the usual physical features with a complex phase-space representing distinct fluvial processes and their relative rates. All classifications have utility as comparative tools, although most suffer from the problem of lacking a strong physical basis that can predict trajectories of channel change given a change in fluvial or sediment variables. A predictive understanding seems to be possible in individual case studies but not in the comprehensive framework of a broad classification. Understanding how the evolution of channel and floodplain changes translates into changes in riparian areas is an even more daunting task.

Future classifications that better emphasize riparian areas will need to incorporate emerging views about interactions between fluvial processes and vegetation. For example, the lateral expansion of channels during floods and the role of riparian areas in modifying flood flows are typically not considered in a standard chan-

continues

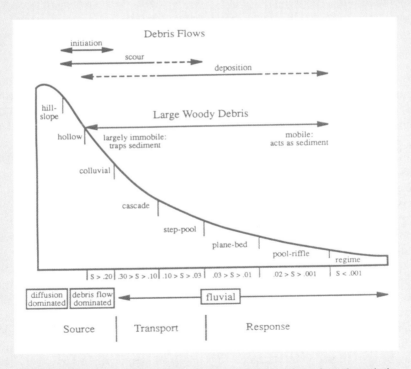

FIGURE 5-4 Illustration of idealized profile from hill tops downslope through the channel network showing general distribution of channel types and controls on channel processes. SOURCE: Montgomery and Buffington (1997).

> nel classification. In addition to the longitudinal and lateral gradients, vertical gradients beneath the channel and floodplain must also be considered (Stanford and Ward, 1993). Subsurface strata of granular material of highly variable texture represent the legacy of deposition and erosion in past times. River water penetrates those bed sediments and mixes with groundwater, creating distinct zones of biogeochemical reactions and habitats for certain aquatic insects and other organisms that live in subsurface environments. Riparian area classifications must therefore be extended to include a number of new variables, such as the size and hydraulic properties of alluvial aquifers, that are typically not considered in standard channel classifications.

themselves may become "standards" to which other projects can be compared, especially those that have matured and succeeded in responding to natural perturbation regimes. For the same reason, much can be learned from unsuccessful restoration projects. Both contribute to a system of reference sites in ways that would be lacking if only unaltered sites were utilized.

Information Needs for Riparian Restoration

History of Resource Development. Because many changes to riparian systems occurred prior to the current generation of landowners and managers, understanding historical trends at both local (e.g., stream reach, valley, watershed) and regional (e.g., across large watersheds or ecoregions) scales can be critical for developing restoration plans. Such information is essential (1) for understanding the present status and trends of existing riparian systems, (2) for identifying possible management practices or forms of resource development that have contributed to existing riparian conditions or have prevented recovery, and (3) for developing effective restoration strategies.

If the historical causes of riparian degradation are not known or have been incorrectly assessed, attempts at restoration may be ineffective or misdirected and opportunities for riparian improvement lost. For example, although the planting of willows on sites for which they are *not* adapted provides temporary satisfaction in that a revegetation effort was undertaken, such planting is likely to have little effect (either positive or negative) on the long-term recovery of a particular riparian system. In other instances, willows may have been planted on appropriate sites, but if continued ungulate grazing (the original cause of willow extirpation) has not been modified, the opportunity for successful regeneration may be lost. In yet other instances, logs and boulders may be added to channels and streambanks in an attempt to replace lost structural elements. Although such an approach may have some basis in forested riparian systems, its application to streams in many meadow systems of the western United States or prairie systems in the Midwest represents a misinterpretation of restoration needs. In these examples, the approach taken may not alter the causes of degradation—even though they may entail a major expenditure—and are likely to provide little prospect of improvement. Hence, understanding historical patterns of resource development and causes of degradation is extremely important where ecological recovery of riparian systems is the primary goal.

Hydrologic Regime. Understanding the characteristics of natural flow patterns—flow frequency, magnitude, duration, and timing—associated with specific riparian areas can be a crucial component of restoration where such flow regimes have been previously modified (e.g., because of dams, other water resources development, or extensive land modification). An important restoration goal may be reestablishment of a streamflow regime that emulates the temporal dynamics of an unaltered system and provides hydrologic connectivity to remaining floodplains and riparian landforms (Hill et al., 1991; Whittaker et al., 1993; Rood et al., 1999; Rood and Mahoney, 2000). For many floodplains, an understanding of subsurface hydrology and geologic stratigraphy is also critical (Jones and Mulholland, 2000; Woessner, 2000).

In some cases reestablishing the hydrologic regime will not be sufficient to restore degraded riparian areas, necessitating a better understanding of the links between flow regime, sediment dynamics, and vegetative growth. This is the case where hillslope or channel erosion processes remain altered in spite of attempts to return natural flows. For example, if accelerated surface erosion or landslides are occurring on upslope areas and the resultant sediments are transported to riparian areas, simply maintaining a natural hydrologic regime may be insufficient to restore a riparian system.

Channel incision and widening (as a result of a variety of land-use practices) and dams that have reduced the magnitude and frequency of high flows can curtail overbank flows, which typically ensures the loss or decline of riparian vegetation. Information on historical conditions of overbank flood events is needed to make decisions about whether healthy riparian plant communities can be reestablished and whether a long-term process of bank-building and channel aggradation may be an achievable restoration goal. Where channel incision or widening has been relatively large, these effects may not be easily reversible.

Soils and Landforms. Soils and landforms can provide important insights into the historical condition of their associated riparian areas. For example, floodplain soils that have developed from overbank flows over the millennia provide a long-term ledger of past hydrologic disturbance regimes and their resultant riparian systems. Even where vegetation has been largely modified, removed, or replaced by various land uses, residual soil properties (e.g., mottles, gleying, organic matter content, soil texture and structure, redox potential) can provide important clues regarding soil development processes and conditions that were prevalent prior to the effects of Euro-American land uses. Soil texture is particularly important as it tends to indicate the prevailing hydraulic conditions at the time of deposition: clay particles are indicative of ponded water, silts of slow-moving water often across a vegetated surface, and sands/gravels of a relatively high-energy environment. Such information affords insight into how these systems may have formed and functioned in the past, the degree to which they have been changed by human activities, and the potential for restoration of degraded systems.

Vegetation. Although there is a great deal of information on the ecological roles of riparian vegetation (Brinson et al., 1981; Salo and Cundy, 1987; Williams et al., 1997; Koehler and Thomas, 2000; Verry et al., 2000; Wigington and Beschta, 2000), there is limited information in the scientific literature on holistically restoring degraded riparian vegetation. Many restoration efforts are simply agronomic projects that consist of planting selected species. Projects that ignore fundamental changes in hydrologic disturbance regimes or other factors that have altered site conditions are unlikely to lead to ecological improvements. Understanding not only the functions that specific species and groups of species perform, but also the hydrologic and edaphic requirements for their successful establishment and growth (e.g., Kovalchik, 1987; Law et al., 2000), is fundamental to any restoration project targeting riparian plant communities. In addition, the underlying causes of vegetation loss must be addressed. For example, attempts to restore native shrubs and other woody species in riparian areas are not likely to be successful if the natural hydrology is not restored or the area continues to experience heavy browsing pressure from domestic or wild ungulates.

Large-Scale Frameworks

Restoration ecologists have become increasingly aware of the need to conduct restoration activities within a context larger than the restoration project itself. This emphasis is particularly vital for riparian areas because they are so well integrated into the landscape by connecting uplands with aquatic ecosystems and creating corridors between high- and low-order streams. Factors that occur beyond the boundary of the site that are relevant to site-specific restoration include the nature and intensity of human activities in the watershed (Kershner, 1997), the potential for biological invasions, and the number of ecosystem types and landforms, to name a few (Aronson and LeFloc'h, 1996).

Two methods are available for organizing large-scale information—Hydrogeologic Equivalence and the Synoptic Approach—that deal explicitly with these landscape-level properties of wetlands and riparian areas. Both assessments were developed from the recognition that the condition and functioning of wetlands and riparian areas are in many cases driven by conditions upslope or upstream. Neither approach is an assessment of specific sites, but rather is a landscape-scale assessment that would include riparian areas. Both approaches are particularly relevant for riparian management because they force a big-picture view that is helpful for planning, prioritizing, and funding restoration efforts. Consequently, they should be conducted prior to implementing smaller-scale assessments.

Hydrogeologic Equivalence

As discussed in Chapter 2, riparian areas vary in their primary source of water, underlying lithology, and inundation frequency. For example, those associated with ephemeral streams receive principally overland flow from adjacent uplands, while those associated with higher-order perennial streams also receive substantial water from groundwater discharge and overbank flow from upstream sources. Given the substantial variation that occurs within a watershed, Bedford (1996) introduced the concept of Hydrogeologic Equivalence [adapted from Winter and Woo (1990) and Winter (1992)]. The approach provides a framework for evaluating the distribution of wetlands at the landscape scale and for gaining insight into how landscape properties control hydrology and water chemistry. The same approach can be used for riparian areas. An assumption of maintaining Hydrogeologic Equivalence is that sources and flow paths of water determine the geographic distribution of riparian areas and wetlands at large scales—an assumption that appears to be well grounded in science (Winter and Woo, 1990). It is a logical extension of the reference concept, but applied at landscape scales rather than to individual sites.

The Hydrogeologic Equivalence concept recognizes that landscapes have developed and maintained different frequency distributions of wetlands and riparian areas with particular "hydrogeologic settings." Such settings may represent desirable endpoints for ecological restoration at a watershed scale. These hydrogeologic settings reflect not only regional climate, but also surface relief, slope, hydrologic properties of soil, and underlying stratigraphy. Information available at landscape scales is increasingly accessible in the form of topographic maps from which the position of riparian areas can be calculated (e.g., headwaters vs. valley bottoms) and aerial photographs from which surface connections can be identified (Bedford, 1999). Once the hydrogeologic settings of a landscape in its pristine condition are determined, the biological and functional attributes of riparian areas can be inferred from the diversity of hydrogeologic patterns.

The approach provides a basis for determining the large-scale changes in wetland or riparian distribution over time and a template for evaluating mitigation strategies designed to replace wetlands and riparian areas that have been lost. It evaluates whether the restored system will be hydrogeologically equivalent to the original, relatively unaltered system. A study that applied the approach in an urbanizing area found that riverine wetlands (i.e., riparian areas) being lost through various modifications were being replaced through compensatory mitigation by small, isolated, deep depressions (Gwinn et al., 1999). This shift in the distribution of wetland types (see Figure 5-5) was not achieving the goals of Hydrogeologic Equivalence.

From a practical perspective, riparian restoration efforts would be most successful if they use these hydrogeologic settings as guides for deciding what kinds of restoration should occur and where on the landscape they should take place. Still uncertain is how progressive changes in land use constrain the capacity of a watershed to maintain the original distribution of hydrogeologic types.

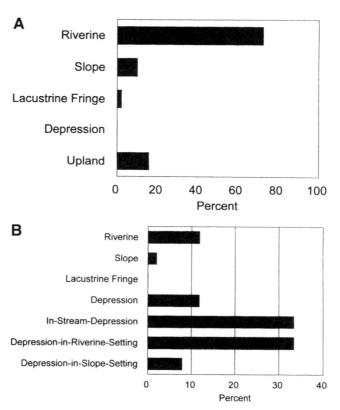

FIGURE 5-5 Comparison of wetland types before (A) and after (B) mitigation construction activities in the urbanizing region of Portland, Oregon, between 1981/1982 and 1993. Riverine wetlands show a reduction from (A) greater than 70 percent to (B) less than 15 percent of the total number of sites (45) in the inventory. SOURCE: Reprinted, with permission, from Gwinn et al. (1999). © 1999 by Dr. Douglas A. Wilcox, Editor-in-Chief, Wetlands.

Synoptic Approach

Originally developed for wetlands, the Synoptic Approach deals with cumulative impacts (Liebowitz et al., 1992). It is intended to provide resource managers with a landscape perspective on wetlands (i.e., the relationship of wetlands and riparian areas to other land forms at regional or statewide scales). The three major steps in a synoptic assessment are the following:

1. Define goals and criteria, prioritize restoration sites, and determine condition by class of wetlands or riparian area.

2. Define synoptic indices and select landscape indicators. Synoptic indices are factors that provide information on the condition of wetlands. Some common categories are nonpoint sources of pollution, stream flow modification, land use, and the condition of habitat. Landscape indicators—the actual measures that estimate the synoptic indices—are then used to estimate how much wetland or riparian types in an area might be affected, including their functions and values, and the significance of altered conditions. For example, an index based on the degree of hydrologic integrity could use as landscape indicators the ratio of waters in natural condition to modified waters. The assumption is that the lower the index (i.e., the more waters have been modified), the greater the amount of effort necessary for restoring waters in the region of analysis.

3. Conduct the assessment and report the results. The assessment team determines how the indices can be combined in a way that best relates to the impacts, both direct and cumulative. Decisions must be made on whether to combine indices through summation, weighting, or multiplication. The assessment itself consists of measurements such as land use (normally done with Geographic Information Systems), analysis of data, and the production of maps to display information and relationships. Accuracy assessments and peer reviews are also components of the program.

This approach is relevant to riparian areas, as evidenced by case studies in Liebowitz et al. (1992) for the Pearl River Basin in Mississippi and Louisiana and for mostly riverine wetlands in Illinois.

As with the Hydrogeologic Equivalence approach, the Synoptic Approach is more a perspective or framework for decision-making than a detailed analysis of specific functions. It can take advantage of principles and data from other geographic regions. A major advantage of the approach is that it is relatively inexpensive and rapid so that management strategies can be developed on a larger scale than is normally used. A number of data sources are available at reasonable costs or at only the cost of acquisition. They may consist of stream discharge and water-quality data, soil surveys, human population trends, and land use–land cover data. Care must be taken in applying the same technique to different geographic areas and comparing them, because the quality of the underlying data may differ.

Additional data sets are likely to become available, at little or no direct cost to users, that can be used for conducting these assessments, thus potentially improving their usefulness. The scale at which they are conducted means that results will provide broad categories of information rather than information regarding the condition of individual sites (Figure 5-6). For example, by conducting an assessment at a national scale, regional priorities for restoration could be

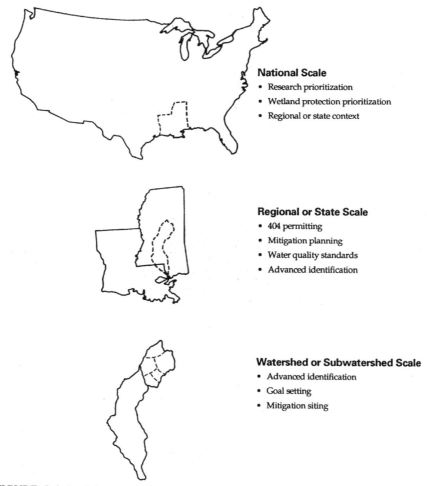

FIGURE 5-6 Applying synoptic assessments at various spatial scales. SOURCE: Liebowitz et al. (1992).

identified and prioritized for funding. At subwatershed scales, priorities could be set for locating restoration projects along specific stream reaches that are in the greatest need of mitigation.

Functional Assessments

As discussed above, the assessment of riparian areas at the scale of individual sites can be done either with functional methods that estimate probabilities that a riparian function exists or with reference-based methods that estimate ecosystem condition. Two functional methods developed in large part for aquatic and wetland ecosystems are the widely used Wetland Evaluation Technique and the Habitat Evaluation Procedure. We discuss both briefly to provide historical context and to acknowledge that some states now using some variation of the Wetland Evaluation Technique may be inclined to adapt it to riparian areas.

Wetland Evaluation Technique

The Wetland Evaluation Technique (WET) is based on the premise that wetlands have an array of functions that can be classed in one of three major categories: hydrologic, water quality, and habitat. Within each of these categories, more specific functions can be identified. Hydrologic functions attributable to riverine wetlands (and, by inference, to riparian areas), for example, include floodwater storage, reduction and desynchronization of downstream flood peaks, and reduction of flow velocity; WET identifies 11 such functions. The method evaluates the probability that a function will occur as high, moderate, or low.

WET is notable for its complexity, its attention to scientific literature, and the degree to which it has been adopted and used. The method was developed for use by the Federal Highway Administration (Adamus, 1983). Although a number of other assessments were in existence at the time (reviewed by Larson and Mazzarese, 1994), federal agencies adopted WET as a non-mandatory regulatory tool. Many permit applications for wetland alterations are accompanied by an assessment of alternatives using WET.

In addition to determining the probability of functionality, WET also evaluates whether the function will be performed (e.g., whether there is a sediment source from land disturbance to be trapped) and the social significance (whether there are people who will benefit from the function). Because WET considers an much broader array of wetland functions than do individual assessors, projects evaluated with WET are subject to greater consistency and comprehensiveness than those evaluated by individual users, who have widely divergent backgrounds and perspectives.

Still, shortcomings of the method would likely remain should WET be modified for use in riparian areas. One of the limitations of WET is the assumption that all wetlands are capable (to some degree) of performing all of the listed functions.

Emphasis is placed on the level or degree of function, not necessarily on whether an unaltered site would normally perform the function at low, but sustainable, levels. As a result, most functional assessments assign high scores (e.g., probabilities of performing the function) to sites that have been modified to perform a certain function, regardless of whether the area would have done so in its unaltered form. If applied to riparian areas, this approach could lead to justification of enhancing specific functions, perhaps with unanticipated or undesired consequences. Extreme examples of enhancing specific functions might be to (1) build reservoirs to reduce downstream flood peaks, (2) convert riparian forest to emergent marsh to encourage waterfowl and wading birds, or (3) divert excessive sediments and other pollutants toward riparian areas to maximize their water-quality functions. In each case, some functions would be sacrificed for the enhancement of others, but the riparian area would depart even further from natural, self-sustaining conditions.

Habitat Evaluation Procedure

The Habitat Evaluation Procedure (HEP) was developed in the early 1970s to evaluate the habitats of aquatic and terrestrial species using certain variables (FWS, 1980). These variables are combined into Habitat Suitability Index (HSI) models constructed to evaluate an "element," such as whitetail deer, wood duck, or hardwood cover type (for gray squirrel, piliated woodpecker, etc.). Once the HSI is determined, it can be multiplied by acreage to determine habitat units. The output from the procedure can be used to make recommendations on projects that would result in a change of HSIs. According to a review of the procedure in 1985 (Whitaker et al., 1985), there were 120 published and 75 draft HSI models.

The scaling of an HSI ranges between 0 and 1.0, with 1.0 representing "the condition...that is needed to support the highest numbers of wildlife species on a regional scale over a long time" (Schroeder et al., 1992). HEP focuses on conditions that are optimal for a species rather than using a reference system as a basis of comparison. The concept has been applied more recently to communities, such as bottomland hardwood forests, rather than species-specific habitats. One of the frequent comments about HSI models is that they are time-consuming in both development and application. Moreover, because the models were developed to predict habitat for species populations, application to riparian condition would require some extrapolation. Finally, the method uses literature reviews and best professional judgment, rather than reference sites, as the primary sources of information for model development.

Reference-based Assessments

The most powerful assessment methods are dependent, to some degree, on knowing the background or reference conditions within a region. Reference-based assessment is premised on the assumption that sites unaltered by human activities represent the best benchmarks or standards for comparison. Some of their advantages are (1) everyone uses the same standard of comparison, thus reducing bias, (2) natural variation, rather than some fixed standard, is recognized as an intrinsic property of riparian areas, and (3) relative comparisons (larger, smaller, equal to) are much more repeatable and are faster to obtain than absolute estimates. Three rapid assessment techniques are presented and compared below: Proper Functioning Condition, which was developed explicitly for riparian areas, and the Hydrogeomorphic Approach and the Index of Biological Integrity, which can be adapted for use in riparian areas.

Proper Functioning Condition

The Bureau of Land Management (BLM), in cooperation with other agencies, developed the Proper Functioning Condition (PFC) assessment for riparian areas and wetlands in the American West, where it has been used widely. PFC refers both to the assessment method and to the condition of a riparian site. As a method, PFC qualitatively assesses how well certain site attributes, mainly physical processes, are working, relying heavily on expert judgment. The method is premised on the assumption that if appropriate physical conditions are restored to a site, the restoration of biological components will follow. PFC is also that condition which will support or allow a riparian area to reach its biological potential. In the late 1980s, BLM set as an agency goal the restoration and maintenance of 75 percent of riparian–wetland systems on BLM land to "proper functioning condition" by 1997 (Prichard et al., 1993, 1998).

Ideally, an interdisciplinary team of specialists in vegetation, soils, and hydrology and of biologists with expertise in fish and wildlife conduct a PFC assessment, filling out a checklist of questions about hydrology, vegetation, and erosion and deposition (as shown in Box 5-4). Sites are then placed into one of four categories (Prichard et al., 1998) described below.

Proper Functioning Condition: Riparian areas are functioning properly when vegetation, landform, and channel characteristics are adequate to dissipate stream energy associated with high water flows (i.e., 25- to 30-year events; Wayne Elmore, BLM, personal communication, 2000), thereby reducing erosion; filtering sediment, capturing bedload, and aiding floodplain development; improving flood-water retention and groundwater recharge; developing root masses that stabilize streambanks against cutting action; developing diverse ponding and channel characteristics to provide the habitat and the water depth, duration, and temperature necessary for fish production, waterfowl breeding, and other uses; and supporting greater biodiversity. The functioning condition of riparian–wetland areas is the result of interaction among geology, soil, water, and vegetation.

> **BOX 5-4**
> **PFC Checklist**
>
> To evaluate a riparian site using PFC, an interdisciplinary group must answer "yes," "no," or "not applicable" to the following items. Asterisks refer to items that lack a "not applicable" choice. For any of the items marked as "no," specific comments must be recorded and discussed by the assessment team.
>
> **Hydrology**
> 1. Floodplain above bank-full is inundated in "relatively frequent" events
> 2. Where beaver dams are present they are active and stable
> 3. Sinuosity, width/depth ratio, and gradient are in balance with the landscape setting (i.e., landform, geology, and bioclimatic region)*
> 4. Riparian-wetland area is widening or has achieved potential extent
> 5. Upland watershed is not contributing to riparian-wetland degradation*
>
> **Vegetation**
> 6. There is diverse age-class distribution of riparian-wetland vegetation (for maintenance/recovery)
> 7. There is diverse composition of riparian-wetland vegetation (for maintenance/recovery)
> 8. Species present indicate maintenance of riparian-wetland soil moisture characteristics
> 9. Streambank vegetation is comprised of those plants or plant communities that have root masses capable of withstanding high-streamflow events
> 10. Riparian-wetland plants exhibit high vigor
> 11. Adequate riparian-wetland vegetative cover is present to protect banks and dissipate energy during high flows
> 12. Plant communities are an adequate source of coarse and/or large woody material (for maintenance/recovery)
>
> **Erosion/Deposition**
> 13. Floodplain and channel characteristics (i.e., rocks, overflow channels, coarse and /or large woody material) are adequate to dissipate energy*
> 14. Point bars are revegetating with riparian-wetland vegetation
> 15. Lateral stream movement is associated with natural sinuosity*
> 16. System is vertically stable*
> 17. Stream is in balance with the water and sediment being supplied by the watershed (i.e., no excessive erosion or deposition)*

Functional-At-Risk: This category includes riparian–wetland areas that are in functioning condition, but existing soil, water, and/or vegetation attributes make them susceptible to degradation.

Nonfunctional: Riparian–wetland areas are nonfunctional when they clearly are not providing adequate vegetation, landform, or large wood to dissipate stream energy associated with high flows and thus are not reducing erosion, improving water quality, etc., as listed above. The absence of certain physical attributes, such as a floodplain where one should be, is an indicator of nonfunctioning conditions.

Unknown: This category includes riparian–wetland areas for which BLM lacks sufficient information to make a determination of the functioning of the area.

If the alterations and stressors that make a site "nonfunctional" are corrected, it is assumed that the site will achieve a rating of "functional-at-risk" or higher as vegetation establishment and growth take place over time. The goal of riparian management is always to achieve proper functioning condition, "because any rating below this would not be sustainable" (Prichard et al., 1998). Twelve PFC publications supplement these basic instructions, addressing methods for measuring vegetation, the characterization of channels, recommendations for grazing practices, interpretation of aerial photography, and use of historical photographs, as well as providing a review of the literature (Prichard et al., 1998). Separate guides have been prepared for lentic areas (non-flowing waterbodies, or lakes and ponds). To improve the condition of sites, the guidebooks suggest that "best management practices need to be set in motion" although this is not a requirement of PFC.

The PFC method was developed principally for riparian areas in the 11 contiguous western states where the majority of BLM's land occurs outside of Alaska. There is nothing about the approach that precludes its use in the eastern part of the United States, although a number of modifications would be necessary to tailor it for that region. For example, PFC is designed to assess physical conditions responsible for maintaining ecosystem structure. It has not been established how well the approach works in low-gradient, vegetation-dominated floodplains of humid climates where the biological components of riparian areas have strong feedback on fluvial geomorphology in terms of stabilizing bank erosion, reducing channel migration, and enhancing sedimentation. In addition, the method does not assess riparian areas along ephemeral streams (because of BLM's definition of "riparian"), although it could be modified to do so.

BLM has established a training and development team with the primary responsibility for instructing local professionals on application of the method in an attempt to improve the effectiveness and efficiency of its application on BLM lands. Because of its relative simplicity and ease of use, PFC is an effective tool for communicating with landowners and others with nontechnical backgrounds. As a result, many stakeholder groups have a greater awareness of the importance of riparian restoration. Acceptance may not have been possible with a more rigorous, quantitative method (Wayne Elmore, BLM, personal communication, 2000). Because of the method's popularity, there is an ongoing effort to enhance data sharing among the agencies using PFC, an effort spearheaded by the Information Center for the Environment at the University of California, Davis.

The method is implicitly a reference-based approach because it identifies the condition of a site relative to one that has a proper functioning condition rating.

However, unlike the Hydrogeomorphic Approach and the Index of Biological Integrity (described below), PFC relies on the judgment of a team of local experts rather than on a group of reference sites that captures the range of natural variation within a geographic region.

It should be noted that the PFC rating is not necessarily the same as the capability or the potential of a site. "Capability" is defined by BLM as the highest ecological status, given political, social, and economic constraints (e.g., presence of dams, ongoing watershed land use, etc.). "Potential" is defined as the highest ecological status without those constraints. These ratings are the result of quantitative data collected using BLM's system for inventory and classification, called an Ecological Site Inventory (Leonard et al., 1992). Ecological Site Inventory is not a component of PFC assessment. Rather, Prichard et al. (1993) recommends that the Ecological Site Inventory document (and others) be reviewed.

Although PFC has been widely used by federal agencies in the West and provides immediate feedback to land managers, it nevertheless has several potential limitations. First, because it is qualitative, PFC is vulnerable to subjective application, which places a great burden on the consistency and skill of the local assessment teams. Consequently, it is difficult to compare assessments over several years to assess progress towards the anticipated condition. Second, emphasis is placed almost exclusively on hydrologic and geomorphic features rather than on biological or ecological functioning. Where vegetation is used (i.e., age class, species composition, plant vigor), it is more an indicator of hydrology and geomorphology than of biology. Virtually no direct attention is given to the terrestrial or wetland habitat functions of riparian areas. If the local assessment team is not familiar with sites in good ecological condition and their characteristic plant communities, its capability to assess the status of a stream is likely to be diminished. Finally, because only the site conditions on a PFC rating form are addressed, the spatial context and connectivity of a given riparian area relative to a watershed setting are not explicitly recognized. Thus, though the PFC approach is an efficient methodology for identifying nonfunctional or functional-at-risk riparian areas—an important issue for many riparian areas in the American West—it has limited capability for quantitatively characterizing the relative level or degree of ecological recovery of riparian areas that have attained a "proper functioning condition" rating.

Prichard et al. (1998) acknowledge that PFC is not a replacement for inventory or monitoring protocols designed for plants and animals dependent on riparian–wetland areas, nor is it a replacement for watershed analysis. However, PFC is a useful tool for prioritizing restoration activities that can reduce the frequency of data collection and labor-intensive inventories by concentrating efforts on the most significant problem areas. A number of the shortcomings discussed above could be addressed with a second-generation, up-to-date method. The approach could be strengthened and the assumptions better understood if studies were conducted to independently validate PFC results.

Hydrogeomorphic Approach

In the early 1990s, the U.S. Army Corps of Engineers' (Corps) Waterways Experiment Station sought an alternative functional assessment method to WET, principally to use in the evaluation of impacts to wetlands, the analysis of alternative sites for development projects that alter or replace wetlands, and the design and evaluation of wetland mitigation projects. The hydrogeomorphic (HGM) approach, developed for this purpose, consists of three components: classification by geomorphic setting, development of a reference system as a benchmark of comparison, and identification and assessment of wetland functions (Smith et al., 1995). The HGM approach compares wetland condition relative to a group of sites that have been minimally impacted so that deflection from the natural variation can be estimated. Functions are used to estimate condition, in contrast to approaches that rely on biota alone. Some functions are based on structural characteristics of a wetland while others rely on species composition.

Seven generic wetland classes (riverine, depression, slope, lacustrine fringe, estuarine fringe, organic soil flats, and mineral soil flats) are used to guide local classifications, or subclasses. Classification is a critical component that allows standards to be developed for and applied to a relatively homogeneous group of wetlands within a biogeographic region (Brinson, 1993). For example, floodplains associated with some first- and second-order streams differ in species composition and hydrology from higher-order streams (Rheinhardt et al., 1999). The purpose of classification is to partition the natural variation among subclasses so that variation due to impacts can be more easily detected.

Measurements are made of structural and biotic variables that relate logically to one of three functional categories: hydrology, biogeochemistry, and habitat. For example, variables relating to nutrient cycling (biogeochemistry) include condition of the riparian buffer, whether or not the stream is channelized, the maturity of the floodplain forest, and the presence of detritus. The nutrient cycling function receives a score based on the status of these variables, each of which is indexed relative to unaltered sites. For example, a site with a channelized stream, a missing buffer, immature forest cover in the floodplain, and suboptimal amounts of detritus would receive a score of 0.4 (relative to 1.0) for the nutrient cycling function (Rheinhardt et al., 1999). Other functions are similarly assessed. Functional indices may be multiplied by surface area in order to compare two or more sites that differ in condition and size.

HGM relies on the development of a guidebook that contains a literature review, an identification of functions, data on variables from reference sites within the biogeographic region of interest, and "models" that relate variables to functions (Smith et al., 1995). Interdisciplinary expertise is incorporated into the development of guidebooks through workshops, field-testing, and peer review. Specific training to use the guidebooks, which helps ensure greater consistency in application, is recommended. This is in contrast to the PFC approach, which requires an interdisciplinary team conducting on-site assessments.

Reference sites are explicitly required in the HGM approach. Terms such as reference standards sites and site potential are provided to distinguish different benchmarks of comparison, recognizing, for example, that restoration of degraded riparian areas in urban landscapes can never achieve natural, unaltered conditions. Given that some recovery in such landscapes may be desirable, the term "site potential" provides an attainable endpoint.

In a typical assessment conducted to estimate the effects of an unavoidable impact, the project site is evaluated in its pre-project condition using the appropriate guidebook. Each variable that contributes to a function is compared to the standard derived from a range of variables at unaltered sites. The condition of the site is then predicted based on alterations expected from the project. Differences between functional indices before and after the project are calculated (upper half of Figure 5-7).

In addition to assessing impacts, the HGM approach was developed to estimate improvement in conditions resulting from restoration (lower half of Figure 5-7). Rather than predicting losses based on project impacts, the method estimates gains in condition resulting from restoration activities. The output of an assessment allows comparisons between changes in condition integrated over surface area. Thus, a restoration of a large but moderately degraded site could potentially compensate for a severe impact to a previously unaltered small site. The details of what is acceptable compensatory mitigation, however, are beyond the intent and capability of the HGM approach. It is important that a policy framework be established to provide guidance necessary for deciding how to interpret and use the results of the assessment.

The HGM method has been adapted for application to selected riparian areas (Hauer and Smith, 1998; Wissmar and Beschta, 1998), even though it was initially developed for evaluating the condition of riverine wetlands. Non-wetland portions of riparian areas are assessed by identifying the "flood-prone width" of a floodplain (Rosgen, 1995). At the time of this writing, only one guidebook for riparian areas has been published (Ainslie et al., 2000), although others are in various stages of development.

Strengths of the method include its use of a reference system, the short time required to conduct an assessment, and the consistency among assessors (Whigham et al., 1999). The reference system involves not only the sites representative of the natural and human-induced variations in a geographic region, but also the body of scientific literature applicable to the riparian or wetland type. Much more data are collected from reference sites than are incorporated in the final assessment tool; the final method used in the field is pared down to include only measurements that are sensitive to activities that alter or degrade the functioning of sites. This background work makes it possible to conduct simple assessments in a matter of hours.

HGM considers components of the flow regime such as the magnitude and frequency of discharge, the duration of specific flow conditions, the predictabil-

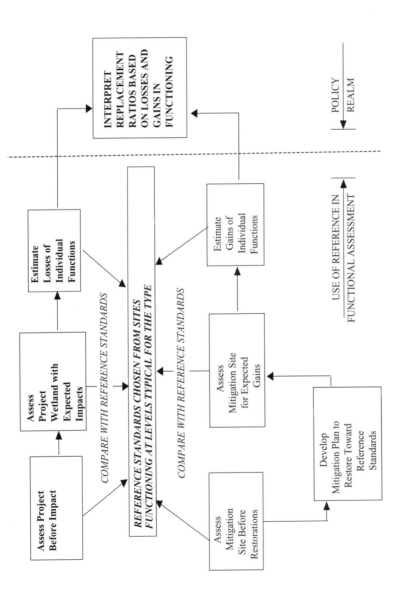

FIGURE 5-7 The HGM approach for assessment, using reference as the basis for comparison of before and after development projects and before and after anticipated restorations. SOURCE: Reprinted, with permission, from Brinson (1996). © 1996 by Environmental Law Institute.

ity of high or low flows, and the rate at which flows change (Poff et al., 1997). Unlike other methods that focus almost entirely on flows within the channel (Peters et al., 1995), HGM can also account for the frequency of overbank flows and their effect on floodplain plant communities. In spite of its "hydrogeomorphic" name, the method takes into account species composition and structure of vegetation as well as functions dealing with animal habitat. Biological variables could be further analyzed using methods developed for biological integrity assessments (see below).

Guidebooks can be updated when new information becomes available. The approach is modular, so changing one portion of a guidebook does not necessarily require revisions in other parts. Assumptions are clearly stated, thus eliminating the "black box" syndrome of WET and PFC's reliance on a team of specialists. An additional benefit of having reference sites is for training and education.

Difficulties surrounding the HGM approach include the lack of high-quality reference sites and the expense of developing the procedure for additional riparian subclasses. Especially in urbanizing areas, streams and their associated riparian areas may already be affected by tree removal, invasive species, gullying caused by changes in runoff patterns, and other alterations. In such generally degraded conditions, there may be a tendency to lower standards for restoration without acknowledging that higher-quality conditions once existed. The development of guidebooks requires a considerable amount of fieldwork and synthesis, an expense that could be considered prohibitive in some riparian management programs.

Indices of Biological Integrity

Biological integrity refers to the ability of a system to support and maintain "a balanced, integrated, adaptive community of organisms having a species composition, diversity, and functional organization comparable to the natural habitat of the region" (Karr and Dudley, 1981). Measures of biological integrity, such as species richness or trophic composition, are used to compare sample site characteristics to that of reference sites or conditions that are minimally influenced by human activities. For example, the Index of Biological Integrity (IBI) (Karr, 1981) explicitly uses reference conditions to ensure that the best available conditions are used as a benchmark for comparison. During the past 20 years, managers have applied a number of biologically based approaches to measure the effects of water pollution and landscape alteration on water quality (Lenat, 1993). These methods principally use aquatic invertebrates or fish communities as the basis for evaluation.

A primary argument for developing these approaches was that reliance on chemical analysis of water failed to account for many of the habitat conditions that are essential to the biological integrity of an aquatic ecosystem (Karr, 1991;

Karr and Chu, 1999). Biota are also viewed as the ultimate integrators of environmental conditions (Karr, 1991).

Although other indices of biological integrity have been developed (see Box 5-5), IBI is the most frequently used and thus is the focus of this section. IBI attempts to integrate the condition of the entire watershed above the point of sampling in a stream, although the actual IBI score reflects a mixture of watershed and site-specific influences (see Roth et al., 1996). A common application of IBI is to compare aquatic insect larvae and other arthropods from streambeds (benthic IBIs) above and below a point source of pollution. In many states, an IBI based on fish is a routine part of water-quality monitoring of streams and provides substantial information for Clean Water Act 305(b) reports. The approach is responsive to larger-scale influences of urbanization, grazing, and agriculture as well as recreation (Karr and Chu, 1999).

Two approaches for analyzing IBI data have been developed: the multimetric procedure (Karr, 1981) and the predictive model method (Hawkins et al., 2000).

BOX 5-5
Floristic Quality Assessment Index

A technique that shares some attributes of biological integrity assessment is the Floristic Quality Assessment Index (FQAI). This index, developed independently of the approaches described above, is based on the method developed for the Chicago region by Wilhelm and Ladd (1988). It was designed to assess the degree of "naturalness" of an area based on the presence of ecologically conservative species. It is thought to reflect the degree of human-caused disturbance to an area by accounting for the presence of non-native and cosmopolitan native species. This index is capable of measuring ecosystem condition because it assigns a repeatable and quantitative value to the plant community.

To calculate the FQAI, a complete species list must be compiled for the site. Each species on the list is then assigned a rating (tolerance values) between 0 and 10 (Andreas and Lichvar, 1995). A rating of 0 is given to opportunistic native invaders and nonnative species. Tolerance values of 1–10 are assigned as follows: values of 1–3 are applied to taxa that are widespread and do not indicate a particular community; values of 4–6 are applied to species that are typical of a successional phase of some native community; values of 7 and 8 are applied to taxa that are typical of stable or "near climax" conditions; and values of 9 and 10 are applied to taxa that exhibit high degrees of fidelity to a narrow set of ecological parameters. The scores are summed to produce an overall score for the community.

Presumably, this approach could be applied to riparian areas with some modification. However, as presently constructed, FQAI is measuring the extent to which a site can support a specialized plant community rather than comparing a site to reference conditions based on minimal human influence. The highest scores, for example, identify unique or rare floristic communities rather than identifying a broader range of community types that are distinguished mainly by having received minimal human influence.

Figure 5-8 shows how the multimetric approach compares the distribution of six invertebrate numeric values for a degraded site and one that is relatively unaltered. The circles represent the position of the attribute (a measurable component of the biological system such as taxa richness) for the unaltered site (E. Fort Cow Creek), and the triangles show the location of the same attributes for the degraded site (Lower Elk Creek), ranging from the relatively unaltered on the right to highly degraded on the left. Attributes are chosen such that they will change (preferably linearly) in value along a gradient of human influence. They should reflect changes in the site from best to highly altered conditions. Thus, taxa are restricted to species compositions sensitive to pollution, but attributes may also include the presence of fish with physical abnormalities. Metrics of 5, 3, and 1 shown at the top of the diagram are assigned to several known relationships between aquatic taxa and environmental conditions. Indices that compile the metrics for several attributes can then be used to rank streams according to their condition. In Figure 5-8, the six metrics shown were combined with five others to yield multimetric scores shown as the benthic IBI. By using these two sites and others that span the range of conditions, criteria can be developed to classify sites as excellent, good, fair, poor, or some other scale that has both biological and regulatory significance. IBIs have also been determined for various metrics such as species richness along a gradient of stream order (Karr and Chu, 1997).

The predictive model approach differs from the multimetric only in the way that data are analyzed, not in the biological components chosen. Predictive models calculate the sum of individual probabilities of finding all taxa of interest relative to a reference site. In addition to sampling the same groups of organisms, additional environmental information (e.g., latitude, elevation, channel slope, alkalinity, etc.) that is likely to be independent of human activity is collected. As a result, it is claimed that predictive models are more effective in regions where streams encounter steep gradients in elevation, temperature, and other factors (Hawkins et al., 2000).

Although IBI was developed for measuring benthic populations in streams, theoretically any biological integrity index can be adapted to riparian areas in one of two ways. One is based on the assumption that riparian area condition and upstream land uses are reflected in a typical analysis of benthic invertebrates or fish. This may be particularly effective in rangelands where grazing by domestic cattle is a principal influence. However, the method may not be able, by itself, to discriminate grazing effects in uplands from the activities of cattle within riparian areas or in the streambed. It should be noted that many riparian areas lack sufficient surface water, seasonally, to use biological assessments based on aquatic macroinvertebrates and fish. Furthermore, the characterization of benthic populations is often an expensive and time-consuming task; thus, results are seldom known immediately. The other approach is to develop metrics for the vegetation and biota of the non-aquatic portion of riparian areas. The specific community of biotic indicators may include soil invertebrates, amphibians, birds, or vegetation.

MANAGEMENT OF RIPARIAN AREAS

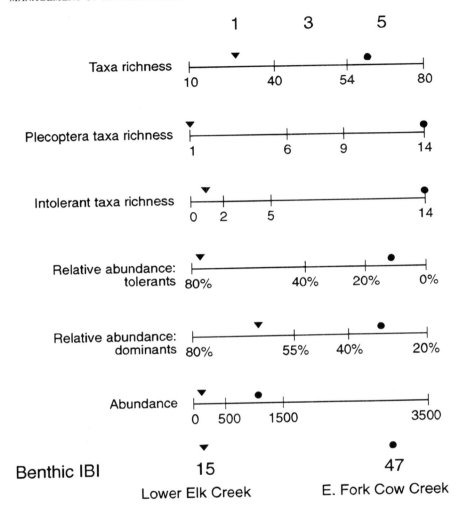

FIGURE 5-8 Benthic IBI scores derived from metrics of invertebrates at a degraded site (Lower Elk Creek) and a relatively unaltered site (E. Fork Cow Creek). Intolerant taxa richness would be the number of species that are sensitive to degradation of water quality in the broadest sense. SOURCES: Reprinted, with permission, from Karr and Chu (1999). © 1999 by Island Press. Reprinted, with permission from Fore et al. (1996). © 1996 by Journal of North American Benthological Society.

In late 1999, EPA led an effort to develop the methodology for biological indicators in wetlands—an approach that should be transferable to riparian areas if appropriate reference conditions and sites are identified.

Vegetation-based biological assessments may be more reliable for riparian areas than animal-based ones because the more persistent vascular plants integrate conditions over longer time periods. Indeed, for wetlands (and presumably for riparian areas) the approach consists of developing biotic indices for an assemblage of wetland plants and combining these with other biological indicators (metrics) into a composite index. Whoever develops and implements the assessments must be skilled in identifying the chosen group of organisms.

Comparison of Methods

The three reference-based assessment methods described above are compared in Table 5-1. All are oriented toward evaluating the condition of ecosystems (or portions of them) by comparing a test or project site with conditions expected in the absence of human activities or in least-disturbed sites. They all use classification as a means of partitioning variation between that due to natural sources (i.e., different wetland classes or different stream orders) and that due to human activities that have degraded the system. Once methods are developed in the form of indices (in the case of biological integrity) or guidebooks (for the HGM approach), their application is relatively rapid, requiring several hours to days depending on the complexity of the assessment. PFC is the most rapid in that the scorecard is filled out in the field and the results are "immediately" known.

PFC does not require a database and thus is highly qualitative and dependent on the knowledge base and judgment of the assessment team. The other two methods are based on data gathered and analyzed for both unaltered and degraded sites, and they use quantitative data to establish the level of variables (for HGM) or indices (for biological assessment) prior to assessor involvement. The collection and analysis of such data require considerable expertise.

For conducting assessments, training is also necessary, but in different ways and to different degrees. For HGM assessments, a science background is expected, and training is oriented toward consistent use of the guidebook. In some cases, taxonomic expertise is required, but usually for a relatively small group of taxa. Field measurements of forest stand structure, percent cover of vegetation, presence of non-native species, types and extent of hydrologic alterations, soil characteristics, and surrounding land uses are typical types of data required for an HGM assessment. For biological integrity assessments, a greater degree of expertise is required for specific taxonomic groups of plants, animals, or both. As in all assessments, experience is a valuable asset in assuring consistency and accuracy. In the PFC approach, the importance of experience and consistency is paramount because of the lack of quantitative information for guidance.

From a practical standpoint, reference-based methods are more easily developed in some areas than others. In landscapes that are highly altered, unaltered reference sites are rare or completely absent. This may require the use of historical records or sites that are geographically separated from the area of interest. In many eastern states, land ownership has fragmented the landscape and made it difficult to gain access to sites for collecting data upon which to build assessment methods. Unlike many western areas where federal ownership is common and access is relatively easy, more populated areas pose significant barriers to access for data collection.

A major impediment to reference-based assessments is the development of standards and metrics for a sufficient number of riparian classes. Both methods that require reference data were developed primarily for wetlands rather than for riparian areas. This is not seen as a barrier because the underlying principles and, to a great extent, the methodologies, would be the same, especially for riparian areas that also contain wetlands. Biological integrity approaches have been in existence about 15 years longer than HGM. Consequently, biological integrity methods have been extensively tested, synthesis volumes have been written on the topic, and a large cadre of technicians has been trained to identify aquatic invertebrates and fishes. This training may not be relevant in riparian systems that lack corresponding aquatic taxa and would have to be developed for the taxa found in riparian areas.

The biological integrity and HGM approaches are not mutually exclusive. In the HGM guidebooks prepared for wetlands, important components are species composition of vegetation, the presence of non-native species, and groups of indicator species sensitive to alteration. For biological integrity methods developed for wetlands, metrics can be developed for hydrophytic plants, much like the Floristic Quality Assessment Index described in Box 5-5. In highly modified landscapes where relatively unaltered conditions are absent and cannot be effectively reconstructed from historical sources, managers are using biological integrity approaches based on an array of plant species that span the range of tolerances to alteration (Lopez and Fennessy, 2002). In either case, methods are directly transferable to riparian areas, especially to riparian classes that are dominated by wetlands.

No single approach will fill all the possible needs for assessing riparian condition. Fortunately, there exist several methods and approaches to choose from to meet the specific needs of restoration, ranging from landscape-level to site-specific, from rapid and qualitative to research-level and model-based, and from those designed to answer ecological questions to those oriented toward socioeconomic issues. However, two features of routine, rapid assessments are essential: (1) availability of information at a scale as large as or larger than the management unit and (2) the availability of reference sites. In spite of the development of these rapid methods, there will always be a need for comprehensive, research-based approaches to generate new ideas and to validate indicators used in the rapid approaches.

TABLE 5-1 Comparison of Environmental Assessment Methods that Are Either Used in Riparian–Wetland–Aquatic Areas or Could Be Modified for These Areas

Attributes	Proper Functioning Condition
Primary purpose for developing	Restore and maintain riparian–wetland systems on BLM lands
Primarily applied to which systems	Riparian–wetland ecosystems on BLM lands, mostly in the arid West
Use of reference	Reference is embedded in the professional judgment of the multidisciplinary team
Use of classification to partition natural variation	No explicit approach to classification is provided; assessment team is responsible for recognizing regional conditions for reference
Use of indicators	
Hydrologic	Physical factors are central to evaluating condition
Geomorphic setting and related attributes	Physical factors are central to evaluating condition
Plant and animal species composition	Species composition of vegetation partly indicates whether physical factors are effective
Physical plant community structure	Vigor and type (shrub, herbaceous) of vegetation indicates whether physical factors are supportive of "proper functioning condition"
Level of effort to develop	Training of regional teams has been conducted; work is continuing
Level of effort to conduct assessment	Requires a multidisciplinary team of experts to visit sites
Expertise of assessor	Several qualified individuals from different disciplines (vegetation, soils, hydrology, fish/wildlife)
Potential challenges in modifying for riparian areas	None; was developed specifically for riparian areas and wetlands, but lacks validation

Hydrogeomorphic Approach	Indices of Biological Integrity
Assess impact and provide compensatory mitigation in the Clean Water Act Section 404 program	Provide a biological method to assess overall condition of ecosystem quality
Developed for wetlands, but encompasses associated riparian areas in arid riverine environments	Developed initially for streams to assess watershed condition; recently being modified for wetlands
Must be developed as the basis for conducting assessments; least-altered sites are the benchmark, but altered sites are needed for scaling condition	Uses lack of human activity for reference. Scores estimate departure of altered sites. The range of conditions is used to categorize indices as high, medium, or low.
Classification is fundamental to developing relatively homogeneous standards for comparison	For wetlands, uses the HGM classification approach
Highly dependent on identifying water sources and flows necessary for evaluating functions Essential both for classification and evaluating condition Plants, especially, used to indicate alteration; non-native species can be used to indicate degraded conditions Essential for evaluating condition	Seldom used; instead, biotic indicators are central to method Not required except for the classification step Biological data are central to the assessment, especially aquatic animals; vegetation indicators are undergoing development Not required
Substantial effort required for field data collection, establishing reference standards, and developing guidebook	Substantial effort required for field sample collection, analysis, and development of multimetrics and predictive models
Requires office preparation (maps, photos, etc.), field visit to collect data, and simple arithmetic computations	Requires field collection and processing of samples
Science background. Although the approach is simplified for the non-specialist, training is required to improve consistency among users	Taxonomic expertise needed for relevant biota (e.g., aquatic macroinvertebrates, fish, native and alien plant species, etc.)
Some; already encompasses flood-prone width for streams; needs little modification for riparian areas that contain wetlands	Need to adapt biological assessments to terrestrial areas

continues

TABLE 5-1 Continued

Attributes	Proper Functioning Condition
Major strengths	Tailored specifically to arid riparian sites; easy for nonprofessionals to understand; a multidisciplinary team of experts must be present at the site
Major weaknesses	Relies on best professional judgment of assessors to establish reference conditions; ignores habitat values
Situations to which the procedure applies	Western riparian situations where a group of specialists can determine whether mostly physical conditions are available to sustain, improve, or further degrade a riparian area

Policy Considerations

None of the approaches described in the preceding section addresses policy interpretations of the results (Kusler and Niering, 1998). Although providing such an analysis would go beyond the scope of this report, it is important to understand the limitations of these assessments when their results are turned over to decision-makers. First, there is a substantial need for policy to bridge the gap between outputs from science-based assessments to management decisions affecting riparian resources. As with the choice of assessment methods for riparian conditions, the policy framework should be scaled to the types of questions being asked. For example, reference-based assessments tend to assign low scores to urban riparian areas in degraded conditions. From this standpoint, one could argue that less overall loss occurs when urban riparian areas are converted to other uses than when rural ones are converted. This could lead, in the absence of a policy framework, to less protection of rare but highly visible urban riparian areas compared to rural areas. In general, concepts of rarity, uniqueness, location (with respect to human uses), and educational value are not taken into account by site-specific assessments of condition. Screening methods such as Hydrogeologic Equivalence and the Synoptic Approach may identify these attributes, if they are calibrated to do so.

There are assessment methods that fall between the science-based assessments described above and the policy realm. "Multicriteria" methods provide ways to optimize environmental management decisions that go beyond assessment of ecological condition. These tend to be project-by-project approaches for siting landfills, highways, etc., and are not specific to riparian areas; Europeans are leaders in this area (Voogd, 1983; Janssen, 1994). Multicriteria approaches

Hydrogeomorphic Approach	Indices of Biological Integrity
Allows detailed restoration planning because of information from the reference system; a simple assessment is rapid	The basic IBI approach has been around for 20 years, allowing many of the problems to be resolved; background literature is available
Has a high front-end cost for developing regional guidebooks	Has been incompletely developed in wetlands where vegetation is the principal biotic variable
Once a reference system is developed for a riparian class, assessment determines current condition of a site, and can project losses or gains due to alterations or restorations, respectively	Once multimetric or predictive models are developed for a riparian class, assessment places the site within a continuum of alteration by human activities

address the need for making decisions that are supported by the public. To better integrate public participation in decision-making on riparian resources, education is required about the condition of riparian areas and implications for social well-being (i.e., "quality of life," internalizing environmental costs, etc.). Unless riparian areas are broadly recognized as contributing to social well-being and policies are put in place to maintain their functions, it is unlikely that decision-makers will have either the tools or the political will to effectively maintain and enhance the functions and values of riparian areas.

Finally, there are many impediments to integrating ecological research into economic development. Foremost are the reluctance of ecologists to recognize that global economies drive environmental decision-making and the conflicting assumptions that exist between ecology and economics (Di Castri, 2000)[1]. The application of integrated approaches to riparian areas is particularly complicated because streams often cross geopolitical and ownership boundaries and interface with vastly distinct land uses. Nonetheless, some of the progress made in valuing wetlands is applicable to riparian areas. For example, scale and landscape setting are attributes that influence the value of both wetlands and riparian areas. Other ideas gleaned from wetlands that might be applicable to riparian areas are that (1) values to individual stakeholders change depending on proximity to the resource, (2) as wetlands become rarer, they become more valuable at the same time that they undergo progressive degradation from human activity, and (3) from strictly economic perspectives, more valuable ecosystems should replace less valuable ones (Mitsch and Gosselink, 2000). Beyond that, assigning monetary values to

---

[1]For example, a tenet of economics is that all resources are "replaceable" or capable of substitution. Ecology does not support this, particularly for species that become extinct.

ecological services of riparian areas can never fully account for true value, and should not be the principal or sole approach to biodiversity and ecosystem services (Gatto and de Leo, 2000). Nevertheless, the assignment of production functions[2] to environmental services of ecosystems can contribute information that may lead to better decisions on the largest net benefits to society (Acharya, 2000). None of the assessment tools described above bridges these science–policy gaps, but they may provide information for economic analyses.

Conclusions and Recommendations

Currently, there are no nationally recognized protocols for assessing the ecological condition of riparian areas. A range of approaches is needed to satisfy assessments at scales ranging from watersheds to individual sites, and from rapid assessment to research-driven analyses and model building. Because riparian areas are the main connectors of landscapes at watershed scales, assessments should begin with large-scale analyses that evaluate the variety, location, and connectivity of riparian areas. Landscape-scale methods that utilize widely available data are useful in determining large-scale restoration needs, protection strategies, and goals for other forms of management. Site-specific assessments have undergone considerable evolution and development over the past 20 years. There are opportunities to tailor many existing wetlands assessment methods to riparian areas.

Tools for riparian management range from assessment approaches that rely on simplistic measures to full watershed analysis, research, and modeling of ecosystem structure and function. Rapid assessments are useful as screening tools to help make decisions where immediate action is necessary, while a more robust science-based approach is required to establish longer-term management actions (including reference-based monitoring and adaptive evaluation of the effectiveness of restoration activities). In either case, the full range of stakeholders must be included in the process to determine desired riparian condition (e.g., ecological restoration), based upon unambiguous articulation of available knowledge of riparian structure and function.

The concepts underlying most assessment tools currently used for wetlands and aquatic ecosystems are transferable to riparian areas, suggesting that these tools can be modified to assess the condition of riparian areas. In

---

[2]A production function is a mathematical description of the relationship between specified inputs, in the present case riparian ecosystem services (water quality, habitat, etc.) and outputs from the production process (cleaner water, fish and wildlife, etc.). Production functions are a useful way to value environmental functions, but data are lacking to make the link to the full range of goods that an ecosystem can and does provide (Acharya, 2000).

some cases, this would require an expansion from the aquatic portion to the floodplain and terrestrial parts of riparian areas. In cases where wetlands are major components of the riparian areas, modifications would be minimal. The extent of riparian area must first be identified so that assessment includes riparian areas that are highly degraded and not easy to recognize.

Proper Functioning Condition, the only method developed specifically for riparian areas, provides a good first-generation framework for riparian assessment. This method can be rapidly applied and may have its greatest utility in quickly identifying riparian areas that have been significantly degraded (i.e., nonfunctional and functional-at-risk). However, there is currently no assessment of biological components because the assessment is built principally by evaluating physical factors. The current version should be refined to increase its capacity to link physical conditions with water quality, instream biota, plant community structure and composition, and terrestrial animal communities. The approach should become more quantitative and rely more on regional reference sites rather than on the exclusive judgment of a team of local experts. Information on reference sites, a component that is currently lacking, would contribute to validation of the method.

The Hydrogeomorphic Approach holds considerable potential for assessing the condition of riparian areas. HGM is a reference-based system, originally developed for wetlands, that provides data useful not only for the assessment of condition, but also for the overall design of regional or watershed-scale restoration efforts. The current HGM methodology should be revised, as needed, for direct use in riparian areas across various geographic regions. This will require the development of guidebooks specifically for ecological restoration of riparian areas.

Biological integrity assessments, which have not yet been used in riparian areas, should be evaluated for their ability to encompass riparian community types. Until recently, most biological assessments have been limited to aquatic ecosystems. However, biological assessment of aquatic systems and riparian plant communities, for example, can be used independently or as a component of the hydrogeomorphic approach. Such assessments are needed to independently validate the biological portion of HGM assessments.

All the methods described above should be expanded to include more types of riparian areas, tested by users, and independently validated with appropriate research. Independent testing and evaluation is a critical need of all assessment methods. This is important to ensure their accuracy, usability, and, perhaps most importantly, their credibility for use across the diverse suite of riparian areas that occur in any given region.

Regardless of the science-based tools used to assess riparian condition, the output from these evaluations must be implemented through policies that take into account both environmental and socioeconomic issues. Decision-makers are usually faced with a number of competing values, of which goods and services derived from riparian areas are only one of many. Unless the output from riparian assessment tools can be placed into this broader context, there is little likelihood that the resulting information will be helpful. Consequently, users of the information should be determined before the assessments are interpreted.

MANAGEMENT STRATEGIES

This section discusses management strategies for restoring the hydrologic regime, geomorphic structure, and vegetation characteristic of riparian areas. Although the scientific basis for restoring riparian areas has rapidly expanded in the last couple of decades, the implementation of restoration practices is in its infancy. In addition, there is much to be learned socially and institutionally about opportunities and limitations for improving these important systems. Nonetheless, some generalities can be made about the appropriateness and effectiveness of certain management strategies. Perhaps most important is that the range of possible restoration activities is broad, from simple activities at a single site to large-scale projects. In many cases, relatively easy things can be done to improve the condition of riparian areas, such as planting vegetation, removing small flood-control structures, or reducing or removing a stressor such as grazing or forestry. Where the objective of restoration is to improve the entire river system, more holistic watershed approaches will be necessary, and management strategies such as removing impediments (e.g., large dams) to the natural hydrologic regime may be required. There are few examples of these larger-scale activities having been conducted with the expressed purpose of restoring riparian areas.

The discussion of management approaches that follows is not meant to be exhaustive, but rather illustrative of how significant ecological improvement of riparian systems might be attained. Some of these strategies will be more passive, some more active, and others a blend of both passive and active approaches. In all examples, successful restoration appears to be based on extensive local knowledge of hydrology and ecology including the range of natural variability, disturbance regimes, soils and landforms, and vegetation; on understanding the history of resource development; and on identifying reference sites. Because restoration is not a deterministic process for which the outcome can necessarily be predicted with high temporal or spatial resolution, it might appropriately be considered a journey involving riparian systems and societal goals, with both evolving over time.

Reestablishing Hydrologic Regime and Geomorphic Structure

Previous chapters have discussed how hydrologic disturbances and sediment dynamics play important roles in maintaining the function of riparian areas. Dams constructed to control floods or generate hydropower, locks, low-water diversion channels, and off-stream storage ponds have precipitated fundamental changes in the flow regimes and the geomorphic character and sediment dynamics of rivers. Many water resources developments have caused concurrent degradation in water quality and fish and wildlife habitats as well as dwindling amounts of riparian areas along lake shores, streambanks, and floodplains. In the opinion of this committee, *repairing the hydrology of the system is the most important element of riparian restoration.* If the flow regime—in terms of magnitude, frequency, timing of peak flows, and other features—is not sufficient to meet the needs of the ecosystem, riparian restoration will ultimately fail (see Poff et al., 1997).

It is important for both stream and riparian restoration for managers to understand the limitations of structural changes in the absence of flow regime changes. For example, typical channel restoration projects meant to improve the complexity of instream fish habitat often involve the addition to channels of gravel or structures with the objective of reversing degradation. However, these projects often fail to consider the importance of hydrologic disturbance, such as the need for peak flows to flush fine sediment from gravels for continued spawning success of salmon (Lisle, 1989; Kondolf and Wolman, 1993). Geomorphic restoration alone cannot bring about the complexity that would result from a fully functioning river corridor with free-flowing exchange of sediment and wood between the channel and riparian area.

Another example is the attempt to increase instream habitat complexity by adding anchored structures on the banks or beds of channels. Even when the added materials are natural in character, anchored structures do not allow the types of adjustments that occur in a system that naturally changes shape in response to floods. Expensive reengineering of a meandering channel with large wood fixed in place has a high probability of substantial reworking during subsequent periods of high flow. Long-term restoration of instream habitat is unlikely to be achieved without full consideration of the flow disturbances on the dynamics of such structures.

Historically, restoration involving changes in flow regime such as dam reregulation has usually targeted fish populations and has not considered riparian objectives. The fact that few hydrologic regimes have been restored expressly for riparian purposes reveals the relative newness of this concept. In addition, most of the available examples of dam reregulation are found on larger rivers, perhaps because there has been greater social and political impetus to makes changes in these systems. For larger rivers, reregulation of dam operations may be one of the only restoration options available because of the limitations imposed by permanent structural modifications. Nonetheless, recreating a more natural hydrologic regime is fundamental to the ecological restoration of rivers of all sizes.

This section discusses how long-term restoration of riparian areas might be achieved if components of the natural hydrologic regime, sediment dynamics, and hydrologic connectivity between rivers and their floodplains were reestablished. Restoring more natural flooding regimes and sediment dynamics in the nation's rivers is likely to be a major theme for environmental science in the twenty-first century (Poff et al., 1997).

Operation, Modification, or Removal of Dams

The vast majority of dams and reservoirs in the United States were completed well before concerns about riparian areas became widely evident. As a result, reservoir release patterns often have done little to support functioning of downstream riparian systems (e.g., Rood and Mahoney, 1990). Although the mandated purposes of a particular dam (e.g., irrigation withdrawals, hydropower generation, flood control) may legally constrain its management, opportunities usually exist to change the storage and release of water to help maintain floodplains and related riparian areas. Restoring the natural flow regime should focus on reestablishing the magnitude, frequency, and duration of peak flows needed to reconnect and periodically reconfigure channel and floodplain habitats (Stanford et al., 1996). In addition, baseflows should be stabilized to revitalize food webs in shallow-water habitats. Another important goal is to reconstitute seasonal temperature patterns (e.g., by construction of depth-selective withdrawal systems on storage dams). These mitigative actions do not reconstitute pristine, pre-alteration conditions but rather normative conditions. Such normative conditions can have measurable benefits to specific attributes such as fish habitat or riparian vegetation if sustained by careful, science-based management of reservoir storage and releases (Stanford et al., 1996).

As increasing numbers of privately owned hydroelectric dams in the United States undergo relicensing by the Federal Energy Regulatory Commission (FERC), regulators should consider modifying flow-release policies to help avoid or mitigate adverse impacts to downstream riparian areas (i.e., to create normative habitat conditions). The Corps and the U.S. Bureau of Reclamation (USBR), in continuing a policy shift toward providing multipurpose benefits in addition to their traditional focus on flood control and irrigation (Whittaker et al., 1993), should consider the maintenance of downstream riparian systems when setting operational policy for federal dams. Box 5-6 considers some of the recent trends in federal dam operations, including the Yakima River in Washington where dam reregulation is being implemented at the watershed scale.

Changes to dam operations have most commonly been motivated by the flow needs of downstream fisheries. In the Pacific Northwest, the proposed removal of the Elwha and Glines Canyon Dams along the northern portion of Olympic

National Park and of four major Snake River dams has been a major topic of discussion because of their detrimental impacts to native salmon runs. A recently completed restoration activity involved removing a dam on the Kennebec River in Maine for fisheries purposes. The Edwards Dam, built on the Kennebec River in 1837 to facilitate upstream navigation and to generate hydropower for sawmills, was removed in 1999 after a protracted legal battle. The actions are expected to restore seven species of migratory fish to the Kennebec River, a goal that required removal of the dam because at least four of the affected fish species were not known to utilize fish ladders of any kind.

Many fewer cases of dam re-regulation involve riparian objectives. Fortunately, research is now demonstrating the essential functions performed by periodic flooding in shaping river channels, building floodplains, creating backwater sloughs, and supporting riparian vegetation; as a result, dam operations are changing in some locations to allow at least some controlled flooding. Prescribed flooding has the potential to become a management tool similar to the use of prescribed burns in managing forests and grasslands. Given the current level of water resources infrastructure, dammed rivers will probably never have flow releases that fully replicate pre-dam flow regimes, and upstream portions of dammed rivers may never be restored. However, in many areas there may be major opportunities for altering flow release patterns so that they are increasingly "friendly" to the hydrologic needs of downstream riparian areas.

An example is the operation of the Corps' dam on the Bill Williams River, which originates in the highlands of western Arizona and flows generally west to its junction with the Colorado River at Lake Havasu. In 1968, the Corps completed construction of Alamo Dam on the Bill Williams River 39 miles upstream from the Colorado. Operation of the dam—primarily for flood control—sharply reduced peak flows, increased flows in some periods, and completely cut off flows in others. Studies in the early 1990s documented the degradation of the riparian habitat along the Bill Williams and attributed that degradation to the change of flow regime resulting from operation of Alamo Dam. In 1996, Congress specifically directed the Corps to modify operation of Alamo Dam. The resulting feasibility report recommended "pulse" releases in the spring and fall, combined with a flooding "event" at least once every 5–10 years, together with base flows varying from 10 to 50 cubic feet per second (USACE, 1998). The Corps based its recommendation on studies demonstrating the effectiveness of different surface and groundwater regimes for supporting native riparian vegetation (Shafroth et al., 1998, 2000).

The cottonwood reestablishment on the Rio Grande from 1993 to 2001 is another example of where simulated flooding has led to the regeneration of riparian vegetation (Crawford et al., 1996; M. C. Molles, University of New Mexico, personal communication, 2001). Other studies have tried to determine the portion of the hydrograph that is essential for driving lateral channel migration and successional changes in riparian vegetation, suggesting that a water

> **BOX 5-6**
> **Changing Operations of Federal Dams—the Yakima River Basin Enhancement Project**
>
> Operating regimes for federally managed water-storage facilities are determined in the first instance by the purposes for which the facilities were authorized for construction and operation. Older facilities tend to be single-purpose, such as for irrigation water, navigation, or flood control. Beginning in the 1930s, federal facilities typically were authorized to serve multiple purposes, and by the 1950s, fish and wildlife generally were included as one of the purposes.
>
> Many federal water facilities—particularly those constructed to provide irrigation water—operate in accordance with water rights obtained from states and from contractual agreements with project beneficiaries who, in return, repay a portion of the cost of constructing and operating the facilities. Water rights, based on beneficial uses, establish the manner and amount of project water storage. Contracts govern the manner and amount of water releases to serve these beneficial uses.
>
> In the Flood Control Act of 1970, Congress authorized the Corps to evaluate modifying its operation of existing facilities when it determined doing so was in the public interest. The 1986 Water Resources Development Act provided the Corps authority (in Section 1135) to undertake environmental enhancements associated with existing projects, including restoration of ecological resources and processes of affected hydrologic regimes.
>
> USBR does not have general authority of the kind possessed by the Corps to alter project operations based on environmental considerations. Nevertheless, USBR has modified operations of many projects—as directed by project-specific congressional enactments, to comply with its obligations under federal environmental law (particularly the Endangered Species Act), or to better serve authorized project purposes. For example, in the 1994 Yakima River Basin Water Enhancement Act, Congress directed USBR to operate the federal Yakima Project so as to provide "target" minimum flows at two key diversion points to facilitate movement

storage and use regime that does not interfere with this portion of the hydrograph could be compatible with riparian ecology (Richter and Richter, 2000). In the northern Great Plains, it has become increasingly recognized that high flows in May and June, when cottonwood seed release commonly occurs, are important for successful cottonwood regeneration (e.g., Rood et al., 1999). These flows not only create mineral seedbeds by their scouring action and deposition of sediment, but as the flows recede they also allow the downward root growth of germinated cottonwood seedlings to track falling water tables. If high flows are curtailed too abruptly, root growth cannot keep up with falling water levels, and establishment will not succeed. Thus, establishing a flow regime downstream of reservoirs that mimics some of the high-flow dynamics of the original river system could serve as a major restoration tool and is important for successfully maintaining gallery forests associated with these river systems. Similarly, experimental high-flow

of salmon (MacDonnell, 1996). The Yakima River in central Washington is a tributary of the Columbia River that historically had robust salmon and steelhead runs. Development of the river basin as a major agricultural production area required extensive regulation of flows to provide irrigation. Regulation of the river, coupled with over-harvest, vastly depleted the anadromous stocks and only remnant populations remain. The USBR-run irrigation system currently depletes average flood peaks, but water storage is insufficient to substantially retard major floods and therefore a fair amount of channel–riparian complexity remains. However, the base flows are substantially altered and some reaches below the major diversion canals are nearly dry during late summer. Moreover, construction of an interstate highway in the riparian corridor of the river coupled with urban and agricultural expansion onto the floodplains has mediated extensive gravel mining and channel revetment.

In 1992, the USBR was charged to determine biologically based flows that would allow restoration of the fisheries. It focused on purchase of water rights for instream flows, water conservation through installation of modern irrigation systems, land acquisition to allow flooding of riparian areas, and research to demonstrate relations between flows, riparian succession, and fish habitat for key floodplain reaches of the river system. The restoration plans call for removal of revetments to naturally restore floodplain function. The project is a good example of an ongoing whole-basin riparian restoration project that has coupled basic research with clear management objectives and stakeholder participation. Implementation of restoration activities, such as acquisition of floodplain land and available water rights from willing sellers, has gone on simultaneously with research that attempts to determine the normative flow regime for the river. The project involves private landowners, the Yakima Indian Nation, federal and state lands and management entities, and university researchers. Normative flows in this river system are expected to be implemented within the next few years and a monitoring plan for evaluating success of the project is in place. For more information see www.umt.edu/biology/flbs.

releases from Flaming Gorge Reservoir on the Green River in Wyoming have been evaluated not only for their benefits for endangered species of fish, but also for their effects on downstream riparian areas and their vegetation (Andrews, 1986; Merritt and Cooper, 2000). As discussed in Box 5-7, releases of water from Stampede Reservoir to the Truckee River in the 1990s in support of the spawning needs of an endangered fish have been associated with cottonwood regeneration along riparian lands in the lower portion of the river. Other prominent examples have occurred on the Upper Colorado River and the Napa River.

Recent decisions regarding the management of dams on the upper Missouri River reflect changing attitudes toward incorporating environmental considerations into dam operations (NRC, 2002). In November 2000, the U.S. Fish and Wildlife Service (FWS) released a biological opinion for the revised management of six dams on the Missouri River in eastern Montana and the Dakotas. The

> **BOX 5-7**
> **Return of Cottonwood Forests During Flow**
> **Regime Reestablishment for the Lower Truckee River**
> (Sources: Rood and Gourley, 1996; Gourley, 1997)
>
> For most of the 1900s, the Truckee River from Lake Tahoe in California to Pyramid Lake in Nevada has been heavily regulated. Starting in 1902, dams along the 140-mile length were constructed for irrigation, domestic, and industrial purposes, with the result that roughly half of the flow from the lower Truckee River was diverted. In addition to the flow changes, the Corps widened and deepened the river from Lake Tahoe to about 3,200 feet downstream to protect the city of Reno from floodwaters. These modifications made the river channel more susceptible to erosive forces, as evidenced by damage to the channel and riparian area following floods in 1963 and 1986.
>
> Prior to 1902, the Truckee River and Lake Tahoe had been home to the Lahontan cutthroat trout, while the lower Truckee and Pyramid Lake supported large populations of cui-ui. Because of alterations to the streamflow that curtailed spawning runs, the native Lahontan cutthroat became extinct in the 1940s; the cui-ui is now close to extinction (and is a federally listed endangered species). Other cutthroat populations have been reintroduced into the system and are now listed as "threatened" under the Endangered Species Act. During the same period (first half of twentieth century), cottonwood-dominated forests in the riparian areas along the river declined to a fraction of their historic extent.
>
> To help stimulate spawning and the eventual recovery of the cui-ui, the FWS began managing the Stampede Reservoir in the 1970s by changing release flows and creating artificial fish passages. The flow modifications consisted primarily of increasing outflow from the reservoir from April to June. By 1992, the adult cui-ui population had rebounded significantly from its low point, with over 1 million fish counted. The success of the recovery was also attributed to several very wet years that complemented the flow modifications.
>
> The change in management coincidentally produced conditions favorable for cottonwood germination and sapling survival. Prior to the flow changes, the river level was so low during summer months that cottonwood recruitment was negligible. In years during which outflow from Stampede was increased, and in several naturally wet years, cottonwood recruitment peaked. Survival was found to correlate not only with flow levels, but also with a slow rate of decline in river levels over time—confirming that the timing, frequency, and duration of flow are all important factors in the lifecycle of riparian vegetation. Parallel with the recovery of cottonwoods, songbird populations have also returned along the Lower Truckee. The FWS has promoted future management of the river's dams to support both fish populations and cottonwood recruitment.

opinion calls for operation of the dams to create higher spring flows and lower summer flows compared with flows under past management. The purpose of increased spring flows is to provide reproductive cues for the federally endangered palled sturgeon and other species, and to build sandbars that would be used by nesting terns and plovers during the time of exposure of the bars in summer.

Lower summer flows would also provide shallow water for young fish. Although not explicitly discussed, this proposed change in management might improve cottonwood recruitment in riparian areas. The plan also calls for adaptive management with monitoring, which would allow dam operations to be adjusted in the future as new information becomes available about changing environmental conditions. FWS's opinion about dam management on the upper Missouri would increase slightly the risk of flooding during spring and have some negative impacts on recreational navigation at low flow; it is hoped that these impacts would be offset by positive effects on fishing, canoeing, and camping. The Corps will make its final decision regarding the biological opinion at the conclusion of the national environmental policy process for the Missouri River master manual.

Complications for River Sediment Dynamics. Manipulating sediment dynamics is not usually a principal restoration activity. However, one consequence of river restoration projects may be changes in the sediment transport regimes that could affect interactions between channels and riparian areas. This is especially true when restoring a river's natural flow regime by manipulating dam operations to create variable flows or by removing dams altogether. Either strategy could change rates of sediment transport and deposition in ways that affect the size, morphology, and disturbance frequency of sediment patches that are suitable for establishment of riparian plants (Friedman et al., 1996; Scott et al., 1996).

An example of a river restoration project that affected sediment dynamics was the controlled flood in the Grand Canyon in 1996. In this case, the principal goal was redistributing sediment from the channel to build exposed, sandy platforms such as point bars (locally called "beaches") that are valued as riparian habitat and as campsites for river rafters (Schmidt et al., 1999). Researchers are still debating how floods should be controlled in order to rebuild and maintain beaches and to keep the beaches free of tree seedlings and saplings. However, the effectiveness of controlled floods in building beaches is strongly related to upstream sediment sources in the river (Topping et al., 2000); these sources were significantly reduced by the construction of Glen Canyon Dam. It may be difficult or impossible to maintain beaches over the long term in such sediment-starved systems.

Sediment dynamics is a major issue for the proposed restoration of the Ventura River within the Los Padres National Forest, CA. The Matilija Dam was built in 1947 for flood control and irrigation water supply. The dam blocked steelhead from much of their upriver habitat. The Ventura River steelhead, which is genetically distinct from populations further north in California, was listed as an endangered species in August 1997. Sedimentation behind Matilija Dam has vastly reduced its flood-control capacity; an estimated 6–11 million cubic yards of clay and silt are located behind the dam. Whether the sediment could be released to the channel below the dam in a staged deconstruction, or whether the sediment would have to be disposed of elsewhere, is one of the major uncertainties associated with any proposal for dam removal. A preliminary analysis by the Bureau of Reclamation suggested that the cost of removal could be as much as $180 million depending on the plan for sediment disposal.

Another California dam, Engelbright Dam on the Yuba River, also poses significant sediment problems. The dam has trapped tailings from hydraulic mining, which are likely to contain significant quantities of toxic metals. Any plans that would involve the release of trapped sediments to the downstream river would have to address the fate of those contaminants. Some important variables that could affect the redistribution of contaminants include (1) association of contaminants with the different grain sizes and differential transport of those grain size fractions, (2) interactions between the reestablished flow regime and the newly released sediment, and (3) relative importance of downstream sediment fluxes versus lateral (between channel and floodplain) sediment fluxes. These factors will determine how the morphology of the channel and the surrounding floodplain changes, the extent to which contaminants are redistributed between them, and the time period that contaminants are likely to be stored in new deposits.

Modification or Removal of Levees and Other Flow Containment Structures

There is considerable potential for restoring riparian areas by altering levees, dikes, and other structures designed to impede the movement of water away from a channel. Where new levees are proposed to reduce or prevent flooding of streamside areas, setting them back some distance from the edge of a river or stream would at least ensure the maintenance of near-channel riparian areas. For areas where levees are already in place, gaps in existing levees could be created that would allow inundation of former floodplains and reestablishment of riparian vegetation. If undertaken in conjunction with setback levees, such an approach would continue to protect specific lands or structural developments from flooding at high flows while allowing overbank flows on some portions of former floodplains. Although such an approach is physically feasible, the costs of creating gaps, the economic effects of flooding former floodplains, and the costs of constructing additional setback levees might preclude the implementation of this approach in many areas. However, where the ecological and social values associated with reestablishing periodic flooding (e.g., increased detention storage of flood waters, increased riparian functions, improved habitat and food web support for aquatic organisms and wildlife) are high and the economic costs are low, breaching or creating gaps in levees and dikes may be a relatively straightforward approach to reclaiming significant amounts of historical riparian area. Complete removal of existing levees or replacement with levees further from the river can be expensive, often necessitating not only the movement of fill, but also substantial changes in existing land uses.

In order to promote successful restoration in "high-intensity use" riparian areas, compromises in the planning process are likely, as illustrated by the restoration of the Napa River riparian area in north-central California. The City of Napa's Living River Plan was developed by a coalition of the Corps, state and local governments, resource agencies, businesses, and environmental organizations in response to a problem created by overdevelopment in the river corridor. The area has experienced 27 floods since the 1850s, causing over $540 million in damages since 1960. Initial plans by the Corps called for deepening the river channel, raising the levees, lining levees with concrete, and frequent dredging. In contrast, the Living River Plan includes terraced marshes and broad wetlands in place of the levees, which were moved farther from the river channel. Rather than deepening the river channel, the river corridor was widened by hydrologically reconnecting it with much of its former floodplain; this required removing from the river's banks 16 houses, 25 mobile homes, eight commercial buildings, and 13 warehouses. The $190 million needed for the project came from Congress and a sales tax in Napa County. In winning the support of a majority of county residents, the coalition successfully argued that if the floodplain restoration reduced flood losses, it would save about $20 million a year, even after considering project maintenance costs.

In urban and residential areas, where buildings, parking lots, and other structures fall into disuse, financial incentives and other resources could be used to assist landowners in removing physical features from near-stream environments. If such policies were adopted at regional or national scales, over time many riparian systems would improve as individual structures were removed and replaced by riparian vegetation.

Many streams and rivers in the United States are crossed by city, county, state, and/or federal highway bridges. Although these bridges are typically designed for passing high flows in a relatively unhindered fashion, the highway fills and embankments leading up to many bridges effectively prevent high flows from accessing historical floodplains and they hydrologically disconnect old channels. When these roadways and bridges are scheduled for updating, reconstruction, or other improvements, reconnecting former floodplains and side channels should be considered and implemented wherever possible.

Streambank Stabilization

Bank stabilization, utilizing a variety of structural approaches, has been undertaken along many streams and rivers throughout the United States. In general, these approaches have been implemented to prevent or retard streambank erosion—often with little recognition of the contributing causes. Even when the causes are known (e.g., increased runoff or unstable watershed conditions caused by loss of streambank vegetation due to harvesting of trees, grazing, or conversion to agricultural crops), a structural approach to stabilization has often been implemented. Streambank stabilization is often relatively easy to design and inexpensive, and it provides relatively immediate relief to local landowners. However, it often precludes the slow and incremental channel adjustments that normally occur in streams. For example, natural flow variation causes streambanks to experience locally alternating periods of erosion and deposition. On the outside of meander bends, a stream may slowly erode the bank only to deposit other sediments on the interior bank (point bar). This natural evolution of fluvial systems and the riparian plant communities they support is truncated when streambanks are structurally stabilized.

Although it is unlikely that many streambank stabilization projects in heavily developed areas are likely to be dismantled in order to improve riparian functions, there is a real need to "soften" the impact of these structures. Particularly important would be their modification to allow and encourage the establishment of native riparian plants. To what extent this can be accomplished by plantings, by adding soil into the large pore spaces associated with many rock structures, by partial removal of a structure and replacement with soil that can support riparian vegetation, or by other approaches is largely unknown. Assessing the potential and practicality of revegetating structural streambank stabilization projects represents a pressing research need. Only recently has there been a systematic assessment of the hydraulic effects of riparian vegetation and an evaluation of flow-resistance equations for vegetated channels and floodplains (Fischenich, 1997; Syndi et al., 1998).

Discouraging Development in Floodplains and Meandering Streams

Chapter 2 defined meander belts as broad swaths of land surrounding sinuous channels that incrementally shift to reflect changing hydrologic, sedimentary, or riparian conditions (see Figure 2-3). Chapter 3 discussed how a variety of activities, including levee construction and urban development, can disrupt and sever the hydrologic linkages across meander belts in a variety of ways, impacting the spatial extent and sustainability of many floodplain riparian areas. Bank stabilization projects and other structural alterations in active floodplains also tend to curtail the natural dynamics, large-wood recruitment, and other riparian functions common to meander belts.

Where the general morphology of a meandering stream or river system remains intact—i.e., it is not significantly altered by human development or land uses—prevention of structural alterations that impact how the river and its riparian areas function is fundamental to ensuring the sustainability of these important systems. For example, 100-year floodplains could serve as an initial screening process regarding proposed development projects: developments proposed within the boundaries of the 100-year floodplain would be subject to a higher degree of scrutiny than similar projects outside the 100-year floodplain. In addition, should such a project go forward, a significant effort at mitigating its potential effects upon riparian and aquatic systems would be required. Structures located relatively close to a stream or river should be more closely evaluated regarding their potential to affect river and riparian processes; whenever possible, structural features and developments (including vegetation removal) should be located farther from rather than closer to the river.

A complex example of restoring the hydrologic regime to a river and riparian area is the reengineering of Florida's Kissimmee River (see Box 5-8). The Kissimmee River was once a broad meandering river and was intimately connected with its floodplain. However, in the 1960s the river was channelized, with attendant effects on wetlands, other fish and wildlife habitats, and water quality. The goals of the restoration effort are to recapture many of these ecological and social values. Unlike many other water resources restoration efforts, the restoration plan utilizes multiple strategies to alter the hydrologic regime, including dechannelization, changes in water release, and removal of several canals and water-control structures.

Conclusions and Recommendations on Hydrologic Regime and Geomorphic Structure

Strategies that focus on returning the hydrologic regime to a more natural state have the greatest potential for restoring riparian functioning. Riparian vegetation has evolved with and adapted to the patterns of changing flows associated with stream and river environments. Furthermore, floodplain structure and functioning are dependent upon periodic inundation. Thus, changing dam operations, removing levees, and otherwise re-creating a more natural flow regime and associated sediment dynamics is of fundamental importance for recovering riparian vegetation and the functions that it provides.

Unless changes are made to return the hydrologic regime to a more natural state, other geomorphic and structural restoration activities are likely to fail. The temporal dynamics of the flow regime are fundamental to the structure of riparian plant communities and the functions they perform. Thus, simply attempting geomorphic or structural modifications is unlikely to meet ecological restoration goals.

Dam operations should be modified where possible to help restore downstream riparian areas. Compared to restoring stream flows for fisheries purposes, in very few cases have riparian objectives been the goal of dam re-regulation. There is an increasing need to send greater flows down long segments of rivers to improve riparian plant communities; specific riparian objectives could involve the amount of vegetation recovered and vegetation structure. The effects on downstream riparian areas of manipulating dam discharges should be moni-

> **BOX 5-8**
> **Restoration of the Central Kissimmee River Corridor**
> Sources: Dahm et al. (1995), Toth et al. (1998), Warne et al. (2000)
>
> The Kissimmee River in south central Florida connects several major lake systems by meandering across a broad floodplain called the Kissimmee Valley. Between 1962 and 1971, the central channel of the Kissimmee River was deepened to 30 feet and compartmentalized into a series of five relatively stagnant pools by flood-control structures. As part of the Central and Southern Florida Flood Control Project, the channelization of the Kissimmee had the main purpose to provide an outlet for floodwaters from the upper basin. The project had substantial unanticipated effects, including the loss of 30,000–35,000 acres of wetlands, a reduction in wading bird and waterfowl usage, and a continuing long-term decline in game fish populations. These impacts spawned numerous state and federally mandated scientific studies that culminated in an overall restoration plan that was authorized as a state–federal partnership in the 1992 Water Resources Development Act.
>
> Restoration of the Kissimmee floodplain mainly involves filling 22 out of 56 miles of flood-control canal and removing two of the five water-control structures. Dechannelization will restore the wider, slower flow-way and increase hydrologic exchange with the adjacent floodplain. Another aspect of restoration is a change in the regulation of water releases from upstream lakes, which will reestablish continuous inflows and allow a more natural seasonal pattern of high and low flows in the river.
>
> Some concerns were raised about the restoration project's impact on water supply and navigation. Although restoration of the river's floodplain wetlands is likely to result in greater evapotranspiration losses of water during wet periods, the project is not thought to be likely to affect regional water supply, even during droughts. During extremely dry periods, navigation potential may be impeded through some sections of the restored river; however, it is expected that navigable depths of three feet or more will be maintained at least 90 percent of the time.

tored to help identify those practices that show potential for aiding riparian restoration. In few cases will it be possible to reinstate pre-dam conditions, but it may be possible to create a smaller, more natural stream that mimics many characteristics of the historical one. Impediments to changing dam operations include both legal and socioeconomic factors.

Future structural development on floodplains should occur as far away from streams, rivers, and other waterbodies as possible to help reduce its impacts on riparian areas, and existing human uses of floodplains should be modified where possible to allow periodic flooding of riparian areas. Structural developments typically have significant and persistent effects on the size, character, and function of many riparian areas. Thus, preventing unnecessary structural development in near-stream areas should be a high priority at local,

The cost of the project, to be shared equally by Florida and the federal government, is expected to reach $414 million (1997 dollars). A considerable portion of the state's costs will be in land acquisition, while federal expenditures will be in construction and maintenance. The plan calls for fee acquisition of floodplain lands up to the five-year flood line and acquisition of flowage easements on lands up to the 100-year flood line. Acquiring only flowage easements on most land, and leasing grazing rights on the newly acquired land, will help maintain county tax revenues. Increased use of the river corridor for recreation, including hunting and fishing, will also be possible through the land ownership changes; this is expected to bring significant economic benefits to the local and regional economies.

Before channelization    After channelization

Photos courtesy of the South Florida Water Management District

regional, and national levels. In addition, acquisition of land through conservation easements can be used to retain currently undeveloped land within floodplains in a more natural state. Communities and municipalities can, for example, use the area between a river and its 100-year floodplain boundary to delineate those areas where significant structural development would not be allowed and where existing structures might be removed when opportunities avail themselves.

Vegetation Management

Because of the fundamental importance of vegetation to the ecological functioning of riparian areas, where such vegetation has been degraded or removed, its recovery is a necessary part of any restoration effort. In many instances, recovery of riparian vegetation can be attained simply by discontinuing those land- or water-use practices that caused degradation. For a variety of reasons, however, eliminating these practices can be a major challenge. Decision-makers and landowners often want to continue the resource use or physical alteration that led to the decline of riparian areas in the first place. But attempts to actively restore altered systems without reversing the cause of decline are not likely to achieve functional riparian vegetation.

The power of passive restoration for achieving functioning riparian vegetation cannot be overstated. Throughout many portions of the United States, streamside forests have been harvested for timber production, causing a multitude of effects (see Chapter 3). On these lands, riparian forests may recover if future tree harvesting is simply excluded from riparian areas. This approach assumes that native riparian plants remained as part of the post-harvest vegetation composition and were not replaced by a single forest species or by exotic species. Some functions will recover rapidly, while others may take considerable time. For example, the reestablishment of forest cover along a small stream might be accomplished within a decade or less, yet significant natural recruitment of large wood may be unlikely for 50–100 years or longer (see Figure 5-2).

With regard to historically harvested riparian forests, there may be opportunities to combine passive and active restoration approaches. The protection of riparian vegetation from future harvest would be a passive approach. Active restoration approaches include planting native trees to encourage the more rapid development of late-successional stages through intermediate harvests and augmenting large wood in streams to meet other ecological goals. In all these situations, the long-term goal would be establishment of a self-sustaining riparian forest.

For overgrazed riparian areas, the passive restoration approach is simply to exclude domestic livestock from riparian areas via fencing, herd management, or other approaches. Grazing strategies that alter the traditional season of use, stocking levels, or duration of use may allow recovery in some instances, but careful herd management is usually required on a year-to-year basis. Although grazing strategies other than full exclusion may promote restoration, they are likely to proceed more slowly and run a greater risk of failure.

In riparian areas that support agricultural crops, the reestablishment of native vegetation requires that cropping practices be altered or curtailed. Once this has been accomplished, natural revegetation may occur sufficiently rapidly that additional efforts at replanting may not be needed. In other instances, the reintroduction of specific plants may be needed. In many agricultural areas, the long-term loss of native plants and the widespread occurrence of exotic plants increase the difficulty of accomplishing restoration goals.

Factors such as water availability, flow duration, flood disturbance, channel and floodplain geomorphic change, soil chemistry, fire disturbance, and competition with exotic species directly influence the regeneration of particular species or communities and must be taken into account during restoration. This section focuses on several broad areas of restoration—management of forested riparian buffers, reintroduction of large wood into streams, creation and maintenance of riparian buffers for water quality and habitat protection in agricultural areas, and grazing management.

Forestry in Riparian Areas

The management of forested riparian areas and its impacts on water quality and near-stream habitats have been an increasingly important issue over the last two decades, both in the Pacific Northwest (Salo and Cundy, 1987; Meehan, 1991; Murphy, 1995; Spence et al., 1996) and in other portions of the nation (e.g., Williams et al., 1997; Verry et al., 2000). In 1993, a multiagency task force (FEMAT) issued *Forest Ecosystem Management: An Ecological, Economic, and Social Assessment*, which summarized the scientific underpinnings for the Northwest Forest Plan and significantly changed how riparian areas on federal forest lands in the Northwest and elsewhere in the nation would be managed (FEMAT, 1993). The report highlighted the many roles of riparian areas, noting that the capability of a riparian area to provide a particular function depends on the distance from a waterbody (i.e., the width of the riparian buffer). For example, as shown in Figure 5-9, the effect of a riparian forest upon root strength (and its associated role in maintaining streambank integrity and channel stability) is greatest relatively close to the edge of a channel. In contrast, riparian forests provide large wood and shade to aquatic ecosystems potentially out to one site potential tree height from the channel, although most of the influence typically occurs within half a tree height (McDade et al., 1990; Murphy, 1995). Figure 5-9 also illustrates that where forested riparian buffers are of sufficient width to provide high levels of large wood and shade protection, functions related to bank stability and litter inputs are also generally satisfied.

FEMAT (1993) indicated that riparian areas were an important component of a four-part aquatic conservation strategy aimed at restoring and maintaining the ecological health of watersheds within the national forests of the Pacific Northwest. These components consisted of riparian reserves, key watershed protection, watershed analysis, and watershed restoration. Site-potential tree heights and slope distances were used as the "ecologically appropriate metrics with which to establish riparian reserve widths." For watersheds with high drainage densities, the proportion of a given watershed that fell within the riparian reserve designation could be relatively high (e.g., in excess of 50 percent). The widths of riparian reserves identified in FEMAT (1993) were considered an interim strategy and it was expected that they would be changed upon completion of watershed analyses. However, these dimensions have become widely accepted within USFS and BLM as the widths of no-harvest riparian areas and were incorporated into the Northwest Forest Plan (see Box 4-4).

Because forested riparian areas often comprise a diversity of plant communi-

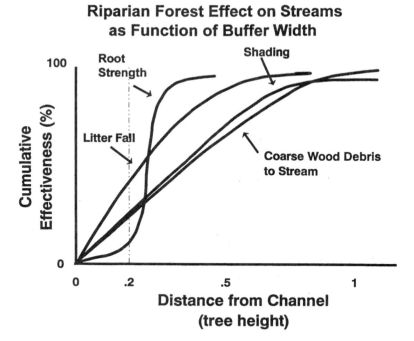

FIGURE 5-9 Generalized relationships indicating percentage of riparian ecological processes and functions occurring within indicated distances from the channel. SOURCE: FEMAT (1993).

ties and provide for a variety of ecosystem processes, functions, and values, it could be argued that specific forest management options are not necessarily well suited for meeting broad ecological goals. Thus, a passive approach to management (no-harvest) may be warranted where broad ecological goals have a high priority and where restoration or sustainability of functioning riparian systems via a no-harvest approach is likely to succeed. However, as indicated by Palik et al. (2000), the designation of riparian reserves or other riparian management areas as no-harvest buffers may preclude other management options and opportunities. These include not only the management of riparian forests for specific species and commercial timber products, but also the enhancement or active restoration of riparian functions. Active restoration may be particularly needed on industrial forestlands and land under other nonfederal ownership where previous management actions have greatly altered the composition and structure of native riparian forests. In such areas, even though a no-harvest option might recover desired ecological functions, the time required may be many decades or longer. (The recruitment of large wood from a previously harvested riparian forest may not occur for many decades; to grow large trees simply requires time.)

A no-harvest option would preclude silvicultural prescriptions that reduce the amount of time needed to develop a late-successional riparian forest, e.g., thinning from below. Similarly excluded would be forest operations that might be used to increase the amount of large wood recruited to a stream via the use of techniques such as directional felling (Garland, 1987).

Given that riparian areas function as ecotones between upland and aquatic ecosystems, Palik et al. (2000) suggests that the degree of harvest or other silvicultural treatments in managed riparian areas should vary with distance from a stream where the protection of aquatic resources is a high priority. According to this paradigm, the initial step for integrating functional objectives into silvicultural practice is to delineate the ecological boundaries of a riparian management area. The second step is to prescribe site-specific silvicultural practices that protect or enhance riparian functions along the riparian ecotone while meeting other management objectives; this prescription would outline the silvicultural system designed for the management area as well as a method for monitoring results over time. "A major purpose of the prescription is to insure that all activities are complementary and based on current knowledge and technology. In other words, the practice of silviculture in riparian forests should anticipate the future and prevent problems rather than respond to problems as they develop" (Palik et al., 2000). A final consideration in developing silvicultural prescriptions is to minimize the cumulative effects of individual activities. Figure 5-10 indicates some types of harvest options that might be considered depending upon ecological and landowner goals.

Once a decision is made to harvest trees within a riparian area, a number of operational considerations can have a major influence on the potential impacts, or lack thereof, associated with timber removal (Garland, 1987; Mattson et al., 2000). For example, ground-based felling, bunching, and yarding systems have the greatest potential for site disturbance and associated degradation of riparian resources, yet they are commonly used in many riparian areas because of their productive and economic advantages over other yarding methods. Thus, selection and use of ground-based logging equipment with specific features, such as wide or dual tires, flexible tracks, or double-axel bogie wheel assemblies, are important for minimizing impacts to soils, to residual vegetation, and to aquatic and riparian habitats and for maintaining riparian functions. Other opportunities for reducing potential onsite impacts might include undertaking yarding operations when soils are dry or frozen and establishing a designated skid trail system (Adams, 1983; Garland, 1983). The directional felling of harvest trees away from a body of water and pulling winch line into the riparian area can reduce or minimize many of the potential adverse effects of ground-based skidding. Where cable yarding systems are used, directional felling away from the stream, intermediate supports, and the use of skyline corridors may reduce or minimize many potential impacts. In other instances, the use of helicopters may be an environmentally sensitive and economically efficient means of yarding trees from ripar-

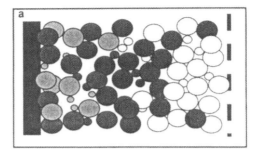
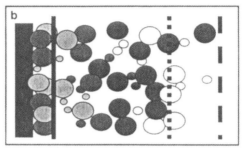
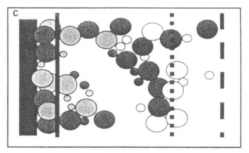

FIGURE 5-10 A riparian forest showing differing harvest patterns. The dark band to the left is a stream or other body of water. (A) The uncut forest showing the riparian management area within the functional ecotone boundary (dashed line on right). (B) The riparian forest is harvested along a gradient. Most of the trees are removed on the right, fewer numbers are removed in the middle such that the residual basal area is dispersed by cutting many small gaps, and an uncut forest remains nearest the stream. (C) Trees are harvested as in (B) except that the residual basal area in the middle part of the riparian area is clumped to open a single large gap. In both (B) and (C), mature stand structure and riparian functions are less impacted by harvest nearest the stream. SOURCE: Reprinted, with permission, Palik et al. (2000). © 2000 by CRC Press, LLC via Copyright Clearance Center.

ian systems. Regardless of which operational approach is taken to remove timber from a riparian area, recognition of the ecological functions and values of these areas should be incorporated into forest practice decisions.

In addition to the riparian areas in national, state, or commercial forest ownership, there are many miles of forested riparian areas (particularly in the eastern United States) that are not managed from a forestry perspective. Many of these riparian systems have experienced significant alteration over time, not only because of the impacts described in Chapter 3 but also simply because landowners may have removed trees for firewood, fence posts, or commercial sale or for aesthetic reasons. Such gallery forests, in various states of disrepair, are in need of the management philosophy embraced in the preceding discussion. The development of management prescriptions that will successfully reestablish various ecological functions over a wide variety of stream and forest types, and which are accepted from both practical and social perspectives, represents an important challenge for individual landowners and the nation.

Conclusion on Forestry

The use of buffer strips is important for maintaining and restoring both aquatic and riparian habitats associated with forest ecosystems. Along with the management of upslope forest practices, functional riparian buffers represent a major component of attaining clean water goals and must be designed to incorporate a strong ecological basis that adequately addresses short- and long-term goals. The dimensions (i.e., width) of riparian buffers and their application throughout a drainage basin (e.g., on intermittent streams, non fish-bearing perennial streams, fish-bearing perennial streams) are likely to vary depending upon ownership and management goals.

Introducing Large Wood

The role of large wood in aquatic ecosystems was first recognized in the early 1970s; thus, management of wood has been a relatively recent subject in riparian ecology and management. Nevertheless, research results have been adopted by many river and forest managers concerned with the protection and restoration of biodiversity, fisheries productivity, and nature conservation. Wood is now deliberately placed in stream channels, and active management of streamside forests is encouraged to supply large wood to the river network. Such practices are particularly popular in the Pacific Northwest, where restoration of stream habitats and riparian areas has sometimes exclusively focused on restoration of large wood. In some cases, management agencies have considered the geomorphic and hydrologic factors that determine appropriate locations, sizes, and amounts of wood before actions were taken. In other cases, agencies have rushed to reintroduce large wood without consideration of restoring the riparian forest over the long term. Wood has also been widely added to streams across the Pacific Northwest that never contained appreciable amounts historically (e.g., meadows). Furthermore, concerns over liability for damage resulting from introduced wood have led many agencies to cable wood in place rather than promote the natural redistribution of wood and channel formation. As a result, assessing the success of restoring large wood has focused on whether the wood remains where it was installed. A more ecologically meaningful standard for success would be to determine whether wood is functioning in the stream system to create the habitat conditions and ecological processes identified as the goal of the restoration effort.

Because most of the pioneering research on large wood has been conducted in certain geographic regions, questions exist about transferring results across differing ecoregions of North America. Models of wood dynamics in streams have been developed over the last decade, with several more comprehensive models of stand dynamics, input, decomposition, and transport being developed currently. Such models offer a broad conceptual framework for transferring results of research on wood dynamics to different regions and forest types.

A final consideration that must accompany the use of wood as a management tool is the societal perception of this strategy. Floods transport and bring into streams large amounts of wood, which can pose threats to bridges and roads. Accumulation of wood in recreational rivers can create problems for boating and concerns about safety. Thus, large wood has traditionally been removed from stream systems, not reintroduced. Especially on private lands, it will be difficult to convince landowners, who are not aware of the ecological role of large wood, that its benefits outweigh the potential damage to, for example, their farming operations. The challenges are equally great for federal and state agencies that must balance multiple objectives, such as restoring biodiversity and reducing the costs of floods. In all situations, the best decision-making will require that large wood be viewed as an agent of restoration and not just as an impediment to flow and a source of damage during floods. As discussed later, programs that both educate the public and help identify locally acceptable compromises (that attain community objectives but alter the dynamics of large wood and riparian areas to the least extent possible) will be important to overcoming these perceptions.

Conclusion on Large Wood

Introducing large wood into streams draining forested watersheds can be an important short-term practice for assisting the recovery of instream habitats, particularly where large wood has been depleted by historical land uses and the remaining riparian forests are unable to provide sufficient large wood for many decades. Even in such situations, the restoration of riparian forests to provide for long-term large wood recruitment must remain a high priority. Where there is no historical or ecological basis for the introduction of large wood (e.g., meadow and prairie systems dominated by graminoids and non-tree woody species), this management practice may cause further degradation to channels and riparian areas.

Buffers for Water-Quality Protection in Agricultural Areas

Buffer zones,[1] both within and upslope from riparian areas, are currently being promoted as management measures for water quality protection throughout the world, particularly in the United States and Europe. As discussed in Chapter 4, major riparian buffer initiatives in the United States include the federal Conservation Buffer Initiative and state and regional restoration programs such as the Conservation Reserve Enhancement Program (CREP) and the Chesapeake Bay Riparian Forest Buffer Initiative. Many of the state CREP programs focus exclusively on riparian area restoration. For example, the CREP goals in Virginia and Maryland are to restore 12,300 and 40,500 hectares (30,500 and 100,000 acres) of riparian habitat, respectively. The 1996 Chesapeake Bay Riparian Forest Buffer Initiative, an agreement between the District of Columbia, Maryland, Pennsylvania, Virginia, and the EPA, seeks to restore 2,010 miles of forested riparian buffers within the Chesapeake Bay watershed by 2010. These are but a few of the hundreds of federal, regional, state, and local buffer programs that are currently restoring riparian areas across the United States.

There are a number of reasons why these programs are so widespread and popular, particularly in agricultural areas. Many of these programs are voluntary, and some compensate landowners and farmers for buffer establishment and maintenance. They generally focus on a narrow band of land along streams, and thus do not affect a large portion of the agriculturally productive landscape. (In many cases, these lands are marginally productive and landowners can make more money enrolling them in buffer programs, while in other instances taking them out of production may not be cost effective.) Furthermore, if properly installed and maintained, buffers can have a high capacity to remove nonpoint source pollutants from upslope activities—as much as 50 percent of the nutrients and pesticides in surface water runoff, 60 percent of certain pathogens, and 75 percent of the sediment load (NRCS, 2000a). Consequently, one might view riparian buffers as a panacea for nonpoint source pollution problems, particularly in agricultural areas, deducing that improved management of upland areas is not necessary for water-quality protection. However, this is not the belief of most conservation professionals, who suggest and/or require that buffers be used as part of a larger conservation management system. This section summarizes the current state of knowledge concerning the effectiveness of buffers for water-quality protection and suggests approaches that can be used to improve their effectiveness.

---

[1]When used as a management tool for water quality protection, riparian areas are referred to as "buffers," "buffer zones," or "riparian buffers" in this report.

Buffer Design. Buffer zones are typically used as best management practices (BMPs) along lower-order streams for enhancement of water quality, protection of fish and wildlife habitat, and possibly production of timber/biomass. To meet such diverse objectives, riparian zones must remove sediment from overland flow, remove and sequester nutrients and other pollutants from overland and shallow subsurface flows, and provide habitat values in the form of streamside shading, generation of coarse and fine particulate matter, and food and cover for wildlife.

Three principal types of buffers are being promoted in the United States for water quality protection: grassed filter strips, the multi-species riparian buffer system, and the three-zone riparian forest buffer. Grassed filter strips are the simplest type of buffer, defined by the NRCS as a strip of herbaceous vegetation situated between cropland, grazing land, or disturbed land (including forest land) and environmentally sensitive areas. Their stated purpose is to reduce sediment and adsorbed and dissolved contaminates in hillslope runoff and to restore, create or enhance herbaceous habitat for wildlife. The filter strip consists of permanent herbaceous vegetation consisting of a single species or a mixture of grasses, legumes and/or forbs adapted to site conditions. The minimum width required for a grassed filter strip is 20 feet (USDA-NRCS, 1999a), but much wider strips—50 to 150 ft—are generally required for participation in the federal CRP and CREP programs (USDA-NRCS, 2000b).

The multi-species riparian buffer system (MSRBS) was developed in the Midwest and is particularly adapted to the American prairie region where trees may not have been a major component of natural riparian areas (Schultz et al., 1995; Iowa State University, 1997). As discussed in detail in Box 5-9, the MSRBS consists of three zones: the first zone of trees to increase bank stability and to produce high value timber, the second zone of shrubs to provide diversity to the ecosystem and help slow flood water, and the third a strip of native warm-season grasses next to the cropland for pollutant removal functions. The MSRBS is flexible in that the tree zone can be replaced by shrubs and/or both the tree and shrub zones can be replaced by grasses. Elimination of the tree zone and expansion of the grass and shrub zones is fairly common in intensively channelized and tile-drained watersheds in the Midwest, partly because some land owners object to trees along the stream that may lead to channel blockage (Schultz et al., 2000). However, the combination of trees, shrubs and grasses helps protect stream quality more than a single species buffer (Iowa State University, 1997).

In the humid eastern portion of the United States, a three-zone riparian forest buffer approach is being promoted to satisfy water quality and limited habitat values in agricultural areas (Lowrance et al., 1985; Inamdar, 1991; Welsh, 1991; Schultz et al., 2000). As described below and in Box 5-10, each of the subzones has specific functional roles. Depending on the management objectives at a site, not all three zones may be required, or the width of less critical zones may be greatly reduced. According to the NRCS, the minimum width for a riparian forest buffer is 100 ft or 30 percent of the floodplain width, whichever is less. However, state CREP and CRP programs may have different (usually wider) requirements.

The *runoff control zone* is located at the upland edge of the riparian zone and has many functions. Because it is composed of densely growing herbaceous vegetation, usually grass, it offers high resistance to shallow overland flow and reduces runoff velocity and sediment transport capacity. It also reduces runoff volume and transport of dissolved pollutants because its vegetative cover promotes infiltration.

This zone should be designed and maintained so that it converts entering concentrated flow into sheet flow in order to improve the effectiveness of the adjacent managed forest zone in trapping pollutants (Dillaha et al., 1989). Under shallow, sheet flow conditions, the runoff control zone will account for most of the sediment trapping in the three-zone buffer. Over time, excessive sediment loading may lead to the formation of sediment deposits and berms, which can hinder further inflow into the buffer and promote concentrated flow. Hence, periodic grading and removal of the accumulated sediment may be required to maintain buffer efficiency. Excess sediment can be moved back upslope to the area it eroded from. Periodic burning, mowing, or harvesting of grass, if permitted, are required to promote vigorous dense growth, to control weeds, and to remove assimilated nutrients.

The runoff control zone is typically a minimum of 20 ft (USDA-NRCS, 2000c) and should be composed of perennial cool season grasses such as brome, orchard grass, fescue, and bermuda grass or warm season grasses such as switch grass (this is region specific). Native species are almost always preferred over non-native species. Buffer grasses should have dense vegetation with stiff, upright stems at ground level. Species that form sods are preferred over bunchgrasses because they provide more uniform coverage and are usually more dense at ground level. Because infiltration is an important pollutant-removal processes, species with deeper roots may also be more effective (USDA-NRCS, 2000a).

The *managed forest zone*, located downslope of the runoff control zone, consists of tree and shrub species. Its main purpose is to remove and sequester dissolved pollutants (especially nutrients) from overland and shallow subsurface flow. Pollutant removal is due mainly to infiltration, plant uptake, and denitrification in the case of nitrate. For the managed forest zone to be effective, it is essential that the shallow subsurface flow move through the biologically active root zone, or there will be little attenuation of nitrate and other dissolved pollutant loads (Lowrance et al., 1995). To encourage high nutrient removal rates, vigorous tree growth is encouraged by periodic harvesting of plant biomass. During harvest, it may be possible to do limited grading to reduce concentrated flow paths through the buffer, but this is rarely practical because of the presence of trees and shrubs that would interfere with grading. The managed forest zone is typically 45- to 75-ft wide and is composed of tree and shrub species (preferably native) adapted for riparian conditions.

> **BOX 5-9**
> **Multiple-Species Riparian Buffer System**
>
> One popular approach to designing riparian buffer zones is the multiple-species riparian buffer system (MSRBS) proposed by Schultz et al. (1995, 2000) for application to agricultural lands in Midwestern states (Figure 5-11). This artificial riparian system consists of two or three zones. In the 3-zone system (Schultz et al., 1995), Zone 1 consists of four or five rows of fast-growing trees planted next to the stream for streambank stabilization, wildlife habitat, stream shading, nutrient removal, and selective timber harvest every 8–10 years. Some slower growing hardwood trees may be interspersed with the fast growing trees. A minimum width of 30 ft is recommended. Zone 2 generally consists of one or two rows of native shrubs with a minimum width of 12 ft. The purpose of Zone 2 is to add diversity and wildlife habitat to the ecosystem and to slow floodwaters when the stream leaves its channel. Zone 3 has a minimum width of 20–24 ft. and is composed primarily of warm season grasses. Warm season grasses, particularly switchgrass, are preferred because their dense, stiff stems slow the overland flow of water, allowing water to infiltrate and sediment to be deposited in the buffer area. Native forbs and grasses may also be mixed with the switchgrass; however, there should always be a 10-ft switchgrass strip at the edge of the field (Iowa State University, 1997). The MSRBS may also contain constructed wetlands for the treatment of tile drainage water and streambank bioengineering for streambank stabilization. Figure 5-11 shows recommended widths of the MSRBS for various functions.
>
> The two-zone MSRBS was developed in response to landowner objections concerning trees along channelized stream in tile drained watersheds (Schultz et al., 2000). In this model, shrubs are planted as the first two or three rows next to the channel rather than trees. On some sites, the whole first zone is planted to shrubs, especially along the upper reaches of first order perennial or intermittent streams where channel incision is minimal. Shrubs are still often recommended next to the grass filter to provide diversity and withstand the pressures of fire management that are part of native grass filter maintenance. In addition to switchgrass, the two-zone MSRBS recommends mixtures of other native grasses and forbs.
>
> For both systems, maintenance includes a combination of mowing grass between the tree and shrub rows once or twice during the growing season and if possible burning the grass each spring for the first five years until grasses are well established, followed by less intensive maintenance in the years after. Fast growing trees may be harvested every 8 to 12 years to remove nutrients and chemicals (Iowa State University, 1997).
>
> Little published information is available on the effectiveness of MSRBS for water-quality protection. One study reported that the MSRBS reduced nitrate-N concentrations in shallow subsurface flow from 12 mg L^{-1} to less than 2 mg L^{-1} (Schultz et al., 1995). Another study (Isenhart et al., 1997) reported that a five-year-old MSRBS in Iowa reduced sediment losses by 80 percent to 90 percent within the first 4 m of the native

The *undisturbed forest zone* is situated immediately next to the stream and consists of unmanaged native trees and shrubs. Harvesting of trees and shrubs in this zone generally requires special permission. The main purpose of this zone is to provide habitat for terrestrial wildlife and aquatic organisms that are dependent on the riparian system. The various direct and indirect functions of this zone

grass strip and reduced nitrate and atrazine loads in shallow groundwater by >90 percent. An additional study reported that a 6-m-wide switchgrass zone (5 percent of the field source area) removed 46, 42, 52, and 43 percent of the influent total-N, nitrate, total-P, and orthophosphorus, respectively, over the short term. A 3-m-wide strip (2.5 percent of the field source area) was somewhat less effective and removed 28, 25, 37, and 34 percent of the influent total-N, nitrate, total-P, and orthophosphorus, respectively (Lee et al., 1999).

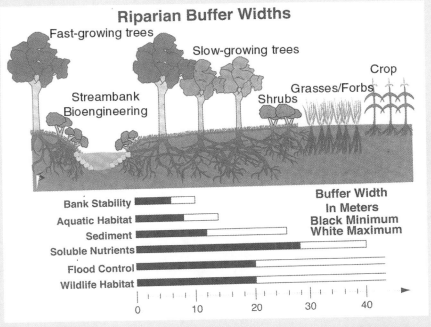

FIGURE 5-11 Three-zone multiple-species riparian buffer strip. SOURCE: Reprinted, with permission, Schultz et al. (2000). © 2000 by American Society of Agronomy.

include regulation of stream temperatures through the canopy effect, streambank stabilization due to tree roots, provision of leaf litter and large wood, and provision of an undisturbed area for wildlife. Additional pollutant removal also occurs in this zone. The unmanaged forest zone is typically 15- to 30-ft wide, but a minimum width equal to mature tree height may be a more effective width for

BOX 5-10
USDA's Three-Zone Approach to Riparian Buffer Design

An approach similar to the multiple-species riparian buffer system (Box 5-9) is the three-zone forest buffer proposed by Welsh (1991), as shown in Figure 5-12. Welsh suggests that the buffer area should be 20 percent of the contributing nonpoint source pollutant source area. Zone 1 is a permanent and undisturbed forested zone immediately adjacent to the stream. Zone 2 is a managed forest zone, just upslope of Zone 1, in which timber is periodically harvested. Zone 3 is the runoff control zone—a managed herbaceous strip, usually grasses, just upslope of Zone 2 that is used to control runoff. The three-zone forest buffers are specified for habitat and water-quality protection of waterbodies adjacent to cropland, pastures, and urban areas that are sources of diffuse pollution. Applicable waterbodies include perennial and intermittent streams, lakes, ponds, wetlands, and groundwater recharge areas.

Recommended widths for all three zones range from a minimum of 100 ft to 150 ft, depending on soil type and land use (Welsh, 1991). There are no published studies on the overall effectiveness of the three-zone forest buffer design for water-quality protection. However, information on the individual effectiveness of forest and grass buffers is summarized later in this report. It is important to recognize that the MSRBS and the USDA's three-zone buffers are not natural systems. They are engineered to approximate the functioning of natural riparian areas and achieve site-specific water-quality and habitat goals. As noted previously, this natural riparian buffer system may not be applicable in many areas of the Midwest because such landscapes are highly modified from their natural state.

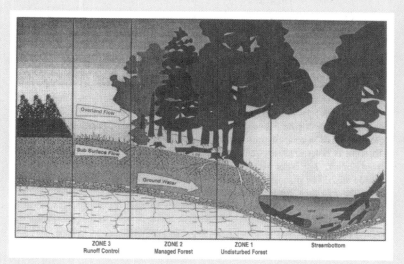

FIGURE 5-12 Three-zone forest buffer. SOURCE: Reprinted, with permission, Lowrance et al. (1985). © 1985 by Soil and Water Conservation Society.

restoring habitat functions. The NRCS requires no less than 35 feet for the undisturbed forest zone and the managed forest zone combined.

Effectiveness of Buffers. Numerous studies have confirmed the role of upland grass buffers and riparian buffer zones for controlling nonpoint source pollution from agricultural and urban areas. Some of these findings are presented in Table 5-2 for various pollutants in overland flow. In reality, most buffers and riparian areas achieve only a fraction of their reported pollutant trapping potential. Most of the trapping studies reported by researchers have been short term in nature and were conducted under very controlled conditions that minimized the influence of factors—such as concentrated flow and long-term accumulation of pollutants—that influence riparian zone performance in nature. In addition, most of these studies report on riparian zone effectiveness for water-quality protection only in the first few years after establishment. These studies are probably not good indicators of the long-term performance of riparian buffers with respect to water-quality protection.

A few researchers have investigated the performance of forested buffers that have been in place below cropland for decades. Lowrance et al. (1983) studied existing forested riparian zones (mixed hardwood and pine) in a subwatershed of the Little River watershed near Tifton, Georgia and reported that the riparian zones reduced nitrogen and phosphorus loading in subsurface flow by 67 and 25 percent, respectively. Reductions in surface loadings were not reported. The authors recommended periodic harvesting of riparian vegetation to maintain nutrient removal efficiencies. In a second watershed-scale study, Cooper et al. (1987) used ^{137}Cs data and sediment–soil morphology to estimate sediment trapping in two riparian areas in the Coastal Plain of North Carolina over a 20-year period. The results indicated that the riparian areas trapped 84 percent to 90 percent of the sediment lost from upland cropland. These studies may be more indicative of the long-term effectiveness of forested buffer zones for water-quality protection. However, because they both were conducted in low-gradient Coastal Plain watersheds, they may not be representative of riparian buffers in other physiographic regions.

It should be noted that neither the three-zone forest buffer nor the multispecies riparian buffer system are effective in removing dissolved contaminates from groundwater in agricultural watersheds where tile drains transport groundwater through the riparian buffer. To reduce pollutant loadings to receiving waters from tile drains, riparian buffers must include constructed wetlands or similar systems to treat tile effluents, as discussed in Box 5-11.

Managing for Success. The conditions necessary for effective riparian buffer performance can be achieved through management. It should be noted that many of the management measures recommended for water-quality protection may not be desirable for enhancing other riparian area functions. For example, periodic

TABLE 5-2 Reported Effectiveness of Buffer Zones for Water Quality Protection

Citation	State	Width (m)	Buffer Type	Reported reductions[a]
Young et al., 1980	MN	25	Grass	Sediment 92%
Horner and Mar, 1982		61	Grass	Sediment 80%
Dillaha et al., 1989	VA	4–9	Grass	Sediment 84%, phosphorus 79%, nitrogen 73%
Magette et al., 1989	MD	5–9	Grass	Nutrients <50%
Schwer and Clausen, 1989	VT	26	Grass	Sediment 45%, phosphorus 78%, total Kedall N 76%, ammonia 2%
Ghaffarzadeh et al., 1992		9	Grass	Sediment 85%
Madison et al., 1992		5	Grass	Nitrate and orthophosphorus 90%
Schellinger and Clausen, 1992	VT	23	Grass	Fecal coliform 30%
Chaubey, 1994	AR	24	Grass	Nitrate 96%, phosphorus 88%, sediment 80%, bacteria 0%
Mickelson et al., 1995	IA	5–9	Grass	Herbicides 28–72%
Arora et al., 1996	IA	20	Grass	Herbicides 8–100%, sediment 40–100%
Daniels and Gilliam, 1996	NC	6–18	Grass	Sediment 30–60%, total Kedall N 35–50%, ammonia 20–50%, nitrate 50–90%, phosphorus 60%, orthophosphorus 50%
Nichols et al., 1998	AR	18	Grass	Estrogen 98%
Lee et al., 1999	IA	3–6	Grass	Sediment 66–77%, total-N 28–42%, nitrate 25–42%, total-P 37–52%, orthophosphorus 34–43%
Lee et al., 2000	IA	7–16	Mixed	Sediment 70–90%, total-N 50–80%, nitrate 41–92%, total-P 46–93%, orthophosphorus 28–85%
Lynch et al., 1985		30	Forest	Sediment 75–80%
Shisler et al., 1987	MD	19	Forest	Nitrogen 89%, phosphorus 80%
Lowrance, 1992	GA	7	Forest	Nitrate (groundwater) 100%

[a]Note that reported reductions were not necessarily adequate to meet water-quality goals. These studies simply quantified the experimental reductions measured.

> **BOX 5-11**
> **Using Constructed Wetlands in Conjunction with Riparian Buffer Zones**
>
> Over 50 percent of the agricultural land in the central Cornbelt area of the United States is drained by surface and subsurface tiles, resulting in spring nitrate levels in surface waters that often exceed 10 ppm. For example, in the Embarras River of east central Illinois, tiles drain 70 percent to 85 percent of the cropland, and total N losses average 39 kg/ha/yr (David et al., 1997). Effluent from tile drainage accounts for 68 percent to 91 percent of the total N load (Kovacic et al., 2000) and 46 percent to 59 percent of the phosphorus load to the Embarras River (Xue et al., 1998). Because of the extensive tile drainage and bypassing of riparian biological processing, buffer zones will be significantly less effective in reducing nitrate levels in these areas, although they still have many other benefits. To reduce pollutant loadings in tile-drained watersheds, riparian buffers should include constructed wetlands or similar pollutant treatment systems.
>
> Constructed wetlands are built specifically to receive tile-drained water prior to its delivery to the stream network. Typically, the constructed wetland is a 0.5- to 1-m-deep depression, or it is created through construction of an earth berm. It is designed to maximize water contact with the soil substrate and plants through a series of baffles or other structures to guide the water (Kovacic et al., 2000; Schultz et al., 2000). The constructed wetland is typically located adjacent to the stream, thus reducing impacts on cropland and intercepting the greatest quantity of water. Guidelines for wetland size and the ratio of wetland area to acres drained vary from 1:100 (Schultz et al., 2000) to 1:15–20 (Kovacic et al., 2000) and depend upon design, precipitation, and other factors.
>
> Initial investigations into the ability of constructed wetlands to reduce pollutant loads are promising. During a three-year evaluation in Illinois, constructed wetlands removed 37 percent of total N; when coupled with a 15.3-m buffer between the wetland and the stream, an additional 9 percent was removed (Kovacic et al., 2000). Future research is needed to develop a better understanding of wetland performance under varying hydrologic and soil conditions as well as an improved understanding of optimal siting of wetlands in relation to the drainage network. Large-scale implementation of this strategy will require an economic evaluation that must take into account the availability of funding, the value of agricultural cropland in areas optimal for wetlands, and the potential beneficial uses of the wetlands for wildlife habitat, aesthetics, and recreation.

burning, mowing, and harvesting of grasses and biomass for nutrient removal and buffer maintenance can affect habitat values.

First, riparian buffers are most effective at pollutant removal when overland and shallow subsurface flow are distributed uniformly across the riparian zone as sheet flow. Areas where flows concentrate have shorter detention times, and pollutant-removal mechanisms in these areas can be overwhelmed. Unfortunately, the majority of overland flow and a significant portion of the shallow subsurface flow from contributing upland areas naturally concentrates before reaching the riparian area. In agricultural areas, overland flow tends to concentrate within 100 meters (Dillaha et al., 1989), while in urban areas, stormwater concentrates into channelized flow within as few as 25 m (75 ft) of its source (Whipple, 1993; Schueler, 1996). In some cases, water-spreading systems can be used to disperse highly concentrated flows across a riparian area. Installation of water bars at intervals across the riparian zone (perpendicular to the slope) can force overland flow to flow across rather than parallel to the riparian area. The runoff control zone of riparian buffers should be maintained by grading out rills, gullies, and excessive sediment deposits to encourage shallow sheet flow.

Second, high infiltration rates will reduce runoff volumes and velocities and the transport of dissolved and adsorbed pollutants associated with overland flow. For this reason, dense herbaceous vegetation or litter layers, which offer high resistance to overland flow, are preferred. Regular mowing of herbaceous cover (cool season grasses) in the runoff control zone 2–4 times per year will encourage thick growth at ground level and high resistance to overland flow. Periodic burning of warm season grasses in the runoff control zone is required for similar reasons. Herbaceous vegetated buffers that have accumulated excessive sediment should be plowed, disked, and graded, if necessary, and re-seeded to reestablish shallow sheet flow conditions. This is practical only in the runoff control zone where trees and shrubs are not present.

Third, both adsorption of dissolved pollutants and microorganisms to soil and plant surfaces and assimilation of dissolved pollutants, particularly nutrients, by plants and soil microorganisms are enhanced as contact times increase. Long-term nutrient removal is the result of nutrient uptake and storage in woody biomass that is not lost at the end of the growing season.

Finally, if water quality protection is the primary objective, priority should be given to installing and maintaining buffers along smaller streams (first- and second-order) rather than higher-order streams (USDA-NRCS, 2000a). This is because only a small portion of the flow in higher-order streams actually flows through their adjacent riparian buffers. Table 5-3 illustrates the relationships between stream order, number of streams, and length of streams. If hydraulic inflow and nonpoint source pollutant loading to streams are assumed to be proportional to stream length, then ephemeral, first-, and second-order streams account for approximately 90 percent of both total stream length and total pollutant loading (Table 5-3). If riparian areas are not functioning along ephemeral drainageways, then approximately 63 percent of the average annual stream loading will enter the riparian areas of higher-order streams as channel flow, with little opportunity for pollutant attenuation by riparian processes. Brinson (1993) reached the same conclusion—that wetlands are the most effective along lower-order streams, where they have been proposed for water-quality protection.

Research Needs. A number of critical questions hold the key to a more holistic utilization of riparian buffers as a landscape feature for both habitat

TABLE 5-3 Relationship Between Stream Order and Length for Any Area Assuming One 10th-Order Stream

Stream Order	Number of Streams	Average Length (km)	Total Length (km)	Percentage of Cumulative Length
Ephemeral[a]	6,271,000	0.7	9,000,000	63
1	1,570,000	1.6	2,526,100	18
2	350,000	3.7	1,295,250	9
3	80,000	8.5	682,200	5
4	18,000	19	347,550	2
5	4,200	45	189,200	1
6	950	103	97,800	1
7	200	237	47,300	<1
8	41	544	22,300	<1
9	8	1,250	10,000	<1
10	1	2,896	2,900	<1
Cumulative	—	—	14,204,164	100

[a]Values for ephemeral streams extrapolated from relationships of 1st- to 10th-order streams ($R^2=0.99$)

SOURCE: Adapted from Leopold et al. (1964).

enhancement and water-quality protection. For example, given the enormous variability in reported best management practice effectiveness for pollutant removal (Table 5-2), what riparian zone width is required to meet site-specific pollutant reduction goals? What are the width requirements for each of the riparian subzones (grass/forb, managed forest or shrub, unmanaged forest/forest)? Under what conditions is each of those subzones required? What type of vegetation should be used for each of the subzones, and how will the vegetation composition affect pollutant/nutrient sequestration and habitat values? How can riparian buffer zones be managed to maximize long-term nutrient/pollutant removal? Are habitat protection goals compatible with pollutant reduction goals? An even more detailed list of buffer research needs were recently identified by a panel of 51 stakeholders (researchers, program administrators, and agricultural and conservation organization representatives) at the National Conservation Buffer Workshop (SWCS, 2001).

Answering these questions will require long-term and expensive experimental studies. Fortunately, it may be possible to answer many of the crucial questions in the near future using process-based riparian zone models such as the Riparian Ecosystems Management Model (REMM) (Inamdar et al., 1999; Lowrance et al., 2000), which is currently under development. REMM simulates long-term sediment transport, plant uptake, nutrient transport, and denitrification in riparian ecosystems and may provide a means of selecting vegetative species and riparian widths required to meet site-specific water-quality goals.

Infield Conservation Practices. Many types of buffer systems are used in agricultural and urban landscapes. Some are located in riparian areas, but most are situated upslope of riparian areas and are designed to keep sediment, agricultural chemicals, and organic matter in the field where they are viewed as valuable natural resources rather than pollutants. Many in the agricultural conservation community prefer to focus on infield BMPs and view riparian buffer zones as a BMP of last resort for water-quality protection.

Infield practices are similar to individual components of the three-zone riparian buffer systems described earlier. For example, a common practice is to place vegetative barriers (USDA-NRCS, 2001) on the contour between strips of row crops. These filter strips are primarily used to prevent sediment loss, but they also trap other potential pollutants. If located within the field, they help maintain agricultural productivity by conserving sediment and nutrients. If located at the lower edge of fields, their principal purpose is for water-quality protection. A similar function is provided by field borders (USDA-NRCS, 1999b). Although their primary purpose is for use as a turn row for agricultural machinery, if located on the downslope edges of fields, they can trap pollutants. Field borders are also useful in providing a buffer to reduce aerial drift of ground-applied pesticides. Contour buffer strips consist of alternating strips of row crops and close-growing herbaceous crops (USDA-NRCS, 1999c). Both are generally harvested. The herbaceous strip is used to detain and utilize sediment and nutrients leaving the more erosion-prone row-crop strips. Every two to four years, the herbaceous and row-crop strips are interchanged.

Windbreaks or shelterbelts (USDA-NRCS, 2000d) and herbaceous wind barriers (USDA-NRCS, 1994) can be designed to prevent wind erosion and to protect crops, livestock, and structures from wind-related damage. They are planted perpendicular to the prevailing wind direction and are usually composed of trees, but shrubs and tall herbaceous species are also used. Because of their effectiveness in trapping wind-blown sediment, however, they usually evolve into terraces that may interfere with overland flow and promote concentrated flow. Although windbreaks can be narrow (less than 20 feet wide), they do provide limited habitat for wildlife.

Grassed waterways (USDA-NRCS, 2000e) are an infield practice designed to prevent gully formation by transporting concentrated runoff downslope in a non-erosive manner. They are constructed by grading ephemeral drainageways to more stable shapes and then vegetating them with an erosion-resistant grass. They are typically designed to convey the runoff from extreme storms (10-year return interval, 24-hour duration) without damage and to maintain velocities that will minimize sediment deposition. In smaller storms, however, they do trap some sediment.

Finally, hedges composed of warm season grasses such as switchgrass (Schultz et al., 2000) can be used to form living terraces that function much like silt fences on construction sites to temporarily detain overland flow and induce deposition of larger sediment particles. They also disperse runoff and provide habitat for some species. Grass hedges are generally narrow, 2–10 feet wide, and are often used in conjunction with lower-growing herbaceous buffer strips.

Although these practices are located upslope, they are extremely beneficial to riparian area functioning, because they reduce pollutant loadings and prevent overloading of pollutant-removal processes in riparian areas.

Conclusions and Recommendations on Riparian Buffers for Water Quality Protection

Engineered and constructed buffer zones are a valuable conservation practice with many important water-quality functions. Under proper conditions, these buffers are highly effective in removing a variety of pollutants from overland and shallow subsurface flow. They are most effective for water-quality improvement when hillslope runoff passes through the riparian zone slowly and uniformly and along lower-order streams where more of the flow transverses riparian areas before reaching the stream channel.

Riparian buffer zones should not be relied upon as the sole BMP for water-quality improvement. Instead, they should be viewed as a secondary practice or BMP of last resort that assists infield and upland conservation practices and "polishes" the hillslope runoff from an upland area.

Riparian buffer zones must be designed using a multiobjective approach that considers all their ecological functions. Even when they are marginally effective for pollutant removal, riparian buffers are still valuable because of the numerous habitat (see below), flood control, groundwater recharge, and other environmental services they provide. Unless new evaluation procedures are developed that consider both the water quality and ecological functions of riparian areas, it is unlikely that riparian zone size (width and length) and composition (vegetation types, other features) will be determined in a way that optimizes their potential for environmental protection.

Cattle Exclusion and Grazing Systems

Several methods have been advanced for managing livestock, particularly cattle, to restore and protect riparian areas:[2]

---

[2]Because sheep and other classes of livestock are less common than cattle and do not concentrate in riparian areas as cattle do, they do not pose the same risk of environmental damage as cattle. Thus, this section focuses primarily on cattle management.

- fencing to exclude cattle,
- extended (e.g., five years or longer) rest of the entire grazing unit (pasture or allotment),
- specialized grazing systems,
- herding practices,
- changing the season of use, class of livestock, or stocking intensity,
- attracting livestock away from riparian areas with upland water sources or mineral or feed supplements and shade,
- culling herds or breeding animals to eliminate "riparian loafers,"
- eliminating grazing from the entire grazing unit,
- revegetating with woody species, and
- constructing drift fences.

Of the strategies listed above, exclusion is by far the most effective means of restoring riparian areas damaged by cattle. The efficacy of each method depends in part on the site potential, the area's grazing history and state of depletion, and the management goals and timeframes for achieving those goals. Some of these management tools are BMPs, which have been adopted by states or recommended by EPA for addressing livestock-related nonpoint source pollution of surface waters (EPA, 1993; Mosley et al., 1997; Sheffield et al., 1997).

Exclusion. Excluding livestock from streams can yield significant, even dramatic, benefits to riparian areas and has consistently resulted in more rapid restoration of riparian areas than other management practices (Ohmart and Anderson, 1986; Elmore and Kauffman, 1994; Fleischner, 1994). Numerous studies have shown a marked difference between the riparian vegetation (and often bank and soil stability) in ungrazed areas compared to grazed areas—with ungrazed areas uniformly being healthier (Gunderson, 1968; Rinne, 1988; Huber et al., 1995; extensive review in Belsky et al., 1999; McInnis and McIver, 2001). For example, Schulz and Leininger (1990) compared areas from which livestock had been excluded for 30 years to study plots where grazing had continued but at a stocking rate reduced by more than two-thirds from the 1939 level. The researchers found that vegetative cover (excluding forbs) was significantly greater in the exclosures, while grazed areas had four times more bare ground. Willow densities were similar between the two areas, but plants were shorter in grazed areas. Even along intermittent streams, excluding cattle can lead to development of significant riparian vegetation (Anderson, 1993). Many researchers have concluded that riparian areas need to be managed separately from upland portions of a grazing unit and that restoring riparian areas will depend on excluding cattle by fencing (Olson and Armour, 1979; Platts and Wagstaff, 1984; Bromberg and Funk, 1998). Exclusion may be the only option in areas where landform limits cattle use to the riparian area and immediately adjacent uplands (e.g., in narrow, steep sided canyons) or where riparian soils remain saturated throughout the year (e.g., seeps and bogs).

Although nearly all researchers seem to agree that improvements in ecological conditions will result following exclusion, they disagree as to the rate, extent, and predictability of recovery. Improved riparian and aquatic conditions may occur within 4–10 years after protection by fencing; this includes reestablishment of shrubs and trees even in heavily damaged areas (Rickard and Cushing, 1982; Skovlin, 1984). Clary and Webster (1990) estimated recovery times of 1–15 years or longer, while Belsky et al. (1999) suggested that *initiation* of recovery alone might take 2–15 years. Estimates of benefits achieved within five years range from 15 percent to 75 percent of the potential for a given site (Platts and Raleigh, 1984; Skovlin, 1984).

Although riparian vegetation may respond quickly to livestock exclusion, stream morphology usually improves more slowly, and fish populations may not improve at all (at least initially) (Platts and Wagstaff, 1984). Schulz and Leininger (1990) observed that trout biomass and fishing opportunities were greater in exclosures than in grazed areas. Similar results were obtained by Hubert et al. (1985) who documented improved habitat for brook trout as a result of livestock exclusion and stocking rate reductions. However, other studies have observed no relationship between grazing pressures and fish populations (Rinne, 1988). Rinne (1999) warns that there is a lack of peer-reviewed literature containing sound data on grazing–fish relationships, suggesting that this is an important area of future research.

In the West, excluding livestock has few if any adverse ecological consequences for riparian areas, although there are other disadvantages. In some more mesic areas, excluding livestock might result in rank growth of vegetation or undesirable thatch accumulation. As Box 5-12 reveals, perhaps the biggest drawback to fencing riparian areas is cost. Most public land ranching operations do not turn a profit, and ranchers oppose any management changes that will increase costs (Wilkinson, 2001). Suggestions to fence cattle out of streamside areas have been labeled impractical except in rare circumstances, and opponents contend that intensive livestock management, for instance by specialized grazing systems, can restore streams at a lower cost (Swan, 1979). Finally, while potential benefits of exclusionary fencing are significant, some can be difficult to quantify, making it hard to balance them against the costs of construction and maintenance, lost forage, and possible negative impacts to wildlife and recreational users. In part for these reasons, several western researchers have urged more research into innovative grazing management strategies that would not require fencing and permanent exclusion of livestock (Olson and Armour, 1979; Platts and Wagstaff, 1984).

Where exclusion of cattle from riparian areas is not an option, changes in grazing system, season of use, or stocking rates, alone or in combination with fencing and with utilization monitoring, are methods for reducing livestock-related impacts. Reports of the effectiveness of changes in grazing management

> **BOX 5-12**
> **Cost of Exclusionary Fencing**
>
> Fencing is the most expensive grazing management strategy in terms of initial capital expenditure and annual upkeep. Reported capital and annualized costs (in 1991 dollars, annualized at 8% interest over 10 years) for permanent fence construction range from $2,640 and $324 per mile, respectively, in the Great Lakes region to $4,015 and $598 per mile, respectively, in Alabama (EPA, 1993). The annualized cost of net wire fences in Alabama exceeded $877 per mile. The Iowa State University and Purdue University Extension Services report the following construction and maintenance costs (per mile per year in 1998 dollars) for straight perimeter fencing (excluding gates): $964 for woven wire, $748 for barbed wire, $520 for non-electric high tensile, and $392 for electric high tensile (Mayer, 1999). Fence construction costs in the interior West have been estimated at $2,000–6,000 per linear mile, with maintenance costs estimated at $25–250 per mile per year. Sheep (woven wire) fencing or "lay-down" fencing is much more expensive than the typical 4-wire cattle fence (PCL Foundation, 1999). In smaller, managed pastures in the East, it is more feasible to protect riparian areas using much cheaper one- or two-strand electric fences (although Alabama reported capital and annualized costs for electric fencing of $2,676 and $399 per mile) (EPA, 1993). Costs of conventional perimeter fencing depend on terrain, contour, number of corners, soil, vegetation type, etc. Providing alternate stock watering facilities entails additional costs—e.g., for developing springs, building pipelines or troughs, or installing tanks.
>
> Olson and Armour (1979) estimated that fencing 9,000 (of an estimated 19,000) miles of streams on BLM lands would cost $45.6 million. The researchers assumed an average cost of $2,400 per mile of fence, and 19,000 miles of fence. They further estimated that trout in the protected stream segments could increase by 300–400 percent. This translated to a conservative $78.2 million increase in the value of sport fishing on BLM lands, yielding a "simplistic first-year benefit–cost ratio of 1.66"—i.e, a "return of $1.66 for each dollar invested." Platts and Wagstaff (1984) estimated that an increase of 47 fishing days per mile per year would be necessary to offset the cost of fencing. Although fencing costs often exceed the gains in fishery value, fencing may be necessary if both grazing and fishing are to be continued, they concluded, because few alternatives are sociologically or ecologically acceptable.
>
> A potential cost of fencing cattle out of riparian areas is the cost of providing alternate livestock forage. The amount of forage foregone will vary with numerous factors, including the geographic region, climate, elevation, and vegetative community. Platts and Wagstaff (1984) estimated that a 100-foot-wide corridor, encompassing about 12 acres per mile, would contain 12 AUMs in the West. (Western public land allotments as a whole typically provide 1 AUM per 13–16 acres.) The cost to national livestock production of excluding livestock from public land riparian areas would be insignificant, given that all public lands contribute only about 2 percent of the nation's livestock feed.

are highly variable, however, and efficacy is difficult to evaluate without reference to site-specific conditions and unless studies are continued over a sufficient time period. Changes in streamside vegetation will occur sooner than changes in hydrology or sediment loads. Grazing management may bring about positive changes in the former, but not in the latter; thus, a short-term (less than 5-year)

The cost of forage foregone would be reduced if riparian areas were fenced and grazed separately from the rest of the grazing unit, rather than put entirely off limits to grazing.

It is useful to compare riparian fencing costs in the West to the current federal grazing fee of $1.35 per AUM. Given the estimate above of 1 AUM per 13–16 acres (or 0.06–0.08 AUM per acre), the grazing fee generates an average of 8–10 cents per acre. (The actual forage contribution of riparian areas is higher because of their productivity and attractiveness to cattle.) Even at the low fencing estimate of $2,000 per mile, fencing $1/_4$-mile-wide riparian corridors would cost more than $31 per acre enclosed. The narrower the riparian enclosure, the greater the disparity between fencing cost per acre and grazing fee revenue.

Additional potential costs of fencing include the visual impact on some landscapes, potential disruption of native ungulate movement patterns, fragmentation of habitat, interference with recreational access, increased runoff resulting from bare ground brought about by cattle trailing along fences, and the contribution of cleared fencelines to the introduction and spread of exotic plant species (Wuerthner, 1990; Anderson, 1993; EPA, 1993; BLM and USFS, 1994; Noss and Cooperrider, 1994; Donahue, 1999).

A sophisticated economic assessment of the efficacy of riparian fencing would compare all foreseeable costs to expected benefits (compare Platts and Wagstaff, 1984). Potential benefits include enhanced riparian functioning, enhanced habitat values, increased native species populations and recreation opportunities, reduced erosion, and improved water quality (Olson and Armour, 1979; Kauffman et al., 1983; Platts and Wagstaff, 1984; Wuerthner, 1990). An additional benefit can be improved cattle herd health, resulting from reduced exposure to waterborne bacteria and fewer leg injuries caused by crumbling streambanks (Bromberg and Funk, 1998). Some benefits (e.g., increased fishing or hunting use) may be easier to quantify than others (e.g., improved water quality or aesthetics).

Fencing may be more appropriate and economical for larger blocks (e.g., 30-40 acres) of riparian meadows, although smaller, more homogenous riparian pastures may be more effective in achieving livestock management and restoration objectives (Mosley et al., 1997). Subdividing pastures into smaller squares, rather than wedge-shaped units, will reduce the amount of fence needed and the number of cattle trails, while allowing more efficient reduction of the distance to water (Hart et al., 1993). The possibility of future grazing of restored riparian meadows "will help amortize fencing costs" (Skovlin, 1984). Fencing can be economically feasible on streams with high fisheries potential. Studies in Idaho showed that trout were 1.5–4.5 times more abundant along ungrazed than along grazed areas (Keller and Burnham, 1982). The economic value of improved trout fishing has been estimated and can be significant (Dalton et al., 1998). When fencing or exclusion of livestock is necessary "for maintaining productive riparian and fishery habitats, the cost of special management pastures may not seem exorbitant" (Platts and Nelson, 1985a).

study could misinterpret the overall effectiveness of the new management system. Livestock manager commitment is crucial, as improper implementation or lack of enforcement will reduce the effectiveness of any grazing management strategy (Platts and Nelson, 1985b; Clary and Webster, 1990; GAO, 1990; Ehrhart and Hansen, 1997).

Grazing Systems. Specialized grazing systems are "systematically recurring periods of grazing and deferment for two or more pastures or management units" (Skovlin, 1984). They include deferred, rotation, rest-rotation, and deferred rotation systems. Three-pasture rest-rotation is probably the most commonly used system (Masters et al., 1996). These systems contrast with continuous livestock use, which can be either season-long or yearlong.

The range management literature contains little if any evidence that these grazing systems, which were designed chiefly to maintain or improve forage species and optimize livestock production, will benefit riparian areas. Nearly all researchers agree that rest-rotation grazing alone will not restore or maintain riparian conditions (Hughes, 1979; Olson and Armour, 1979; Skovlin, 1984; Hart et al., 1993; but see Masters et al., 1996). Rest rotation reportedly works well when precipitation exceeds 15–20 inches per year and is predictably distributed (Busby, 1979), but few western rangelands meet these criteria. [In fact, 95 percent of BLM lands receive less than 15 inches of annual precipitation (Foss, 1960).] Grazing impacts on riparian areas can be minor if the period of use is sufficiently short. Mosley et al. (1997) recommend three weeks or less, though the appropriate period will depend on soil moisture conditions, timing during growing season, number of animals, etc. Suggested periods of use on the order of two to three days are likely to be too abbreviated to be practical for many operators (Davis and Marlow, 1990).

At least in the West, riparian area vegetation will be utilized under any stocking rate or grazing system because of cattle's tendency to spend a disproportionate time in riparian areas (Bryant, 1982; Gillen et al., 1985; Howery et al., 1996). Indeed, riparian forage is often over-utilized, even when upland vegetation use meets the grazing system's management prescription (Platts and Nelson 1985a,b,c). The smaller the fraction of the grazing unit occupied by the riparian area and the drier the surrounding uplands, the more likely and severe are the impacts.

Many researchers have recommended that more attention be given to developing grazing systems designed to improve and maintain riparian conditions. For example, Platts and Nelson (1985c) suggest that "longer rest periods (such as double rest-rotation) or deferred grazing that allows a protective vegetation mat to be maintained on the streambank during critical periods" are promising strategies that might avoid the need to exclude livestock completely. Mosley et al. (1997) contend that "adjusting [the] timing, frequency, and intensity of grazing in individual pasture units is more important than adopting a formalized grazing system." They recommend that riparian areas not be grazed during "critical" periods (usually between late spring and early fall) more often than once every three or four years, and that annual grazing of riparian areas could occur during noncritical periods. Still others opine that uniform grazing can be achieved more economically by subdividing pastures and providing additional water than by implementing rotation grazing (Hart et al., 1993).

Season of Use. Differing viewpoints exist regarding the effect of season of use on maintaining or improving riparian area conditions (compare Platts and Raleigh, 1984; Platts and Nelson, 1985c; Elmore and Beschta, 1987; Clary and Webster, 1990; Masters et al., 1996; Phillips et al., 1999). The tendency of cattle to use riparian areas most heavily during hot periods (summer) apparently relates to water availability, forage availability and quality, and microclimate (temperature and humidity) (Enrenreich and Bjugstad, 1966; Bryant, 1982). For these reasons, grazing at these times or at the peak of the growing season is generally not recommended. But views regarding the effects of late-season grazing vary widely (Bryant, 1982; Kauffman et al., 1983; Platts and Nelson, 1985a; Conroy and Svejcar, 1991). According to one Montana study, use of riparian areas is related more to the timing and amount of precipitation, forage production and quality, and daily weather changes and insects (all or some of which can be difficult to predict) than to season alone (Marlow and Pogacnik, 1986). Regardless of the season, utilization should be monitored and kept within prescribed levels, and grazing may need to be ended early during drought conditions.

Stocking Intensity. Stocking rate, or grazing intensity, refers to the density of animals on a given area, and is usually described as light, moderate, or heavy.[3] The failure of many studies to document stocking rates or to define terms such as heavy and moderate stocking makes it difficult to evaluate claims that reduced stocking intensity can improve riparian conditions. Even very short periods of heavy stocking may have hydrologic consequences (Branson, 1984), while little information is available on the hydrologic impacts of light to moderate grazing intensity (Skovlin, 1984). Clary (1990) observed relatively similar improvements in bank stability, willow height, and cover when historically heavy stocking rates were reduced to no, light, or moderate stocking, and concluded that light to medium spring cattle use was compatible with these riparian habitats (in Idaho). Gifford and Hawkins (1978) concluded that infiltration and sediment loss rates for lightly grazed, moderately grazed, and ungrazed areas were not statistically different. But Blackburn (1984) cautioned that if an area has been severely overgrazed, reducing stocking to moderate levels may not reduce sediment loss from the watershed. In some circumstances, reducing both livestock numbers and the length of the grazing season may not be sufficient to achieve management objectives (Chaney et al., 1990).

---

[3]Although the terms are seldom defined in particular studies, "light" indicates a utilization rate of 20–25% (Clary, 1999), also described as a "conservative amount of forage" use (Skovlin, 1984). "Moderate" grazing indicates 35–50% utilization (Clary, 1999); this is the "maximum amount of forage [use] that will still maintain forage, soil, and watershed conditions" (Skovlin, 1984). "Heavy grazing," or more than 50% utilization, "exceeds the capacity of the rangeland system" (Skovlin, 1984). In contrast, Mosley et al. (1997) view herbaceous utilization levels of less than 65% and shrub use not exceeding 50–60% as "usually appropriate."

Measuring Utilization. Increasingly, public land managers are using stubble heights of the most palatable herbaceous species as an indication of acceptable grazing use of riparian areas. Effective use of this indicator requires frequent monitoring and the ability to remove cattle quickly when conditions so require. Although no standards have been established, EPA recommended to BLM that a resource management plan for an area in Colorado limit autumn utilization of streamside vegetation to 30 percent with 4–6 inches of stubble remaining at the end of the grazing season, and it recommended that the plan require stubble heights greater than 6 inches in critical fishery habitats (EPA, 1992). A national forest proposed to adopt a criterion of 6-inch stubble height or 30 percent utilization (whichever occurs first) for forage utilization directly adjacent to the stream on Curlew National Grassland, Idaho (Federal Register, 1999).

At stubble heights of less than 3 inches, continued livestock use can cause damage to riparian areas (including shrub use and bank breakage) within a few days. Hall and Bryant (1995) recommend that livestock be moved when stubble height of preferred species approaches 3 inches. Stubble heights greater than 6 inches may be required to protect critical fisheries or easily eroded streambanks (Clary and Webster, 1990). Clary (1990) reported that nearly all variables indicative of favorable salmonid fisheries habitat improved when grasses in riparian pastures were grazed to not less than approximately 5–6 inches.

The remaining methods for managing grazing (of those listed on page 386) are of generally less utility for protecting riparian areas. Culling and breeding strategies intended to develop a herd that avoids riparian areas are apparently not widely used. Changing from cattle to sheep can lessen the impacts of grazing on riparian areas because sheep are more easily herded, can negotiate steeper terrain, and tend to spend less time near water or in valley bottoms (Busby, 1979). But because of market factors and individual operator preferences, this strategy has only limited application. Herding and "frequent riding," if constant and assiduous, can be useful to keep cattle away from riparian areas (Platts, 1990). This is most easily achieved where pastures are relatively small and/or animals are monitored closely. Herding is of limited practical usefulness in the West, where cattle are often grazed over extremely large pastures and/or are simply turned out at the start of the grazing season and are rounded up at the end.

Attracting livestock away from riparian areas with mineral or feed supplements and alternate water sources can help limit cattle use of riparian areas (Skovlin, 1984; McInnis and McIver, 2001), but the efficacy of this approach depends significantly on climate, weather factors, terrain, and animal behavior. Off-stream livestock watering combined with fencing is one of the most common strategies for protecting riparian areas in more mesic parts of the country. But it is much less effective in the arid West, especially in steep terrain, where there is little other shade, and/or during hot weather (Clawson, 1993). Supplying alternate water sources in arid, steep terrain can be logistically difficult. Moreover, because cattle avoid steep slopes (> 35 percent) where possible, the steeper and drier the grazing land adjacent to the riparian area, the less likely it is that these tools will be effective (Cope, 1979; Platts, 1990).

Conclusions and Recommendations on Grazing

Excluding cattle from riparian areas is the most effective tool for restoring and maintaining water quality and hydrologic function, vegetative cover and composition, and native species habitats. If ecological restoration is the primary management objective, and if cost is not an obstacle, excluding livestock (or extended rest from grazing) will likely be the preferred management strategy. Excluding livestock permanently from riparian areas may be desirable or even necessary because of the relative value or importance of non-livestock resources (such as recreation, endangered species habitat, or water quality) or because of the degraded condition of the area. Once ecological and hydrologic functions are restored, grazing in some cases could have minor impacts if well managed.

Even where grazing in riparian areas is excluded or properly managed, grazing also must be managed on uplands to protect riparian areas. Riparian conditions are a function not only of activities within the riparian area, but of upland conditions and activities as well. Any upland activities that contribute to excessive soil erosion or runoff can negatively impact riparian area condition and functioning.

Where cattle are not excluded from riparian areas degraded by livestock grazing, conditions will not improve without changes in grazing management. Changing the season of use, reducing the stocking rate or grazing period, resting the area from livestock use for several seasons, and/or implementing a different grazing system can lead to improvements in riparian condition and functioning. Improvements, e.g., in vegetation conditions, soil and water quality, and/or streambank stability, will depend in part on site conditions and potential.

Further research is needed concerning effective grazing management strategies in both the interior/arid West and more mesic areas of the country. As long as riparian lands continue to be used to produce livestock, research should address grazing systems or other strategies that could lessen the ecological impacts of livestock use. To be effective, grazing strategies must be site-specific. Management methods will also vary according to land ownership and land-use goals.

Riparian Buffers—Management for Habitat and Movement Corridors

Managing riparian lands for a diversity of plants and animals involves balancing multiple factors, all of which are accentuated in intensively managed or developed landscapes. These factors include width and vegetative composition of the riparian area, activities within the riparian area and on adjacent lands, connectivity with other patches on the landscape, and potential negative edge effects such as those associated with modified microclimates, exotic plants, or increased predation or parasitism. These factors come into play for both riparian areas left in their natural state (for example, no-harvest buffers in managed forests) as well as engineered buffers (such as those designed for water-quality concerns). Given the multiple factors that must be considered when managing riparian buffers as habitat for plants and animals, it is no surprise that no single management prescription can optimize diversity in all situations.

Noss (1991) advocates an approach that seeks to optimize the width (distance from the water body) and variety of natural habitats in order to accommodate the full spectrum of native species. But it is unclear how a management plan can take into account the many different taxa that use riparian areas. To be truly comprehensive, management techniques—for rare sedges, neotropical migrant birds, and large carnivores, for example—would have to be similar. Or a particular species or suite of species would have to serve as an umbrella in the management of certain riparian areas. Vital to all situations are the identification and use of reference sites (described earlier) that can provide guidance in establishing and accomplishing management goals. Reference sites can provide an understanding of the potential functioning of riparian areas as habitat and movement corridors.

Habitat. Historically, emphasis in wildlife management has been placed on a single species—an organism-centered approach. With a single-species approach, however, management runs the danger of neglecting the interactions of organisms. As an alternative, Naiman and Rogers (1997) advocate examining influences via functional groups, a particularly useful construct for riparian areas. In this scheme, animals are grouped by primary activities associated with movement, dwelling, and feeding, as well as the habitat modifications produced by these activities. Instead of species-focused management, the authors advocate managing for spatiotemporal variability in populations as a way to perpetuate resilience in riparian areas. Similarly, Ilhardt et al. (2000) argue for a functional approach, where function is defined as a process that moves material between the terrestrial and aquatic portions of the riparian area and the width is designed to perpetuate selected functions (including animal habitat needs). A functional approach is more likely to foster consideration of a wider diversity of plants and animals than would be found only in the aquatic ecosystem and its immediate surroundings.

Buffer width required for habitat is almost always the first issue confronted by managers and is one of the most difficult to address. Substantial variability in required width has been observed in numerous studies. For example, strips at least 60 m wide are needed to maintain breeding habitat for forest-dwelling birds adjacent to a clearcut in the Laurentian Mountains of Quebec (Darveau et al.,

1995). A Vermont study in a mountainous landscape advocated widths of 75–175 m to optimize bird diversity (Spackman and Hughes, 1995), while a South Carolina study in bottomland hardwoods described the need for riparian areas 500 m wide (Kilgo et al., 1998). And these examples are drawn from only one taxon: birds! Brinson et al. (1981) graphically summarized the spatial distribution of selected riparian vertebrates in relation to streams, with distances ranging from a few meters in the case of salamanders to several kilometers in the case of herons (Figure 5-13). Far-ranging species such as cougar or bear might require widths that are measured in kilometers (Smith, 1993). Several tables of documented habitat widths for a variety of taxa are found in Verry et al. (2000). Increasingly, such information on widths, combined with data on reference sites, can be used by resource managers in their design of leave areas or engineered buffer strips, although the wide disparity in desired riparian widths associated with various taxa presents an ongoing challenge.

Any question of width must also include concern about negative effects associated with fragmentation of habitats, often loosely grouped under the term "edge effects." Edge effects can result from reduction in habitat area as well from increases in isolation of habitat patches, predation and parasitism, and disturbance from adjacent lands (Noss, 1983; Robbins et al., 1989; Robinson, 1992). The linear nature of riparian areas, and the abrupt transitions between some uplands and riparian lands, make them particularly susceptible to such effects. Riparian areas severely affected by fragmentation and edge effects may cease to function as breeding habitat—particularly for birds (e.g., Robinson, 1992; Trine, 1998). Work in Maine conservatively estimated that negative edge effects adjacent to a clearcut extended 25–35 m into the adjacent forest (Demaynadier and Hunter, 1998). A bird community will persist, even in the narrowest riparian fringe, but often it is not the community that would have typified the riparian area in a less degraded condition. Clearly, a landscape approach is needed to address edge-effect issues when managing riparian areas for optimal biodiversity.

Management decisions in riparian areas must consider the potential simplification of habitat that can characterize degraded riparian areas. A loss of plant species, a reduction in understory diversity, the elimination of flooding or other disturbances such as fire, a reduction in amount of snags and woody debris, and disruptions of ecosystem continuity by roads, trails, or recreational facilities can lead to a parallel decline in animal diversity. In disturbed riparian areas, often only the most tolerant species remain, as was shown in a study from an Iowa agricultural landscape (Stauffer and Best, 1980). Likewise, in a comparative study of grazed and ungrazed riparian areas in Colorado, mammal and bird species needing more complex vertical structure and lush herbaceous understory were displaced in grazed areas by more tolerant species such as American robin and deer mouse (Schulz and Leininger, 1991). Similar findings of lower bird diversity and species richness, coupled with an increase in a few common spe-

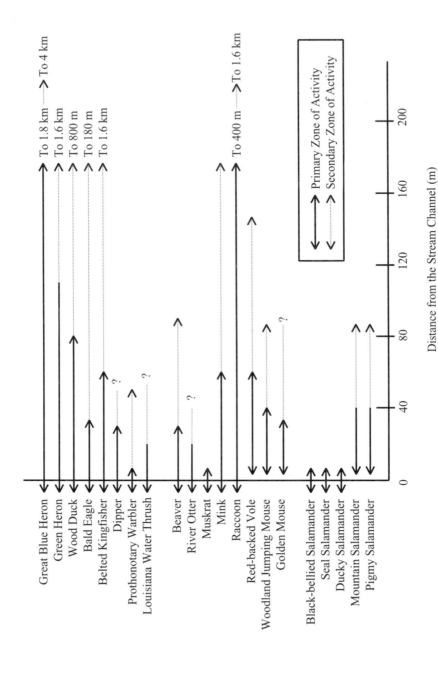

FIGURE 5-13 Distribution of some vertebrate species in relation to streams. SOURCE: Brinson et al. (1981).

cies, were reported for the citrus orchards that have replaced cottonwood forests along the lower Colorado River (Wells et al., 1979).

Many of the engineered buffer zones employed in eastern and Midwestern agricultural landscapes exhibit a simplified vegetative diversity and structure. Thus, efforts have been made to increase habitat diversity with the planting of trees and shrubs to create a more varied agricultural field edge. Identifying native species for this purpose through the use of reference sites can be a formidable task where agricultural practices have so altered soil characteristics that they are no longer suitable for some native vegetation. The Plant Materials Program of the NRCS (www.nhq.nrcs.usda.gov) has begun to assemble information on the use of native plant species for conservation as well as to identify seed sources. In agricultural areas that were originally a mosaic of uplands and prairie sloughs, the establishment of trees and many shrubs represents creation of habitat that was not present in pre-settlement times. Nevertheless, these species can provide valuable habitat adjacent to intensively managed farmland, consistent with the management paradigm of naturalization (Rhoads and Herricks, 1996) whereby managers strive for diversity, stability, and self-regulation within the framework and constraints of human utilization of the natural resources. It should be noted that the use of such trees or plants could affect the trophic structure of the aquatic ecosystem by changing instream temperatures and acting as a source of carbon and energy inputs.

Managing riparian areas for wildlife may be as simple as identifying and eliminating those practices that render the area unsuitable as habitat. For example, timber harvest in forested riparian areas, even selective harvesting, produces changes in the forest and ground cover structure that often result in reduced habitat complexity and losses in species richness and abundance (Conner et al., 1975; Niemi and Hanowski, 1984), as well as changes in microclimates that may extend as much as 1,000 feet from a harvest (Brosofske et al., 1997). Before timber harvest is proposed for a riparian area, management objectives for plant and animal diversity need to be identified. For the most part, current silviculture in riparian areas employs the same techniques used on uplands—techniques that are based on a paradigm of relatively homogeneous forest patches and that fail to address the spatial and temporal diversity inherent in riparian areas. Some foresters, however, are working to develop approaches that are more compatible with the heterogeneity of the riparian ecotone (Ilhardt et al., 2000). For example Short (1985) advocates managing for optimized vertical structure of habitat within a riparian buffer. One might also consider ways to perpetuate the horizontal patchiness of vegetative types that typifies many riparian areas with carefully targeted selective harvesting. Substantial innovations in silvicultural and agricultural protocols are needed before most vegetative management will be effective at perpetuating diverse natural riparian communities.

Corridors for Movement. Riparian areas may provide the best opportunity for restoring linkages between larger patches of more natural ecosystems and promoting gene flow between populations (MacClintock et al., 1977; Noss and Harris, 1986; Mackintosh, 1989). This is particularly evident in developed landscapes where often the only remaining natural or semi-natural land exists along floodplains, or in agricultural landscapes where buffer zones planted for water-quality protection also serve a wildlife habitat function (Forman and Baudry, 1984; Barrett and Bohlen, 1991). Such riparian remnants frequently include wetlands where anthropogenic alterations have been limited.

Theoretical work on corridors (based on the discipline of biogeography) often involves modeling the probabilities of extinction in isolated patches versus interconnected patches (Simberloff and Wilson, 1969; Turner, 1989). Using this theory as a springboard, much attention has been paid to designing for the movements of large carnivores between reserves (Harris and Atkins, 1991; Grumbine, 1992; Noss et al., 1996). Applicable work has also drawn from percolation theory, which models different combinations of population dispersal characteristics and patterns of habitat boundaries to predict patterns of population growth and habitat utilization (Gardner et al., 1991). This type of analysis is useful in understanding how human disturbance disrupts movement of animals at the landscape scale (O'Neill et al., 1988).

With this theoretical underpinning, practical examples of corridor design and the benefits of corridors have multiplied. Of importance in any design is a consideration of quality of habitat within the corridor, as well as width and connectivity (Noss, 1993). Beier (1993) advocated the use of radiotelemetry data combined with geographical information system (GIS) mapping to identify the movement range for a far-ranging carnivore (such as the cougar) and there are several recent examples of this approach. For example, a GIS modeling analysis of movement corridors in the northern Rockies focused on a suite of species that included a carnivore (cougar), omnivore (grizzly bear), and ungulate (elk), under the hypothesis that a diverse combination of far-ranging species would provide a protective umbrella for a greater number of other species (Walker and Craighead, 1997). The Yellowstone to Yukon initiative, a conservation proposal emerging from a coalition of more than 140 environmental groups, seeks to preserve and create a series of wildlife corridors that will link populations of bear, wolves, and other large predators from Yellowstone National Park to Canada's Yukon Territory. These corridors are based fundamentally on riparian areas and mountain ranges (The Wildlands Project, *http://www.twp.org*). Similarly, in Florida a GIS analysis was conducted to link significant reserve areas with the most appropriate and effective corridors; identification of linkages relied heavily on inland and coastal riparian areas (Hoctor et al., 2000). "Rewilding" is the term that has come to be applied to this approach of creating and preserving connectivity among large wild reserves, with a focus on the roles of species—frequently large carnivores—whose influence on ecosystem function and diversity is disproportionate to their numerical abundance (Soulæ and Noss, 1998). Riparian areas inevitably figure in the rewilding approach to landscape-scale corridor design.

Skeptics have accused managed riparian corridors of facilitating the spread of pests and diseases (Forman, 1995), and they have complained of a perceived deficit of studies regarding the effects of riparian zone design and management on biodiversity (Wigley, 1996). However, a considerable body of data exists for a variety of taxonomic levels from diverse ecoregions (Karr, 1996). Assessment of corridors is a significant challenge that continues to be addressed using both economic and scientific frameworks (Diamond et al., 1976; Simberloff et al., 1992). Often, corridors used by organisms have been designed by humans for other functions, which confounds any assessment of their effectiveness.

Solutions for providing movement corridors can be costly. The cost of one bridge that allows animal movement along a river under a road is an estimated 13 times greater than that of the usual bridge (Simberloff and Cox, 1987). Increasingly, movement corridors are being considered during major highway planning (Smith, 1999). Nevertheless, ongoing study is needed to assess the effectiveness of such created crossings. Studies from Banff National Park in Canada indicate that the effectiveness of constructed wildlife crossings can be compromised by recreational use of the same areas (Clevenger and Waltho, 2000). The time scale over which corridors may function can also make it difficult to assess effectiveness. For example, riparian areas that function as refugia for rare plants may act as migration corridors over hundreds of years. All these factors make it difficult to design and manage corridors and to assess their cost-effectiveness (Hunter, 1996). In spite of gaps in scientific and economic knowledge, there is little question that reestablishing connectivity of riparian areas as a means of counteracting habitat fragmentation is crucial for long-term species survival and perpetuation of biodiversity (Noss, 1993; Walker and Craighead, 1997; Soulæ and Noss, 1998).

Conclusions and Recommendations on Riparian Buffers for Habitat

Riparian areas—both natural reserves and managed buffer zones—provide some of society's best opportunities for restoring habitat connectivity on the landscape. Identification, mapping, and assessment of these areas are needed. Management of riparian areas in ways that optimize their value as habitat and movement corridors for plants and animals will require planning and action at both site-specific and landscape scales.

Much riparian buffer zone management suffers from focusing on a single species or taxon. Integrated management that uses a functional approach and seeks to optimize habitats for a variety of native species is needed. Integrated management should include the use of reference sites and monitoring and should focus on the rehabilitation of native vegetation. Management of riparian areas should shift from focusing on stable populations of individual species to managing these species for variability as well as for their interactive roles in the ecosystem.

Current silvicultural and agricultural management approaches do not adequately address the habitat values of riparian areas. Most forestry buffers, fenced riparian exclosures, and agricultural buffers have the protection of water quality (or sometimes fisheries) as their main focus. These water-quality protection buffers are usually considerably less diverse structurally and vegetatively than they would be if wildlife habitat were actively considered in their planning.

Controlling Exotic Species

Because of their great impact on biodiversity and ecosystem function, as well as the economic burdens they exact, the control of exotic species is often a high priority in the management of riparian areas. Approaches to exotic species control include hand removal, mechanical removal, herbicide applications, controlled burns, controlled flooding, and biological controls. Biological controls essentially involve reestablishing the natural control mechanisms exerted by herbivores and pathogens in the native ranges from which exotic species come. Though very effective control has been achieved in many circumstances, there are no universal prescriptions for control of exotic species in riparian areas. The effectiveness of the various control methods will depend on the growth habit of the particular exotic species and its reproductive strategy, the extent of the infestation including whether it is urban versus rural or near critical habitat for endangered species, and the state of knowledge and of testing concerning potential biological control agents.

Control strategies must also be tailored to the relative costs and benefits, which vary among species and with land use conditions. Because these costs and benefits can be difficult to quantify, there is considerable debate in both the scientific and political arenas about whether certain exotic species can and should be eradicated from riparian areas. For example, there is no effective control agent for Chinese privet. Other species, such as saltcedar, are so widespread (1 million acres in the USA) that eradication is impossible. In addition, eradication of saltcedar is complicated by the discovery of its use as nesting habitat for the southwestern willow flycatcher, a federally listed endangered species. Each situation must be assessed in terms of the forgoing constraints, as well as whether the removal activity and associated habitat disturbance, including the application of herbicides, will result in more good than harm. Specific examples of control strategies for three exotic plant species in riparian areas are described in Box 3-4.

Managing Other Activities

Human activities and land uses have resulted in the widespread alteration of riparian areas along rivers and other waterbodies across the United States. By changing these activities, many of these riparian areas can be ecologically restored and improved. In some cases the types and extent of needed change will be minor and relatively easy to implement, as in designating leave areas exempt from traditional agriculture, grazing, or forestry. For other activities, restoration will require a new understanding of why riparian areas are important. Through continued research, educational programs, tax incentives, awards, regulations, legislation, and perhaps other approaches, the ecological importance and intrinsic values associated with these lands may be better balanced against the competing wants and needs of a modern society.

Recreation

Although it may seem insignificant compared to other land uses, recreation can impair riparian area functioning to a substantial degree in many areas and must be part of riparian management plans (e.g., Loeks, 1985). Managers of recreation must consider not only its impacts on the aquatic ecosystem—which has been the usual focus of water-based recreation (Field et al., 1985)—but also on the riparian area itself (see Chapter 3). Fortunately, the public tends to place a high value on the natural habitat present along streams and drainages (Black et al., 1985).

Some of the most relevant research regarding management of recreational activities in riparian areas comes from work on greenways. A greenway is a linear open space that is more natural than the surrounding area (Smith and Hellmund, 1993); it is typical of many recreational lands along rivers, lakes, and coasts. The challenge in the management of greenways is to preserve natural functions while still allowing for human enjoyment of these areas.

Management of recreational activities in riparian areas involves a combination of careful design, limitation of use, and public education (Cole, 1993). A frequent concern is the disturbance of soil, plants, and animals by placement and use of trails and roads. Conservation-oriented trail design suggests making use of existing roads and trails (unless they are degrading the area) rather than cutting new paths. Ideally, before designing public access to a riparian area, surveys should be conducted for rare or sensitive plants and animals or culturally sensitive sites that might be disturbed by human use, so that impacts can be minimized. Rather than placing the trail entirely within the riparian area, trails should allow access and view points in discrete locations (Trails and Wildlife Task Force, 1998). Durable areas should also be sought out as locations for placement of recreational facilities (Cole, 1993). Placement should also consider the inevitable negative edge effects on surrounding natural areas and should seek to mitigate these with buffer zones or screening. Having well-identified, highly developed riparian recreation sites such as marinas, overlooks, and picnic grounds can also channel a high percentage of users into discrete areas with the capacity to accommodate them, and thus limit impacts on the larger riparian area. Careful design and placement of such high-intensity use areas (and of other facilities such as outhouses, flush toilets, and litter receptacles) can go a long way to minimizing impacts on riparian areas.

In any riparian area, recreational impacts can be reduced by choosing a style of development that seeks to maintain as many ecological functions of the area as possible. In surfacing trails and roads, one needs to consider impacts of runoff to adjacent waterbodies; this suggests that permeable materials should be used whenever possible (Cole, 1993). Perpetuation of natural vegetation, whether preserved or restored, should be a high priority in any riparian recreational area. Where natural vegetation cannot be maintained because of unavoidable heavy use, durable, non-native species may serve some vegetative functions (Binford and Buchenau, 1993).

Certain construction techniques can alleviate impacts to sensitive areas such as spring seeps, wetlands, cliffs, or outcrops. A boardwalk can be used to route foot traffic through a fragile riparian wetland or dune. Carefully placed water bars on trails can lessen runoff to adjacent streams and lakes.

Limitations on human access to riparian areas can take many forms. Knight and Temple (1995) describe spatial, temporal, behavioral, and visual restrictions on human use. Any prohibitions require a combination of education and enforcement to be effective. Examples include bans on bicycles or horses on erosion-prone trails and bans on motorized vehicles in areas where noise is an issue because of disturbance of animals. Wilderness areas often have setbacks from waterbodies for camping or horse use to protect the riparian area and aquatic ecosystem; these range from 20 to 200 feet (Cole et al., 1987). In some areas, certain activities may need to be excluded entirely if ecological restoration is the goal.

Human access can also be limited by quota or by limiting access to certain times or seasons. Recreational carrying capacity considers both ecological damage and perceptions of crowding that can be used to set such limits (Chilman et al., 1985). An example of exclusion for a discrete time period is the prohibition, by law, of people and pets from designated Great Lakes beaches during the nesting period of the rare piping plover. Similarly, a recreational area along a South Platte River reservoir in Colorado allows no visitor access for two months while herons are courting and building nests and limited access while the birds are laying eggs. In addition, the location of the viewing platform on a bluff creates a buffer between visitor and nesting birds. The mechanism for imposing these restrictions is generally signs and barriers (both artificial and natural) supplemented by educational interpretive materials and enforcement (Larson, 1995).

Finally, adaptive management is particularly appropriate in riparian recreational areas. Monitoring should be used to identify problems that can be rectified early on through restoration, redesign of structures and trails, and changes in human use patterns (Trails and Wildlife Task Force, 1998). To date, however, little effort has been expended to adaptively manage recreation within riparian areas.

Conclusions and Recommendations on Recreation

Management of recreational use in riparian areas needs to combine careful design, limitation of use, and public education. In most cases, all or many of these components are lacking in recreational management plans. The goal of managing recreational activities in riparian areas is to perpetuate natural functions (e.g., water quality, wildlife habitat, etc.) while still allowing human use and enjoyment of these areas.

Most recreational development in riparian areas lacks sound ecological assessment and planning. Recreation planning should include a landscape perspective, and it should involve the local community and other stakeholders. Some recreational uses are incompatible with preservation or rehabilitation of riparian areas and may need to be prohibited. Examples include prohibiting the use of off-road vehicles in fragile riparian wetlands or erosion-prone areas and prohibiting intensive public visitation in a habitat of a rare plant or animal species.

Education

To be effective at improving the ecological functions and integrity of the nation's riparian areas, education must have as a primary goal increasing "riparian literacy" in the general population. In addition, education needs to effectively transfer practical interdisciplinary information to natural resources managers, policy makers, watershed councils, and those more directly affecting riparian areas such as developers and zoning officials. Finally, riparian education must encompass the ongoing training of riparian scientists for the next generation, people who will assuredly face even greater management challenges than those faced today. At its most fundamental level, riparian education should be directed at understanding the effects human activities have had and are continuing to have on these vital and vulnerable areas so that alternatives may be developed to sustain these areas for future generations (Orr, 1990a).

Riparian education is as inherently multifaceted as the ecotone it addresses. A basic education in riparian functions must integrate the physical, natural, and social sciences. A good understanding of riparian science draws from disciplines as diverse as hydrology, geology and geomorphology, soil science, ecology, and limnology. It includes all the fields dealing with organisms inhabiting aquatic and terrestrial ecosystems: ichthyology, malacology, herpetology, entomology, mammalogy, ornithology, and botany. Also critical are disciplines that deal more directly with natural resource management such as wildlife biology, fisheries, forestry, range, and recreation. Of crucial importance is education that addresses methods for statistically sound assessment and monitoring. Riparian education also needs to include information on the legal framework that pertains to human activities in riparian areas.

The content of riparian education is, by necessity, broad and complex, yet the public at large needs to receive information about the ecological services of riparian areas in clearly understandable language and format. Hence, the challenge of educators is to integrate and distill this diverse information so that it can be presented effectively to the citizen body. It is important that the public at large quickly be brought "on board" regarding the importance of riparian areas and their contributions to quality of life. Their proximity to waterbodies, already highly valuable in the public eye (Black et al., 1985), may streamline this task. Thus, it may be easier to cultivate a positive perception of riparian areas than it has been to convince people that wetlands are valuable and deserving of protection.

Formal education about riparian areas should involve students at all levels, from elementary through graduate school. Teachers, particularly in elementary and secondary schools, will need curricula development (lessons and projects) along with training to address such an interdisciplinary and nontraditional topic within the framework of their mandated curriculum. Project-based or site-based education, such as typifies many river or watershed study programs, is an effective way to accomplish riparian education. The benefit of using environment-based education as an integrating context for learning has been well demonstrated in a study of 40 schools (elementary through high school) drawn from 13 states (Lieberman and Hoody, 1998).

Higher education needs to envision riparian science as a truly interdisciplinary field that includes a solid grounding in hydrology, limnology, ecology, conservation biology, experimental design, mapping (GIS), and statistics (Noss, 1997). Most institutions of higher education will need to revitalize their field science courses, making clear their practical connection to solving specific environmental challenges, such as riparian restoration and management, and the larger challenge of creating an ecologically sustainable society. Elder (1999) advocates the development of bioregional curricula that use place-based education to teach subjects in both the arts and sciences. Riparian areas and their waterbodies provide a natural opportunity to emphasize a bioregional perspective. Students in professional programs such as engineering, forestry, range, agriculture, and urban development will need to understand the potential impacts that various practices and land uses can have on riparian areas, the capability to minimize such impacts, and the opportunities for restoration and improvement of riparian systems across diverse landscapes. Making these changes will require the crossing of discipline barriers and cultivating a systems approach to thinking (Orr, 2000).

Natural resources managers and regulators form another crucial audience in need of information on riparian functions as well as up-to-date findings on alternative management strategies. Although it might be assumed that these professionals are adequately prepared to address the challenges associated with riparian areas and their management, the rapidly increasing information base of the last 25 years (see Chapter 1) indicates a need to ensure that they have current information and skills. Few resource managers have the broad, interdisciplinary training necessary for effective riparian management. Although all managers can benefit from information on the latest research and management approaches within their disciplines, perhaps more important to them is knowledge from disciplines outside of their formal training.

Government officials at all levels (e.g., town managers, planning and zoning officials, county commissioners, and state and federal legislators) and citizen governmental councils should also be included as recipients of information on riparian area functions and values. Non-governmental environmental organizations, often in a position to influence policy and frequently in need of solid scientific information, need to update their understanding of riparian areas in ways that help them better understand the broader consequences of their specific interests. (In states such as Arizona, Colorado, Montana, and New Mexico, nonprofit organizations have formed to help educate the public about the importance of riparian areas.) Other potential audiences include private sector entities, such as real estate professionals and developers, likely to have a disproportionate influence on riparian areas. One such educational program has been started by the North American Lake Management Society to train real estate agents selling waterfront property (as described in Box 5-13).

The mechanisms of information transfer will be varied, running the gamut of community involvement projects, printed popular material, targeted seminars, brochures, videos, multimedia computer programs, and peer-reviewed scientific papers. All approaches should strive to present interdisciplinary material in a clearly understandable and "user-friendly" way. Terminology should be clearly defined; jargon and gratuitous use of acronyms should be avoided whenever possible.

There are challenges to providing continuing education on riparian areas, the greatest of which may be effectively reaching wide and diverse audiences. Elementary and high school teachers are often ill equipped to undertake the type of project-based education that can most effectively integrate the diverse disciplines of riparian science. Researchers may find it easier to obtain funding on narrow disciplinary topics than to develop a coordinated research proposal that requires a funding source willing to support broad-based interdisciplinary efforts. Natural resources professionals may see continuing education as impugning the value of their past education or may perceive the blurring of disciplinary boundaries as a

> **BOX 5-13**
> **A Pilot Educational Program for Real Estate Professionals**
>
> *"The waterfront owner has invested in an ecosystem, not just a piece of property."* This is the operating premise of a new program designed to help real estate professionals understand the positive correlation between maintaining or increasing the health of the ecosystem and the value of waterfront property. In 1998, the North American Lake Management Society, with funding from EPA and assistance from the Wisconsin Association of Lakes, launched a pilot program to train waterfront real estate professionals in techniques for promoting the sale and maintenance of healthy waterfront property. The program links ecological understanding with property values and real estate transactions (Premo and Rogers, 1998). A real estate agent is very likely to be involved in the repeated sale of a property, particularly in a waterfront market. Thus, increasing the value of a waterfront property ensures higher commissions with each transaction. In addition, real estate professionals are a critical first point of contact with people who are planning to live near water. More often than not, their customers lack understanding about the land and water as living ecosystems and have no idea how their activities will affect the health and functioning of these areas. Providing real estate agents with solid understanding and information that they convey to customers can foster more sensitive riparian stewardship by new property owners. This training also allows the real estate agent to better match customers with property and thus minimize drastic riparian modifications by new owners.
>
> The program seminar uses a blend of illustrated lectures and activities to convey ecological functions to an audience of non-scientists who are inclined to view wildlife and vegetation as negative shoreline attributes—while seeing pavement, docks, and manicured lawns as desirable. The seminars address riparian ecology, water quality, shoreline law, and the human waterfront community, relating all topics to real estate transactions. Seminars also include an aquatic "zoo" containing macroinvertebrates and wetland plants, which provides an up-close look at riparian organisms. The day-long event concludes with a virtual real estate property selling tour conducted by the attendees using a selection of slides on riparian ecology and waterfront property. The long-range and ambitious goal of this program is to broaden the perspective of real estate agents such that healthy, functioning riparian areas become the waterfront property business standard.

threat to their profession. Layered on this challenge are the conflicts of varied human uses of riparian areas. Educational efforts may easily break down when users perceive them as simply threats to grazing privileges or available wood fiber. Riparian education will have to address a blend of socioeconomic and ecological issues that go beyond the questions scientists usually research—issues that are often neglected by those making management decisions. To be successful, riparian education must also foster a sense of community and responsible stewardship (Orr, 1990b). The unique functions and values of riparian areas must better understood by the layperson if there is to be a shift in society's management of these vitally important areas.

Conclusions and Recommendations on Riparian Education

Riparian education needs to reach broad and diverse audiences if it is to succeed in effecting positive change in riparian management. It needs to include formal educational institutions and reach out directly to policy makers, natural resources personnel, government officials, developers, landowners, and the public at large. Natural resources professionals need to expand their perspectives beyond their formal background and training.

Riparian education should strive to be inclusive. It should avoid using jargon, acronyms, and single-perspective approaches. The public's aesthetic appreciation of waterbodies is already high. This appreciation should be harnessed to further public stewardship of riparian areas.

CONCLUSIONS AND RECOMMENDATIONS

As noted by the Federal Interagency Working Group (1998) in its report on stream corridor restoration, water and other materials, energy, and organisms meet and interact within riparian areas over space and time. Riparian areas provide essential life functions such as maintaining streamflows, cycling nutrients, filtering chemicals and other pollutants, trapping and redistributing sediments, absorbing and detaining floodwaters, maintaining fish and wildlife habitats, and supporting the food web for a wide range of biota. The protection of healthy riparian areas and the restoration of degraded riparian areas relate directly to at least five national policy objectives: protection of water quality, protection of wetlands, protection of threatened and endangered species, reduction of flood damage, and beneficial management of federal public lands. The following conclusions and recommendations are intended to bring national awareness to riparian areas commensurate with their ecological and societal values.

Restoration of riparian functions along America's waterbodies should be a national goal. Over the last several decades, the nation (through both federal and state programs) has increasingly focused on the need for maintaining or improving environmental quality, ensuring the sustainability of species, protecting wetlands, and reducing the negative impacts of high flow events—all of which depend on the existence of functioning riparian areas. Unless an ambitious effort to restore the nation's riparian areas in undertaken, it will be difficult to achieve the goals of the Clean Water Act, the Endangered Species Act, wetland protection, and flood damage control programs. There is a clear need for legal guidance at the federal, state, and local levels that explicitly recognizes the importance of riparian areas and provides a legal framework for their protection, restoration, and sustainability.

Protection should be the goal for riparian areas in the best ecological condition, while restoration is needed for degraded riparian areas. Management of riparian areas should give first priority to protecting those areas in natural or nearly natural condition from future alterations. The restoration of altered or degraded areas could then be prioritized in terms of their relative potential value for providing ecological services and/or the cost effectiveness and likelihood that restoration efforts would succeed. There is only a limited track record of restoring biophysical systems that have been previously degraded. Nevertheless, where degradation has occurred—as it has in many riparian areas throughout the United States—there are vast opportunities for restoring the functions and values of these areas. Because riparian areas perform a disproportionate number of biological functions on a unit area basis (see Chapter 2), efforts focused upon restoring them could have a major influence on improving overall water quality and fish and wildlife habitat.

Patience and persistence in riparian management is needed. The current degraded status of many riparian areas throughout the country represents the cumulative, long-term effects of numerous, persistent, and often incremental impacts from a wide variety of land uses and human alterations. For many sites, substantial time, on the order of years to decades, will be required for a full sequence of high and low flows to occur, for riparian vegetation to establish and plant communities to fully function, for channels to adjust, for water quality to improve, and so on. Restoring riparian areas to fully functional condition may not be possible in those cases where permanent modifications to the hydrologic regime have been made. On the other hand, recovery may be rapid at sites where the impacts are easily reversible, such as where there is only a single stressor and where native vegetation is still present. Regardless of site condition, an adaptive management framework (NRC, 2002)—in which well-understood and relatively simple steps are taken first (such as passive restoration or pilot program initiation) and used to inform later activities (such as more active restoration or landscape-scale and regional programs)—is ideal for approaching riparian restoration.

Many of the impacts that have altered and destroyed riparian areas are the byproducts of local, state, and federal programs designed to develop and utilize land and water resources in a variety of ways and over many decades. In the process, the values and functions of riparian systems were not recognized or considered, were devalued or marginalized, or were considered as obstacles to ongoing management or development. Restoration of the nation's riparian areas will require a newly educated public that understands what riparian functions have been lost and what can be recovered in conjunction with a change in values, a change in institutional perspectives, and most likely a change in laws. Like recovery of the natural system, it will take time for education and outreach programs to generate broad-based public and political support. Thus, while initiating efforts to restore riparian areas is an urgent need, patience and persistence in meeting long-term restoration goals are required.

Although many riparian areas can be restored and managed to provide many of their natural functions, they are not immune to the effects of poor management in adjacent uplands. Because the subject of this report is riparian areas, it might seem that restoration activities need only focus on those areas. Indeed, in many situations this is all that may be needed to achieve certain restoration goals. However, where upslope management practices significantly alter the magnitude and timing of overland flow, the production of sediment, and the quality of water arriving at a downslope riparian area, then simply focusing on the riparian system may be inadequate for achieving restoration goals. In such situations, upslope practices that are contributing to riparian degradation must be addressed in order for long-term success to be achieved. Restoration of riparian areas should be approached with full recognition of the larger physical structure of which it is a part; that is, riparian area management must be a component of good watershed management.

REFERENCES

Acharya, G. 2000. Approaches to valuing the hidden hydrological services of wetland ecosystems. Ecological Economics 35:63–74.

Adams, P. W. 1983. Soil compaction on woodland properties. Oregon State University Extension Circular 1109. Corvallis, OR: Oregon State University. 5 pp.

Adamus, P. R. 1983. A method for wetland functional assessment, volume II. FHWA Assessment Method. Rep No. FHWA-IP-82-24. Washington DC: Federal Highway Administration, U.S. Department of Transportation. 134 pp.

Adler, R. W. 1995. Addressing barriers to watershed protection. Environmental Law 25(4):973–1106.

Ainslie, W. B., R. D. Smith, B. A. Pruitt, T. H. Roberts, E. J. Sparks, L. West, G. L. Godshalk, and M. V. Miller. 2000. A regional guidebook for assessing the functions of low gradient, riverine wetlands in Western Kentucky. Technical Report Number WRP-DE-17. Vicksburg, MS: U.S. Army Corps of Engineers Waterways Experiment Station.

Anderson, S. 1993 (updated 1994). Threats to amphibians: grazing and wildlife: a selected literature review. Internet source: http://ice.ucdavis.edu/Toads/grzaway.html

Andreas, B. K., and Lichvar. 1995. A floristic assessment system for northern Ohio. Wetlands Research Program Technical Report WRP-DE-8. Vicksburg, MS: U.S. Army Corps of Engineers, Waterways Experiment Station.

Andrews, E. D. 1986. Downstream effects of Flaming Gorge Reservoir on the Green River, Colorado and Utah. Geological Society of America Bulletin 97:1012–1023.

Aronson J., and E. LeFloc'h. 1996. Vital landscape attributes: missing tools for restoration ecology. Restoration Ecology 4:377–387.

Arora, K., S. K. Mickelson, J. L. Baker, D. P. Tierney and C. J. Peters. 1996. Herbicide retention by vegetative buffer strips from runoff under natural rainfall. Trans. ASAE 39:2155–2162.

Barrett, G. W., and P. J. Bohlen. 1991. Landscape ecology. Pp 149–161 In: Landscape linkages and biodiversity. W. E. Hudson (ed.). Washington, DC: Island Press.

Bedford, B. L. 1996. The need to define hydrologic equivalence at the landscape scale for freshwater wetland mitigation. Ecological Applications 6:57–68.

Bedford, B. L. 1999. Cumulative effects on wetland landscapes: links to wetland restoration in the United States and southern Canada. Wetlands 19:775–788.

Beier, P. 1993. Determining minimum habitat areas and habitat corridors for cougars. Conservation Biology 7:94–108.

Belsky, A. J., A. Matzke, and S. Uselman. 1999. Survey of livestock influences on stream and riparian ecosystems in the western United States. J. Soil and Water Conservation 54:419–431.

Beschta, R. L., and J. B. Kauffman. 2000. Restoration of riparian systems —taking a broader view. Pp. 323–328 In: Riparian ecology and management in multi-land use watersheds. J. P. J. Wigington, Jr., and R. L. Beschta (eds.). Middleburg, VA: American Water Resources Association. 616 pp.

Beschta, R. L., W. S. Platts, J. B. Kauffman, and M. T. Hill. 1994. Artificial stream restoration— money well spent or an expensive failure? Pp. 76–104 In: Environmental restoration. Carbondale, IL: University Council on Water Resources, University of Southern Illinois.

Binford, M. W., and M. J. Buchenau. 1993. Riparian greenways and water resources. Pp. 69–104 In: Ecology of greenways: design and function of linear conservation areas. D. S. Smith and P. C. Hellmund (eds.). Minneapolis, MN: University of Minnesota Press.

Black, S. P., Broadhurst, J. Hightower, and S. Schauman. 1985. The value of riparian habitat and wildlife to the residents of a rapidly urbanizing community. Pp. 413–416 In: Riparian ecosystems and their management: reconciling conflicting uses. USDA/Forest Service Gen. Tech. Bull. RM-120. Washington, DC: USDA Forest Service.

Blackburn, W. H. 1984. Impacts of grazing intensity and specialized grazing systems on watershed characteristics and responses. Pp. 927–83 In: Developing strategies for rangeland management. Boulder, CO: Westview Press.

Branson, F. A. 1984. Evaluation of impacts of grazing intensity and specialized grazing systems on watershed characteristics and responses. Pp. 985–1000 In: Developing strategies for rangeland management. Boulder, CO: Westview Press.

Brinson, M. M. 1993. A hydrogeomorphic classification for wetlands. Technical Report WRP-DE-4. Vicksburg, MI: Waterways Experiment Station, Army Corps of Engineers.

Brinson, M. M. 1996. Assessing wetland functions using HGM. National Wetlands Newsletter 18(1):10–16.

Brinson, M. M., and R. Rheinhardt. 1996. The role of reference wetlands in functional assessment and mitigation. Ecological Applications 6:69–76.

Brinson, M. M., B. L. Swift, R. C. Plantico, and J. S. Barclay. 1981. Riparian ecosystems: their ecology and status. FWS/OBS-81/17. Washington, DC: U.S. Fish and Wildlife Service, Biological Sciences Program. 155 pp.

Bromberg, M., and T. Funk. 1998. Fencing livestock away from streams reduces erosion and health risks. University of Illinois Cooperative Extension Service. Internet source: www.ag.uiuc.edu/news/articles/898265683.html.

Brooks, K. N., P. F. Ffolliott, H. M. Gregersen, and J. L. Thames. 1991. Hydrology and the management of watersheds. Ames, IA: Iowa State University Press. 392 pp.

Brosofske, K. D., J. Chen, R. J. Naiman, and J. F. Franklin. 1997. Harvesting effects on microclimatic gradients from small streams to uplands in western Washington. Ecol. Applications 7:1188–1200.

Bryant, L. D. 1982. Response of livestock to riparian zone exclusion. J. Range Mgmt. 35:780–85.

Busby, F. E. 1979. Riparian and stream ecosystems, livestock grazing, and multiple-use management. Pp. 6–12 In: Proceedings of the forum: grazing and riparian/stream ecosystems, Nov. 3–4, 1978. O. B. Cope (ed.). Denver, CO: Trout Unlimited.

Cairns, J., Jr. 1993. Is restoration ecology practical? Restoration Ecology 1:3–6.
Chaney, W., W. Elmore, and W. S. Platts. 1990. Livestock grazing on western riparian areas. Eagle, ID: Northwest Resource Information Center, Inc. 45 pp.
Chaubey, I., D. R. Edwards, T. C. Daniel, P. A. Moore Jr., and D. J. Nichols. 1994. Effectiveness of vegetative filter strips in retaining surface-applied swine manure constituents. Trans. ASAE 37:845–850.
Chilman, K. C., D. Foster, and A. Everson. 1985. Using visitor perceptions in river use planning, 1972–1984. Pp. 393–397 In: Riparian ecosystems and their management: reconciling conflicting uses. USDA/Forest Service Gen. Tech. Bull. RM-120. Washington, DC: USDA Forest Service.
Clary, W. P. 1990. Stream channel and vegetation responses to late spring cattle grazing. J. Range Mgmt. 52:218–27.
Clary, W. P., and B. F. Webster. 1990. Riparian grazing guidelines for the intermountain region. Rangelands 12:209–211.
Clary, W. P. 1999. Stream channel and vegetation responses to late spring cattle grazing. Journal of Range Management 52:218–227.
Clawson, J. E. 1993. The use of off-stream water developments and various other gap configurations to modify the watering behavior of grazing cattle. M.S. thesis, Oregon State Univ., Corvallis. 80 pp.
Clevenger, A. P., and N. Waltho. 2000. Factors influencing the effectiveness of wildlife underpasses in Banff National Park, Alberta, Canada. Conservation Biology 14:47–55.
Cole, D. N., M. E. Petersen, and R. C. Lucas. 1987. Managing wilderness recreation use: common problems and potential solutions. USDA/FS Gen. Tech. Report INT-230. Washington, DC: USDA Forest Service.
Cole, D. N. 1993. Minimizing conflict between recreation and nature conservation. Pp. 105–122 In: Ecology of greenways: design and function of linear conservation areas. D. S. Smith and P. C. Hellmund (eds.). Minneapolis, MN: University of Minnesota Press.
Collier, M. P., R. H. Webb, and E. D. Andrews. 1997. Experimental flooding in Grand Canyon. Scientific American 276:82–89.
Commission for Environmental Cooperation. 1998. Sustaining and enhancing riparian migratory bird habitat on the upper San Pedro River.
Committee of Scientists. 1999. Sustaining the people's lands: recommendations for stewardship of the National Forests and Grasslands into the next century. Washington, DC: US Department of Agriculture. 193 pp.
Conner, R. N., R. G. Hooper, H. S. Crawford, and H. S. Mosby. 1975. Woodpecker nesting habitat in cut and uncut woodlands in Virginia. J. Wildl. Manage. 39:144–150.
Conroy, S. D., and T. J. Svejcar. 1991. Willow planting success as influenced by site factors and cattle grazing in northeastern California. J. Range Mgmt. 44:59–63.
Cooper, J. R., J. W. Gilliam, R. B. Daniels, and W. P. Robarge. 1987. Riparian areas as filters for agricultural sediment. J. Soil Sci. Soc. Am. 51(2):416–420.
Cope, O. B. 1979. Proceedings of the forum: grazing and riparian/stream ecosystems. Nov. 3–4, 1978. Denver, CO: Trout Unlimited.
Crawford, C. S., L. M. Ellis, M. C. Molles, Jr, and H. M. Valett. 1996. The potential for implementing partial restoration of the middle Rio Grand ecosystem. Pp. 93–99 In: Desired future conditions for southwestern riparian ecosystems: bringing interests and concerns together. D. W. Shaw, D. M. Finch (eds.). General Technical Report no. RM—GTR—272. Fort Collins, CO: USDA Forest Service.
Cummins, K. W., and C. N. Dahm. 1995. Restoring the Kissimmee. Restoration Ecology 3:147–148.
Dahm, C. N., K. W. Cummins, H. M. Valett, and R. L. Coleman. 1995. An ecosystem view of the restoration of the Kissimmee River. Restoration Ecology 3:225–238.

Dalton, R. S., C. T. Bastian, J. J. Jacobs, and T. A. Wesche. 1998. Estimating the economic value of improved trout fishing on Wyoming streams. N. Amer. J. Fisheries Mgmt. 18:786–97.

Daniels, R. B., and J. W. Gilliam. 1996. Sediment and chemical load reduction by grass and riparian filters. Soil Sci. Soc. Amer. J. 60:246–251.

Darveau, M., Beauchesne, P., Belanger, L., Huet, J., Larue, P. 1995. Riparian forest strips as habitat for breeding birds in boreal forests. J. Wildlife Manage. 59:67–78.

David, M. B., L. E. Gentry, D. A. Kovacic, and K. M. Smith. 1997. Nitrogen balance in and export from an agricultural watershed. Journal of Environmental Quality 26:1038–1048.

Davis, K. C., and C. B. Marlow. 1990. Altering cattle behavior through grazing management. Montana Ag Research 7(1):11–14.

DeLoach, C. J. 2000. Saltcedar biological control: methodology, exploration, laboratory trials, proposals for field releases, and expected environmental effects. http://refuges.fws.gov/nwrsfiles/HabitatMgmt/PestMgmt/SaltcedarWorkshopSep96/deloach.html

Demaynadier, P. G., and M. L. Hunter, Jr. 1998. Effects of silvicultural edges on the distribution and abundance of amphibians in Maine. Conservation Biology 12:340–352.

Di Castri, F. 2000. Ecology in a context of economic globalization. BioScience 50:321–332.

Diamond, J., J. Terborgh, R. F. Whitcomb, J. F. Lynch, P. A. Opler, D. S. Simberloff, L. F Abele. 1976. Island biogeography and conservation: strategy and limitations. Science 193:1027–1032.

Dillaha, T. A., R. B. Reneau, S. Mostaghimi, and V. O. Shanholtz. 1989. Vegetative filter strips for nonpoint source pollution control. Transactions of the ASAE 32(2):491–496.

Donahue, D. L. 1999. The western range revisited: removing livestock from public lands to conserve native biodiversity. Norman, OK: University of Oklahoma Press. 388 pp.

Ehrhart, R. C., and P. L. Hansen. 1997. Effective cattle management in riparian zones: a field survey and literature review. Montana BLM Riparian Technical Bulletin No. 3. Montana State Office, BLM. 92 pp.

Elder, J. 1999. In pursuit of a bioregional curriculum. Orion Afield, Spring:26–28.

Elmore, W., and R. L. Beschta. 1987. Riparian areas: perceptions in management. Rangelands 9: 260–265.

Elmore, W., and R. L. Beschta. 1989. The fallacy of structures and the fortitude of vegetation. General Technical Report PSW-110. Washington, DC: U.S. Department of Agriculture.

Elmore, W., and B. Kaufman. 1994. Riparian and watershed systems: degradation and restoration. Pp. 212–231 In: Ecological implications of livestock herbivory in the West. M. Vavra et al. (eds.). Denver, CO: Society for Range Management.

Enrenreich, J. H., and A. J. Bjugstad. 1966. Cattle grazing time is related to temperature and humidity. Journal of Range Management 19:141–142.

Environmental Protection Agency (EPA). 1992. Federal Register 57:22746.

EPA. 1993. Guidance specifying management measures for sources of nonpoint pollution in coastal waters. EPA-840-B-92-002. Washington, DC: EPA Office of Water.

EPA. 1995. Watershed protection: a statewide approach. EPA-841-R-95-004. Washington, DC: EPA Office of Water.

Federal Interagency Working Group. 1998. Stream corridor restoration. principles, processes, and practices. Washington, DC: National Technical Information Service, U.S. Department of Commerce.

Federal Register 64:23592, May 3, 1999.

FEMAT (Report of the Forest Ecosystem Management Assessment Team). 1993. Forest ecosystem management: an ecological, economic, and social assessment. Washington, DC: USDA Forest Service.

Field, D. R., M. E. Lee, and K. Martinson. 1985. Human behavior and recreation habitats: conceptual issues. Pp. 227–231 In: Riparian ecosystems and their management: reconciling conflicting uses. USDA/Forest Service Gen. Tech. Bull. RM-120. Washington, DC: USDA Forest Service.

Fischenich, J. C. 1997. Hydraulic impacts of riparian vegetation; summary of the literature. Technical Report EL-97-9. Vicksburg, MS: U.S. Army Corps of Engineers, Waterways Experiment Station. 53 pp.

Fleischner, T. L. 1994. Ecological Costs of Livestock Grazing in Western North America. Conservation Biology 8:629–644.

Fore, L. S., J. R. Karr, and R. W. Wisseman. 1996. Assessing invertebrate responses to human activities: evaluating alternative approaches. Journal of the North American Benthological Society 15:212–231.

Forman, R. T. T. 1995. Land mosaics. Cambridge, UK: Cambridge University Press. 632 pp.

Forman, R. T. T., and J. Baudry. 1984. Hedgerows and hedgerow networks in landscape ecology. Environmental Management 8:495–510.

Foss, P. 1960. Politics and grass. Seattle, WA: University of Washington Press.

Friedman, J. M., W. R. Osterkamp, and W. M. Lewis, Jr. 1996. The role of vegetation and bed-level fluctuations in the process of channel narrowing. Geomorphology 14:341–351.

Gardner, R. H., M. G. Turner, R. V. O'Neill, and S. Lavore. 1991. Simulation of the scale-dependent effects of landscape boundaries on species persistence and dispersal. Pp. 76–89 In: Ecotones, the role of landscape boundaries in the management and restoration of changing environments. M. M. Holland, P. G. Risser, and R. J. Naiman (eds.). New York: Chapman and Hall.

Garland, J. J. 1983. Designated skid trails minimize soil compaction. Oregon State University Extension Circular 1110, Corvallis, OR. 6 pp.

Garland, J. J. 1987. Aspects of practical management in the streamside zone. Pp. 277–288 In: Streamside management: forestry and fisheries interactions. E. O. Salo and T. W. Cundy (eds.). Contribution No. 57. Seattle, WA: University of Washington, Institute of Forest Resources. 471 pp.

Gatto, M., and G. A. de Leo. 2000. Pricing biodiversity and ecosystem services: the never-ending story. BioScience 50:347–355.

General Accounting Office (GAO). 1990. Rangeland management: BLM efforts to prevent unauthorized livestock grazing need strengthening. GAO/RCED—91-17. Washington, DC: GAO.

Ghaffarzadeh, M., C. A. Robinson, and R. M. Cruse. 1992. Vegetative filter strip effects on sediment deposition from overland flow. Madison, WI: Agronomy Abstracts, ASA. 324 pp.

Gifford, G. F., and R. H. Hawkins. 1978. Hydrologic impact of grazing on infiltration: a critical review. Water Resources Research 14:305–313.

Gillen, R. L., W. C. Krueger, and R. F. Miller. 1985. Cattle use of riparian meadows in the Blue Mountains of northeastern Oregon. J. Range Mgmt. 38:205–209.

Goodwin, C. N., C. P. Hawkins, and J. L. Kershner. 1997. Riparian restoration in the western United States: overview and perspective. Restoration Ecology 5(4s):4–14.

Gore, J. A., and F. D. Shields, Jr. 1998. Can large rivers be restored? BioScience 45:142–152.

Gourley, C. R. 1997. Restoration of the Lower Truckee River ecosystem: challenges and opportunities. Paper presented at the Wallace Stegner Conference: "To cherish and renew": restoring western ecosystems and communities. April 17–19, 1997.

Grumbine, R. E. 1992. Ghost bears: exploring the biodiversity crisis. Washington, DC: Island Press.

Gunderson, D. R. 1968. Floodplain use related to stream morphology and fish populations. J. Wildl. Manage. 32(3):507–514.

Gwinn, S. E., M. E. Kentula, and P. W. Shaffer. 1999. Evaluating the effects of wetland regulation through hydrogeomorphic classification and landscape profiles. Wetlands 19:477–489.

Hall, F. C., and L. Bryant. 1995. Herbaceous stubble height as a warning of impending cattle grazing damage to riparian areas. Gen. Tech. Rept. PNW-GTR-362. Portland, OR: USDA Forest Service, Pacific Northwest Research Station. 9 pp.

Harris, L. D., and K. Atkins. 1991. Faunal movement corridors in Florida. Pp. 117–148 In: Landscape linkages and biodiversity. W. E. Hudson (ed.). Washington, DC: Island Press.

Hart, R. H., J. Bissio, M. J. Samuel, and J. W. Waggoner, Jr. 1993. Grazing systems, pasture size, and cattle grazing behavior, distribution, and gains. J. Range Mgmt. 46:81–87.

Hauer, F. R., and R. D. Smith. 1998. The hydrogeomorphic approach to functional assessment of riparian wetlands: evaluating impacts and mitigation on river floodplains in the U.S. A. Freshwater Biology 40(3):517–530.

Hawkins, C. P., R. H. Norris, J. N. Hogue, and J. W. Feminella. 2000. Development and evaluation of predictive models for measuring the biological integrity of streams. Ecological Applications 10:1456–1477.

Hendrickson, D. A., and W. L. Minckley. 1984. Cienegas—vanishing climax communities of the American Southwest. Desert Plants 6:131–175.

Hill, M. T., W. S. Platts, and R. L. Beschta. 1991. Ecological and geomorphological concepts for instream and out-of-channel flow requirements. Rivers 2:198–210.

Hoctor, T. S., M. H. Carr, and P. D. Zwick. 2000. Identifying a linked reserve system using a regional landscape approach: the Florida ecological network. Conservation Biology 14:984–1000.

Horner, R. R., and B. W. Mar. 1982. Guide for water quality impact assessment of highway operations and maintenance. Rep. WA-RD-39.14. Olympia, WA: Washington Dept. of Transportation.

Howery, L. D., F. D. Provenza, R. E. Banner, and C. B. Scott. 1996. Differences in home range and habitat use among individuals in a cattle herd. Applied Animal Behavior Sci. 49:305–20.

Huber, S. A., M. B. Judkins, L. J. Krysl, T. J. Svejcar, B. W. Hess and D. W. Holcombe. 1995. Cattle grazing a riparian mountain meadow: effects of low and moderate stocking density on nutrition, behavior, diet selection and plant growth response. J. Anim. Sci. 73:3752–3765.

Hubert, W. A., R. P. Lanka, T. A. Wesche, and F. Stabler. 1985. Grazing management influences on two brook trout streams in Wyoming. U.S. Forest Service General Technical Report RM-120:290–294.

Hughes, L. E. 1979. Rest-rotation grazing vs. season-long grazing on Naval Oil Shale Reserve Allotment in Colorado. Rangelands 1:55–56.

Hunter, C. 1991. Better trout habitat, a guide to stream restoration and management. Washington, DC: Island Press.

Hunter, M. L., Jr. 1996. Fundamentals of conservation biology. Cambridge, MA: Blackwell Science. 482 pp.

Ilhardt, B. L., E. S. Verry, and B. J. Palik. 2000. Defining riparian areas. In: Riparian management in forests of the Continental Eastern United States. E. S. Verry, J. W. Hornbeck, and C. A. Dolloff (eds.). Washington, DC: Lewis Publishers.

Inamdar, S. P., J. M. Sheridan, R. G. Williams, D. D. Bosch, R. R. Lowrance, L. S. Altier and D. L. Thomas. 1999. Riparian ecosystem management model (REMM): I. Testing of the hydrologic component for a Coastal Plain riparian system. Trans. Am. Soc. Agr. Eng. 42:1679–1689.

Inamdar, S. P. 1991. Riparian zone management model for optimized water quality, biomass and wildlife species diversity. Unpublished Independent Study Report, University of Kentucky, Lexington, KY.

Iowa State University. 1997. Stewards of our streams: buffer strip design, establishment, and maintenance. Pub. 1626b. Ames, IA: Iowa State University Extension.

Isenhart, T. M., R. C. Schultz, and J. P. Colletti. 1997. Watershed restoration and agricultural practices in the Midwest: Bear Creek of Iowa. Pp. 318–334 In: Watershed restoration: principles and practices. J. E. Williams, C. A. Wood, and M. P Dombeck (eds.). Bethesda, MD: American Fisheries Society.

Jackson, L. L., N. Lopoukhine, and D. Hillyard. 1995. Ecological restoration: a definition and comments. Restoration Ecology 3:71–75.

Janssen, R. 1994. Multiobjective decision support for environmental management. Dordrect, The Netherlands: Kluwer Academic Publishers. 232 pp.

Jones, J. B., and P. J. Mulholland. 2000. Streams and ground waters. San Diego, CA: Academic Press.

Karr, J. R. 1981. Assessment of biotic integrity using fish communities. Fisheries 6:21–27.
Karr, J. R., and D. R. Dudley. 1981. Ecological perspective on water quality goals. Environmental Management 5:55–68.
Karr, J. 1996. Making meaning of all the science. Pp. 149–152 In: At the water's edge: the science of riparian forestry. Conference Proceedings, January 1996. BU-6637-S. Minneapolis, MN: University of Minnesota Extension Service.
Karr, J. R. 1991. Biological integrity: a long-neglected aspect of water resource management. Ecological Applications 1:66–84.
Karr, J. R., and E. W. Chu. 1999. Restoring life in running waters: better biological monitoring. Washington, DC: Island Press.
Kauffman, J. B, R. L. Beschta, N. Otting, and D. Lytjen. 1997. An ecological perspective of riparian and stream restoration in the western United States. Fisheries 22(5):12–24.
Kauffman, J. B., W. C. Krueger, and M. Vavra. 1983. Effects of late season cattle grazing on riparian plant communities. Journal of Range Management 36:685–91.
Keller, C. R., and K. P. Burnham. 1982. Riparian fencing, grazing, and trout habitat preference on Summit Creek, Idaho. North American Journal of Fisheries Management 2:53–59.
Kenney, D. S., S. T. McAllister, W. H. Caile, and J. S. Peckham. 2000. The *new* watershed source book: a directory and review of watershed initiatives in the western United States. Natural Resources Law Center.
Kershner, J. L. 1997. Setting riparian/aquatic restoration within a watershed context. Restoration Ecology 5(4s):15–24.
Kilgo, J. C., R. A. Sargent, B. R. Chapman, and K. V. Miller. 1998. Effect of stand width and adjacent habitat on breeding bird communities in bottom hardwoods. J. Wildl. Manage. 62:72–83
Knight, R. L., and S. A. Temple. 1995. Wildlife and recreationists: coexistence through management. Pp. 327–333 In: Wildlife and recreationists: coexistence through management and research. R. L. Knight and K. J. Gutzwiller (eds.). Washington, DC: Island Press. 372 pp.
Koehler, D. A., and A. E. Thomas. 2000. Managing for enhancement of riparian and wetland areas of the western United States: an annotated bibliography. General Technical Report RMRS-GTR-54. Washington, DC: USDA Forest Service. 369 pp.
Kondolf, G. M., and M. G. Wolman. 1993. The sizes of salmonoid spawning gravels. Water Resources Research 29:2275–2285.
Kovacic, D. A., M. A. David, L. E. Gentry, K. M. Starks, and R. A. Cooke. 2000. Effectiveness of constructed wetlands in reducing nitrogen and phosphorus export from agricultural tile drainage. Journal of Environmental Quality 29:1262–1274.
Kovalchick, B. L. 1987. Riparian zone associations: Deschutes, Ochoco, Fremont, and Winema National Forests. USDA Forest Service, Pacific Northwest Region, Portland, OR, RG-ECOL-TP-297-87. 171 pp.
Kusler, J. A., and M. E. Kentula, eds. 1990. Wetland creation and restoration; the status of the science. Washington, DC: Island Press. 594 pp.
Kusler, J., and W. Niering. 1998. Wetland assessment: have we lost our way? National Wetlands Newsletter 20(1):9–14.
Larson, J. S., and D. B. Mazzarese. 1994. Rapid assessment of wetlands: history and application to management. Pp. 625–636 In: Global wetlands: old world and new. W. J. Mitsch (ed.). Amsterdam: Elsevier Science B.V.
Larson, R. A. 1995. Balancing wildlife viewing with wildlife impacts: a case study. Pp. 257–270 In: Wildlife and recreationists: coexistence through management and research. R. L. Knight and K. J. Gutzwiller (eds.). Washington, DC: Island Press. 372 pp.
Law, D. J., C. B. Marlow, J. C. Mosley, S. Custer, P. Hook, and B. Leinard. 2000. Water table dynamics and soil texture of three riparian plant communities. Northwest Science 74(3):233–241.

Lee, D., T. A. Dillaha and J. H. Sherrard. 1999. Modeling phosphorous transport in grass buffer strips. J. Environ. Eng. ASCE 115(2):409–427.

Lee, K. H., T. M. Isenhart, R. C. Schultz, and S. K. Mickelson. 2000. J. Environmental Quality 29:1200–1205.

Lenat, D. R. 1993. A biotic index for the southeastern United States: derivation and list of tolerance values, with criteria for assigning water quality ratings. Journal of the North American Benthological Society 12:279–290.

Leon, S. C. 2000. Southwestern Willow Flycatcher. http://ifw2es.fws.gov/swwf/

Leonard, S., G. Staidl, J. Fogg, K. Gebhardt, W. Hagenbuck, and D. Prichard. 1992. Procedures for ecological site inventory – with special reference to riparian-wetland sites. Technical Reference 1737-7. Denver, CO: Bureau of Land Management, National Applied Resource Science Center. 135 pp.

Leopold, L. B., and M. G. Wolman. 1957. River channel patterns: braided, meandering and straight. U.S. Geological Survey Professional Paper 282-B, pp. 39-85.

Leopold, L. B., M. G. Wolman, and J. P. Miller. 1964. Fluvial processes in geomorphology. San Francisco, CA: W. H. Freeman and Company. 522 pp.

Lieberman, G. A., and L. L. Hoody. 1998. Closing the achievement gap using the environment as an integrating context for learning. San Diego, CA: State Education and Environment Roundtable.

Liebowitz, S. G., B. Abbruzzese, P. R. Adamus, L. E. Hughes, and J. T. Irish. 1992. A synoptic approach to cumulative impact assessment: a proposed methodology. Report EPA/600/R-92/167. Corvallis, OR: EPA Environmental Research Laboratory.

Lieurance, F. S., H. M. Valett, C. S. Crawford, and M. C. Molles, Jr. 1994. Experimental flooding of a riparian forest: restoration of ecosystem functioning. Pp. 365–374 In: Proceedings of the second international conference on ground water ecology. J. A. Stanford and H. M. Valett (eds.). Herndon, VA: American Water Resources Association.

Lisle, T. E. 1989. Sediment transport and resulting deposition in spawning gravels, north coastal California. Water Resources Research 25:1303–1319.

Loeks, C. D. 1985. Thinking laterally: strategies for strengthening institutional capacity for integrated management of riparian resources. Pp. 13–20 In: Riparian ecosystems and their management: reconciling conflicting uses. USDA/Forest Service Gen. Tech. Bull. RM-120. Washington, DC: USDA Forest Service.

Lopez, R. D., and M. S. Fennessy. 2002. Testing the floristic quality assessment index as an indicator of wetland condition. Ecological Applications 12(2):487–497.

Lowrance, R., L. S. Altier, R. G. Williams, S. P. Inamdar, D. D. Bosch, R. K. Hubbard, and D. L. Thomas. 2000. REMM: The riparian ecosystem management model. J. Soil & Water Conserv. 55:27–36.

Lowrance, R. R. 1992. Groundwater nitrate and denitrification in a Coastal Plain riparian forest. J. Environ. Qual. 21:401–405.

Lowrance, R. R., R. L. Todd and L. E. Asmussen. 1983. Waterborne nutrient budgets for the riparian zone on an agricultural watershed. Agr. Ecosystems and Environ. 10:371–384.

Lowrance, R., R. Leonard, and J. Sheridan. 1985. Managing riparian ecosystems to control nonpoint pollution. Journal of Soil and Water Conservation 40:87–97.

Lowrance, R., L. S. Altier, J. D. Newbold, R. R. Schnabel, P. M. Groffman, J. M. Denver, D. L. Correll, J. W. Gilliam, J. L. Robinson, R. B. Brinsfield, K. W. Staver, L. Lucas, and A. H. Todd. 1995. Water quality functions of riparian forest buffer systems in the Chesapeake Bay Watershed. Annapolis, MD: U.S. EPA Chesapeake Bay Program. 67 pp.

Lynch, J. A., E. S. Corbett and K. Mussallem. 1985. Best management practices for controlling nonpoint source pollution on forested watersheds. J. Soil and Water Conservation 40:164–167.

MacClintock, L., R. F. Whitcomb, and B. L. Whitcomb. 1977. Evidence for the value of corridors and minimization of isolation in preservation of biotic diversity. Am. Birds 31:6–16.

MacDonnell, L. J. 1996. Managing reclamation facilities for ecosystem benefits. University of Colorado Law Review 67(2):197–257.

Mackintosh, G. ed. 1989. Preserving communities and corridors. Washington, DC: Defenders of Wildlife. 95 pp.

Madison, C. E., R. L. Blevins, W. W. Frye, and B. J. Barfield. 1992. Tillage and grass filter strip effects upon sediment and chemical losses. Madison, WI: Agronomy Abstracts, ASA. p. 331.

Magette, W. L., R. B. Brinsfield, R. E. Palmer, and J. D. Wood. 1989. Nutrient and sediment removal by vegetated filter strips. Transactions of the ASAE 32(2):663–667.

Malanson, G. P. 1993. Riparian landscapes: Cambridge studies in ecology. Cambridge, UK: Cambridge University Press.

Marlow, C. B., and T. M. Pogacnik. 1986. Cattle feeding and resting patterns in a foothills riparian zone. J. Range Mgmt. 39:212–217.

Masters, L., S. Swanson, and W. Burkhardt. 1996. Riparian grazing management that worked: introduction and winter grazing. Rangelands 18:192–195.

Mattson, J. A., J. E. Baumgras, C. R. Blinn, and M. A. Thompson. 2000. Harvesting options for riparian areas. Pp. 255–272 In: Riparian management in forests of the continental United States. E. S. Verry, J. W. Hornbeck, and D. A. Dolloff (eds.). New York: Lewis Publishers. 402 pp.

Mayer, R. 1999. Estimated costs for livestock fencing. Iowa State University Extension Service. FM 1855. Internet source: www.agry.purdue.edu/ext/forages/rotational/fencing/fencing_costs.html.

McDade, M. H., F. J. Swanson, W. A. McKee, J. F. Franklin, and J. Van Sickle. 1990. Source distances for coarse woody debris entering small streams in western Oregon and Washington. Canadian Journal of Forest Resources 20:326–330.

McInnis, M. L., and McIver, J. 2001. Influence of off-stream supplements on streambanks of riparian pastures. J. Range Manage. 54:648–652.

Meehan, W. R, ed. 1991. Influences of forest and rangeland management on salmonid fishes and their habitats. Special Publication 19. Bethesda, MD: American Fisheries Society. 751 pp.

Merritt, D. M., and D. J. Cooper. 2000. Riparian vegetation and channel change in response to river regulation: a comparative study of regulated and unregulated streams in the Green River Basin, USA. Regulated Rivers 16:543–564.

Mickelson, S. K., J. L. Baker, K. Arora, and A. Misra. 1995. A summary report: the effectiveness of buffer strips in reducing herbicide losses. Proc. 50th Annual Mtg. Soil and Water Conservation Society.

Mitsch, W. J., and J. G. Gosselink. 2000. The value of wetlands: importance of scale and landscape setting. Ecological Economics 35:25–33.

Molles, M .C., Jr., C. S. Crawford, and L. M. Ellis. 1995. The effects of an experimental flood on litter dynamics in the Middle Rio Grande riparian ecosystem. Regulated Rivers: Research and Management 11:275–281.

Molles, M. C., Jr., C. S. Crawford, L. M. Ellis, H. M. Valett, and C. N. Dahm. 1998. Managed flooding for riparian ecosystem restoration. BioScience 48:749–756.

Montgomery, D. R., and J. M. Buffington. 1997. Channel classification, prediction of channel response, and assessment of channel condition. Prepared for the Washington State Timber/Fish/Wildlife Committee. Report TFW-SH10-93-002.

Mosley, J. C., P. S. Cook, A. J. Griffis, and J. O'Laughlin. 1997. Guidelines for managing cattle grazing in riparian areas to protect water quality: review of research and best management practices policy. Rept. No. 15. Moscow, ID: University of Idaho, Wildlife and Range Policy Analysis Group.

Murphy, M. L. 1995. Forestry impacts on freshwater habitat of anadromous salmonids in the Pacific Northwest and Alaska—requirements for protection and restoration. U.S. Department of Commerce, National Oceanic and Atmospheric Administration, Decision Analysis Series No. 7, 156 pp.

Naiman, R. J., and K. H. Rogers. 1997. Large animals and system-level characteristics in river corridors. BioScience 47:521–529.
National Research Council (NRC). 1992. Restoration of aquatic ecosystems. Washington, DC: National Academy Press.
NRC. 1995. Wetlands: characteristics and boundaries. Washington, DC: National Academy Press.
NRC. 1996. Upstream: salmon and society in the Pacific Northwest. Washington, DC: National Academy Press.
NRC. 1999. New strategies for America's watersheds. Washington, DC: National Academy Press.
NRC. 2000. Watershed management for potable water supply: assessing the New York City strategy. Washington, DC: National Academy Press.
NRC. 2001. Compensating for wetland losses under the Clean Water Act. Washington, DC: National Academy Press.
NRC. 2002. The Missouri River ecosystem: exploring the prospects for recovery. Washington, DC: National Academy Press.
Natural Resources Law Center. 1996. The watershed source book: watershed-based solutions to natural resource problems. Boulder, CO: University of Colorado.
Nichols, D. J., T. C. Daniel, D. R. Edwards, P. A. Moore, and D. H. Pote. 1998. Use of grass filter strips to reduce 17-ß-estradiol in runoff from fescue applied poultry litter. J. Soil and Water Conservation 53:74–77.
Niemi, G. J., and J. M. Hanowski. 1984. Relationships of breeding birds to habitat characteristics in logged areas. J. Wildl. Manage. 48:438–443.
Noss, R. F., H. B. Quigley, M. G. Hornocker, T. Merrill, and P. C. Paquet. 1996. Conservation biology and carnivore conservation in the Rocky Mountains. Conservation Biology 10:949–963.
Noss, R. 1993. Wildlife corridors. Pp. 43–68 In: Ecology of greenways: design and function of linear conservation areas. D. S. Smith and P. C. Hellmund (eds.). Minneapolis, MN: University of Minnesota Press.
Noss, R. F. 1983. A regional landscape approach to maintain diversity. BioScience 33:700–706.
Noss, R. F. 1997. The failure of universities to produce conservation biologists. Conservation Biology 11:1267–1269.
Noss, R. F., and L. D. Harris. 1986. Nodes, networks, and MUMs: preserving diversity at all scales. Environmental Management 10:299–309.
Noss, R. F., and A. Y. Cooperrider. 1994. Saving nature's legacy. Washington, DC: Island Press.
Noss, R. F. 1991. Landscape connectivity: different functions at different scales. Pp. 27–39 In: Landscape linkages and biodiversity. W. E. Hudson (ed.). Washington, DC: Island Press.
O'Neill, R. V., B. T. Milne, M. G. Turner, and R. H. Gardner. 1988. Resource utilization scales and landscape pattern. Landscape Ecology 2:63–69.
Ohmart, R. D., and B. W. Anderson. 1986. Riparian habitats. Pp. 169–99 In: Inventory and monitoring of wildlife habitat. A. Y. Cooperider, R. J. Boyd, and H. R. Stuart (eds.). USDI-BLM.
Olson, R. W., and C. L. Armour. 1979. Economic considerations for improved livestock management approaches for fish and wildlife in riparian/stream areas. Pp. 67–71 In: Proceedings of the forum: grazing and riparian/stream ecosystems. Nov. 3–4, 1978. O. B. Cope (ed.). Denver, CO: Trout Unlimited.
Orr, D. 1990a. Is conservation education an oxymoron? Conservation Biology 4:119–121.
Orr, D. 1990b. The virtue of conservation education. Conservation Biology 4:219–220.
Orr, D. 2000. Ideasclerosis: part two. Conservation Biology 14:1571–1572.
Palik, B. J., J. C. Zasada, and C. W. Hedman. 2000. Ecological principles for riparian silviculture. Pp. 233–254 In: Riparian management in forests of the continental Eastern United States. E. S. Verry, J. W. Hornbeck, and C. A. Dolloff (eds.). New York: Lewis Publishers. 402 pp.
Peters, M. R., S. R. Abt, C. C. Watson, J. C. Fischenich, and J. M. Nestler. 1995. Assessment of restored riverine habitat using RCHARC. Water Resources Bulletin 31:745–752.

Petersen, R. C., L. B. M. Petersen, and J. Lacoursiere. 1992. A building block model for stream restoration. P. Boon, G. Petts, and P. Calow (eds.). The conservation and management of rivers. New York: John Wiley & Sons.

Phillips, R. L., M. J. Trlica, W. C. Leininger, and W. P. Clary. 1999. Cattle use affects forage quality in a montane riparian ecosystem. J. Range Mgmt. 52:283–289.

Planning and Conservation League (PCL) Foundation. 1999. The benefits of watershed management: water quality and supply. June 23, 1999.

Platts, W. S. 1990. Managing fisheries and wildlife on rangelands grazed by livestock: a guidance and reference document for biologists. Nevada Dept. of Wildlife Rept. 114 pp.

Platts, W. S., and J. N. Rinne. 1985. Riparian and stream enhancement management and research in the Rocky Mountains. North American Journal of Fishery Management 5:115–125.

Platts, W. S., and F. J. Wagstaff. 1984. Fencing to control livestock grazing on riparian habitats along streams: is it a viable alternative? North American Journal of Fisheries Management 4:266–272.

Platts, W. S., and R. F. Raleigh. 1984. Impacts of Grazing on Wetlands and Riparian Habitat. Pp. 1105–17 In: Developing strategies for rangeland management. Boulder, CO: Westview Press.

Platts, W. S., and R. L. Nelson. 1985a. Will the riparian pasture build good streams? Rangelands 7:7–10.

Platts, W. S., and R. L. Nelson. 1985b. Impacts of rest-rotation grazing on stream banks in forested watersheds in Idaho. North American Journal of Fisheries Management 5:547–556.

Platts, W. S., and R. L. Nelson. 1985c. Streamside and upland vegetation use by cattle. Rangelands 7:5–7.

Poff, N. L., J. D. Allan, M. B. Bain, J. R. Karr, K. L. Prestegaard, B. D. Richter, R. E. Sparks, and J. C. Stromberg. 1997. The natural flow regime: a paradigm for river conservation and restoration. BioScience 47(11):769–784.

Premo, D., and E. Rogers. 1998. Promoting healthy waterfront property—a pilot program hits the ground running. Lakeline 18:16–17, 38–39.

Prichard, D., H. Barrett, J. Cagney, R. Clark, J. Fogg, K. Gebhardt, P. L. Hansen, B. Mitchell, and D. Tippy. 1993. Riparian area management: process for assessing proper functioning condition. Technical Reference 1737-9. Denver, CO: Bureau of Land Management.

Prichard, D., J. Anderson, C. Correll, J. Fogg, K. Gebhardt, R. Krapf, S. Leonard, B. Mitchell, and J. Staats. 1998. A user guide to assessing proper functioning condition and supporting science for lotic areas. Technical Reference 1737-15. Denver, CO: Bureau of Land Management, National Applied Resource Science Center. 126 pp.

Quammen, M. L. 1986. Measuring the success of wetlands mitigation. National Wetlands Newsletter 8:6–8.

Rheinhardt, R. R., M. C. Rheinhardt, M. M. Brinson, and K. E. Faser, Jr. 1999. Application of reference data for assessing and restoring headwater ecosystems. Ecological Restoration 7(3):241–251.

Rhoads, B. L., and E. E. Herricks. 1996. Naturalization of headwater streams in Illinois: challenges and possibilities. In: River channel restoration: guiding principles for sustainable projects. A. Brookes and F. D. Shields (eds.). New York: John Wiley & Sons.

Richter, B. D., and H. E. Richter. 2000. Prescribing flood regimes to sustain riparian ecosystems along meandering rivers. Conservation Biology 14:1467–1478.

Rickard, W. H., and C. E. Cushing. 1982. Recovery of streamside woody vegetation after exclusion of livestock grazing. Journal of Range Management 35:360–361.

Rinne, J. N. 1988. Grazing effects on stream habitat and fishes: research design considerations. N. Amer. J. Fisheries Manage. 8:240–247.

Rinne, J. N. 1999. Fish and grazing relationships: the facts and some pleas. Fisheries 24(8):12–21.

Robbins, C. S., D. K. Dawson, and B. A. Dowell. 1989. Habitat area requirements of breeding forest birds of the middle Atlantic states. Wildl. Monogr. 103.

Robinson, S. K. 1992. Population dynamics of breeding neotropical migrants in Illinois. Pp 408–418 In: Ecology and conservation of neotropical migrant landbirds. J. M. Hagan III, and D. W. Johnston (eds.). Washington, DC: Smithsonian Institution Press.

Rood, S. B., and J. M. Mahoney. 1990. Collapse of riparian poplar forests downstream from dams in western prairies: probable causes and prospects for mitigation. Environmental Management 14:451–464.

Rood, S. B., and J. M. Mahoney. 2000. Revised instream flow regulation enables cottonwood recruitment along the St. Mary River, Alberta, Canada. Rivers 7:109–125.

Rood, S. B., K. Taboulchanas, C. E. Bradley, and A. R. Kalischuk. 1999. Influence of flow regulation on channel dynamics and riparian cottonwoods along the Bow River, Alberta. Rivers 7:33–48.

Rood, S., and C. R. Gourley. 1996. Instream flows and the restoration of riparian cottonwoods along the Lower Truckee River, Nevada. Unpublished report to the U.S. Fish and Wildlife Service, Reno, NV.

Rosgen, D. 1995. Applied river morphology. Wildland Hydrology, Pagosa Springs, CO.

Roth, N. E., J. D. Allan, and D. E. Erickson. 1996. Landscape influences on stream biotic integrity assessed at multiple spatial scales. Landscape Ecology 11:141–156.

Salo, E. O., and T. W. Cundy, eds. 1987. Streamside management: forestry and fisheries interactions. Contribution No. 57. Seattle, WA: University of Washington, Institute of Forest Resources. 471 pp.

Schellinger, G. R., and J. C. Clausen. 1992. Vegetative filter treatment of dairy barnyard runoff in cold regions. J. Environ. Qual. 21:40–45.

Schmidt, J. C., E. D. Andrews, D. L. Wegner, D. T. Patten, G. R. Marzolf, and T. O. Moody. 1999. Origins of the 1996 controlled flood in Grand Canyon. Pp. 23–36 In: The controlled flood in the Grand Canyon. American Geophysical Union, Monograph 110. R. H. Webb, J. C. Schmidt, G. R. Marzolf, and R. A. Valdez (eds.). Washington, DC: American Geophysical Union.

Schroeder, R. L., L. J. O'Neil, and T. M. Pullen, Jr. 1992. Wildlife community habitat evaluation: a model for bottomland hardwood forests in the southeastern United States. (draft) Biological Report 92. Washington, DC: U.S. Fish and Wildlife Service.

Schueler, T. 1996. The architecture of urban stream buffers. Watershed Protection Techniques 1(4):155–165.

Schultz, R. C., J. P. Colletti, T. M. Isenhart, W. W. Simpkins, C. W. Mize, and M. L. Thompson. 1995. Design and placement of a multi-species riparian buffer strip system. Agroforestry Systems 29:201–226.

Schultz, R. C., J. P. Colletti, T. M. Isenhart, C. O. Marquez, W. W. Simpkins, C. J. Ball, and O. L. Schultz. 2000. Riparian forest buffer practices. Pp. 189–281 In: North American agroforesty: an integrated science and practice. Madison, WI: American Society of Agronomy.

Schulz, T. T., and W. C. Leininger. 1990. Differences in riparian vegetation structure between grazed areas and exclosures. Journal of Range Management 43:295–299.

Schulz, T. T., and W. C. Leininger. 1991. Nongame wildlife communities in grazed and ungrazed montane riparian sites. Great Basin Naturalist 51:286–292.

Schwer, C. B., and J. C. Clausen. 1989. Vegetative filter treatment of dairy milkhouse wastewater. J. Environmental Quality 18:446–451.

Scott, M. L., J. M. Friedman, and G. T. Auble. 1996. Fluvial process and the establishment of bottomland trees. Geomorphology 14:327–339.

Scurlock, M., and J. Curtis. 2000. Maximizing the effectiveness of watershed councils: policy recommendations from Pacific River Council and Trout Unlimited, downloadable at http://www.pacriver.org/alerts/watershed.html

Seehorn, M. E. 1992. Stream habitat improvement handbook. Atlanta, GA: U.S. Department of Agriculture, Forest Service, Southern Region.

Shafroth, P. B., G. T. Auble, J. Stromberg, and D. T. Patten. 1998. Establishment of woody riparian vegetation in relation to annual patterns of streamflow, Bill Williams River, Arizona. Wetlands 18(40):577–590.

Shafroth, P. B., J. C. Stromberg, and D. T. Patten. 2000. Wood riparian vegetation response to different alluvial water table regimes. Western North American Naturalist 60(1):66–76.

Sheffield, R. E., S. Mostaghimi, D. H. Vaughan, E. R. Collins, Jr. and V. G. Allen. 1997. Off-stream water sources for grazing cattle as a stream bank stabilization and water quality BMP. Transactions of the ASAE 40:595–604.

Shisler, J. K., R. A. Jordan, and R. N. Wargo. 1987. Coastal wetland buffer delineation. Trenton, NJ: New Jersey Department of Environmental Protection.

Short, H. L. 1985. Management goals and habitat structure. Pp. 257–262 In: Riparian ecosystems and their management: reconciling conflicting uses. USDA/Forest Service Gen. Tech. Bull. RM-120. Washington, DC: USDA Forest Service.

Simberloff, D., and J. Cox. 1987. Consequences and costs of conservation corridors. Conservation Biology 1:63–71.

Simberloff, D., J. Farr, J. Cox, D. W. Mehlman. 1992. Movement corridors: conservation bargains or poor investments? Conservation Biology 6:493–504.

Simberloff, D. S., and E. O. Wilson. 1969. Experimental zoogeography of islands: the colonization of empty islands. Ecology 50:278–296.

Skovlin, J. M. 1984. Impacts of grazing on wetlands and riparian habitat: a review of our knowledge. Pp. 1001–1103 In: Developing strategies for rangeland management. Boulder, CO: Westview Press.

Smith, D. J. 1999. Identification and prioritization of ecological interface zones on state highways in Florida. In: Proceedings of the Third International Conference on Wildlife Ecology and Transportation, FL-ER-73-99. Evink, G. L., P. Garrett, and D. Zeigler (eds.). Tallahassee, FL: Florida Department of Transportation. 330 pp.

Smith, D. S. 1993. Greenway case studies. Pp. 161–214 In: Ecology of greenways: design and function of linear conservation areas. Minneapolis, MN: University of Minnesota Press.

Smith, D. S., and P. C. Hellmund, eds. 1993. Ecology of greenways: design and function of linear conservation areas. Minneapolis, MN: University of Minnesota Press.

Smith, R. D., A. Ammann, C. Bartoldus, and M. M. Brinson. 1995. An approach for assessing wetland functions using hydrogeomorphic classification, reference wetlands and functional indices. Technical Report TR-WRP-DE-9. Vicksburg, MI: Waterways Experiment Station, Army Corps of Engineers.

Soil and Water Conservation Society (SWCS). 2001. Realizing the promise of conservation buffer technology. Ankeny, IA: Soil and Water Conservation Society. 32 pp.

Soulæ, M., and R. Noss. 1998. Rewilding and biodiversity: complementary goals for continental conservation. Wild Earth, Fall 1998:1–11.

Spackman, S. C., and J. W. Hughes. 1995. Assessment of minimum stream corridor width for biological conservation: species richness and distribution along mid-order streams in Vermont, USA. Biological Conservation 71:325–332.

Spence, B. C., G. A. Lomnicky, R. M. Hughes, and R. P. Novitzki. 1996. An ecosystem approach to salmonid conservation. Management Technology TR-4501-96-6057, Corvallis, OR, 356 pp.

Stanford, J. A., and J. V. Ward. 1993. An ecosystem perspective of alluvial rivers: connectivity and the hyporheic corridor. Journal of the North American Benthological Society 12:48–60.

Stanford, J. A., J. V. Ward, W. J. Liss, C. A Frissell, R. N. Williams, J. A. Lichatowich, and C. C. Coutant. 1996. A general protocol for restoration of regulated rivers. Regulated Rivers 12(4–5) 391–413.

Stauffer, D. F., and L. B. Best. 1980. Habitat selection by birds of riparian communities: evaluating effects of habitat alterations. J. Wildl. Manage. 44:1–15.

Swan, B. 1979. Riparian habitat: the cattlemen's viewpoint. Pp. 4–6 In: Proceedings of the forum: grazing and riparian/stream ecosystems. Nov. 3–4, 1978. O. B. Cope (ed.). Denver, CO: Trout Unlimited.

Syndi, J. F., S. R. Abt, C. D. Bonham, C. C Watson, and J. C. Fischenich. 1998. Evaluation of flow-resistance equations for vegetated channels and floodplains. Technical Report EL-98-2. Vicksburg, MS: U.S. Army Corps of Engineers Waterways Experiment Station. 34 pp.

Thompson, K. 1961. Riparian forests of the Sacramento River Valley, California. Annals of the Association of American Geographers 51:294–315.

Topping, D. J., D. M. Rubin, and L. E. Vierra. 2000. Colorado river sediment transport Part 1. Natural sediment supply limitation and the influence of Glen Canyon Dam. Water Resources Research 36:515–542.

Toth, L. A., S. L. Melvin, D. A. Arrington, and J. Chamberlain. 1998. Hydrologic manipulations of the channelized Kissimmee River: implications for restoration. BioScience 48:757–764.

Trails and Wildlife Task Force, Colorado State Parks, and Hellmund Associates. 1998. Planning Trails with Wildlife in Mind. Denver, CO. (www.dnr.state.co.us/parks)

Trine, C. L. 1998. Wood thrush population sinks and implications for the scale of regional conservation strategies. Conservation Biology 12:576–585.

Turner, M. G. 1989. Landscape ecology: the effect of pattern on process. Annual Review of Ecology and Systematics 20:171–197.

U.S. Army Corps of Engineers (Corps). 1998. Alamo Dam & Lake Feasibility Report and EIS.

U.S. Fish and Wildlife Service (FWS). 1980. Ecological Services Manual (101-104 ESM). Washington, DC: U.S. Fish and Wildlife Service Division of Ecological Services.

USDA-NRCS. 1994. NRCS Conservation Practice Standard—Herbaceous Wind Barrier Code 422A. Washington, DC: USDA Natural Resources Conservation Service. 4 pp.

USDA-NRCS. 1998. Stream Visual Assessment Protocol. Technical Note 99-1. National Water and Climate Center. Washington, DC: USDA Natural Resources Conservation Service. Found at http://www.ncg.nrcs.usda.gov/pdf/svapfnl.pdf

USDA-NRCS. 1999a. NRCS Conservation Practice Standard—Filter Strip Code 393. Washington, DC: USDA Natural Resources Conservation Service. 4 pp.

USDA-NRCS. 1999b. NRCS Conservation Practice Standard—Field Border Code 386. Washington, DC: USDA Natural Resources Conservation Service. 3 pp.

USDA-NRCS. 1999c. NRCS Conservation Practice Standard—Contour Buffer Strips Code 332. Washington, DC: USDA Natural Resources Conservation Service. 4 pp.

USDA NRCS. 2000a. Conservation buffers to reduce pesticide losses. Washington, DC: USDA National Resources Conservation Service. 21 pp.

USDA-NRCS. 2000b. Filter Strips Conservation Reserve Enhancement Program CREP-CP21. Washington, DC: USDA Natural Resources Conservation Service. 4 pp.

USDA-NRCS. 2000c. NRCS Conservation Practice Standard—Riparian Forest Buffer Code 391. Washington, DC: USDA Natural Resources Conservation Service. 3 pp.

USDA-NRCS. 2000d. NRCS Conservation Practice Standard—Windbreak/Shelterbelt Establishment Code 380. Washington, DC: USDA Natural Resources Conservation Service. 4 pp.

USDA-NRCS. 2000e. NRCS Conservation Practice Standard—Grassed Waterway Code 412. Washington, DC: USDA Natural Resources Conservation Service. 2 pp.

USDA-NRCS. 2001. NRCS Conservation Practice Standard—Vegetative Barrier Code 601. Washington, DC: USDA Natural Resources Conservation Service. 4 pp.

U.S. Department of the Interior Bureau of Land Management (BLM) and U.S. Department of Agriculture Forest Service (USFS). 1994. Rangeland reform '94: draft environmental impact statement executive summary. Washington, DC: US DOI and USDA.

Verry, E. S., J. S. Hornbeck, and D. A. Dolloff. 2000. Riparian management in forests of the eastern United States. New York: Lewis Publishers. 402 pp.

Voogd, H. 1983. Multicriteria evaluation for urban and regional planning. London: Pion Limited. 367 pp.

Walker, R., and L. Craighead. 1997. Analyzing wildlife movement corridors in Montana using GIS. 1997 ESRI International Users Conference, July 8-11, 1997. (www.esri.com)

Warne, A. G., L. A. Toth and W. A. White. 2000. Drainage-basin-scale geomorphic analysis to determine reference conditions for ecologic restoration—Kissimmee River, Florida. Geologic Society of America Bulletin 112(6):884–899.

Wells, D., B. W. Anderson, and R. D. Ohmart. 1979. Comparative avian use of southwestern citrus orchards and riparian communities. Journal of the Arizona-Nevada Academy of Science 14:53–58.

Welsh, D. 1991. Riparian forest buffers—function and design for protection and enhancement of water resources. USDA-FS Pub. No. NA-PR-07-91. Radnor, PA: USDA-FS.

Wesche, T. A. 1985. Stream channel modifications and reclamation structures to enhance fish habitat. Pp. 103–163 In: The restoration of rivers and streams: theories and experience. J. A. Gore (ed.). Stoneham, MA: Butterworth Publishers. 279 pp.

Whigham, D. F., L. C. Lee, M. M. Brinson, R. D. Rheinhardt, M. C. Rains, J. A. Mason, H. Kahn, M. B. Ruhlman, and W. L. Nutter. 1999. Hydrogeomorphic (HGM) assessment—a test of user consistency. Wetlands 19:560–569.

Whipple, W. 1993. Buffer zones around water-supply reservoirs. J. Water Res. Plann. Manage. 119(4):495–499.

Whitaker, G., R. Misso, M. Schamberger, C. Cordes, and G. Hickman. 1985. HEP Review. Unpublished manuscript dated October 7, 1985. Washington, DC: U.S. Fish and Wildlife Service.

Whiting, P. J., and J. B. Bradley. 1993. A process-based classification system for headwater streams. Earth Surface Processes and Landforms 18:603–612.

Whittaker, D., B. Shelby, W. Jackson, and R. Beschta. 1993. Instream flows for recreation: a handbook on concepts and research methods. Anchorage, AK: U.S. Department of the Interior, National Park Service Water Resources Division. 103 pp.

Wigington, P. J., and R. L. Beschta, eds. 2000. International conference on riparian ecology and management in multi-use watersheds. Middleburg, VA: American Water Resources Association. 616 pp.

Wigley, T. B. 1996. Wildlife in streamside management zones: what do we really know? Pp. 85–90 In: At the water's edge: the science of riparian forestry. Conference Proceedings, January 1996. BU-6637-S. Minneapolis, MN: University of Minnesota Extension Service.

Wilhelm, G. S., and D. Ladd. 1988. Natural area assessment in the Chicago region. Pp. 361–375 In: Transactions of the 53rd North American Wildlife and Natural Resources Conference, March 18–23, 1988 in Louisville, Kentucky. Washington, DC: Wildlife Management Institute.

Wilkinson, T. 2001. Open grazing vs. barbed wire in New West. Christian Science Monitor, Jan. 22.

Williams, J. E., C. A. Wood, and M. P. Dombeck (eds.). 1997. Watershed restoration: principles and practices. Bethesda, MD: American Fisheries Society. 561 pp.

Winter, T. C. 1992. A physiographic and climatic framework for hydrologic studies of wetlands. Pp. 127–148 In: Aquatic ecosystems in semi-arid regions: implications for resource management. R. D. Robarts and M. L. Bothwell (eds.). N.H.R.I. symposium Series 7. Saskatoon, Sasketchewan: Environment Canada.

Winter, T. C., and M.-K. Woo. 1990. Hydrology of lakes and wetlands. Pp. 159–187 In: The geology of North America. Vol. 0-1. Surface water hydrology. M. G. Wolman and H. C. Riggs (eds.). Boulder, CO: The Geologic Society of America.

Wissmar, R. C., and R. R. Beschta. 1998. Restoration and management of riparian ecosystems: a catchment perspective. Freshwater Biology 40(3):571–585.

Woessner, W. W. 2000. Stream and fluvial plain ground water interactions: rescaling hydrogeologic thought. Ground Water 38:423–429.

Wuerthner, G. 1990. Grazing the western range: what costs, what benefits? Western Wildlands: 27–29.
Xue, Y., M. B. David, L. E. Gentry, and D. A. Kovacic. 1998. Kinetics and modeling of dissolved phosphorus export from a tile-drained agricultural watershed. Journal of Environmental Quality 27:917–922.
Young, R. A., T. Huntrods, and W. Anderson. 1980. Effectiveness of vegetative buffer strips in controlling pollution from feedlot runoff. J. Environ. Qual. 9:483–487.

Appendix A

Committee Member and Staff Biographies

Mark M. Brinson, *Chair*, is a professor of ecology at East Carolina University. His research focuses on wetland ecosystem structure and function, tidal and nontidal marshes, and riverine and fringe forested wetlands. Dr. Brinson interacts extensively with federal and state government agencies in providing advice on wetland issues, participating in training courses, testifying at congressional hearings, and guiding research priorities and directions. He received his B.S. from Heidelberg College, his M.S. from the University of Michigan, and his Ph.D. in botany from the University of Florida. Dr. Brinson was a member of the NRC Committee on Characterization of Wetlands.

Lawrence J. MacDonnell, *Vice-Chair*, is an environmental and natural resources attorney with the firm of Porzak, Browning & Bushong in Boulder, Colorado. His practice emphasizes water law and the Endangered Species Act. He is active in several nonprofit organizations involved in community-based conservation in Colorado. Between 1983 and 1994, he served as the initial director of the Natural Resources Law Center at the University of Colorado School of Law, where he also taught courses in environmental and natural resources law. His most recent book is *From Reclamation to Sustainability: Water, Agriculture, and the Environment in the American West*. He is a past president of the Colorado Riparian Association. Dr. MacDonnell served on the NRC Committee on the Future of Irrigation. He received his B.A. from the University of Michigan, his J.D. from the University of Denver, and his Ph.D. from the Colorado School of Mines.

Douglas J. Austen is head of the Watershed Management Section for the Illinois Department of Natural Resources. His work involves directing an interagency watershed program that involves planning, implementation, and assessment of a wide variety of land and aquatic management practices in four paired watersheds located throughout Illinois. Dr. Austen also coordinates the assessment of the Illinois River Conservation Reserve Enhancement Program (CREP), he supports ecosystem restoration efforts throughout the state, and he co-developed the Illinois Watershed Academy. He also has extensive experience and has authored publications on fisheries management, natural resources information systems, sampling, and aquatic ecology. Dr. Austen received his B.S. from South Dakota State University, M.S. from Virginia Polytechnic Institute and State University, and Ph.D. from Iowa State University in Animal Ecology.

Robert L. Beschta is an emeritus professor of forest hydrology at Oregon State University. He received his B.S. from Colorado State University, his M.S. from Utah State University, and his Ph.D. in watershed management from the University of Arizona. Dr. Beschta's research interests include forest hydrology, forest-land use and water quality, riparian area management, and channel morphology. In particular, he studies large organic debris and channel morphology; hillslope hydrology; road drainage; the effects of mass soil movements upon channel characteristics; how sediment is stored along stream channels and released during high stream flow events; and interactions of stream and subsurface water in eastern Oregon. Dr. Beschta served on the NRC Committee on the Protection and Management of Pacific Northwest Anadromous Salmonids.

Theo A. Dillaha is a professor of biological systems engineering at the Virginia Polytechnic Institute and State University. His research interests include environmental engineering, functioning of riparian buffer zones, water-quality modeling, nonpoint source pollution control, TMDLs, and water supply and sanitation in developing countries. He has published extensively on the abilities of grass buffer strips to remove nutrients, sediment, and bacteria from overland flow. Dr. Dillaha received his B.E. and M.S. in environmental and water resources engineering from Vanderbilt University and his Ph.D. in agricultural engineering from Purdue University.

Debra L. Donahue is a professor of law at the University of Wyoming. She received her B.S. in wildlife science from Utah State University, her M.S. in wildlife biology from Texas A&M University, and her J.D. from the University of Colorado School of Law. Her formal work on riparian areas began during her graduate studies on river otters in Louisiana. She has subsequently been employed as a seasonal biologist for the Bureau of Land Management surveying stream and riparian habitats; as an area wildlife biologist in Nevada; and as an environmental coordinator for the Freeport Gold Company, where her environmental compliance responsibilities included monitoring stream and spring water quality. She now teaches courses in public land, land use, water pollution, and natural resources law. Her current research interests are in using science and the law to protect riparian areas and the quality of surface waters, particularly in the West.

Stanley V. Gregory has been a professor of fisheries and wildlife at the Oregon State University since 1986. His research focuses on stream ecosystems, including channel dynamics, woody debris, benthic algae, invertebrates, fish, salamanders, and riparian vegetation. He has studied the influence of human activities on ecosystem structure and function, and he is an expert in the historical reconstruction of rivers and riparian forests. Dr. Gregory received his B.S. in zoology from the University of Tennessee and his M.S. and Ph.D. in fisheries from Oregon State University. He has served on the NRC Committee on Environmental Issues in Pacific Northwest Forest Management.

Judson W. Harvey has been a hydrologist with the U.S. Geological Survey's National Research Program in Reston, VA since 1995. His research is focused on hydrologic transport and biogeochemical reactions near the interface between groundwater and surface water. Dr. Harvey has investigated water flow and chemical reactions in streams and wetlands throughout the country, including steep forested watersheds of the Rocky Mountains in Colorado, alluvial floodplains of semi-arid basins in Arizona, and peatlands in the Everglades of south Florida. His training in hydrology was at the Department of Environmental Sciences at the University of Virginia, after which he held an NRC Postdoctoral Fellowship at the USGS in Menlo Park, CA. Dr. Harvey currently serves on the editorial board of *Water Resources Research* and on the Water Quality Committee of the American Geophysical Union.

Manuel C. Molles, Jr. is a professor of biology at the University of New Mexico and curator of ichthyology and arthropods at the Museum of Southwestern Biology. He received his Ph.D. in zoology from the University of Arizona. Dr. Molles' research focuses on riparian ecology, ecology of desert streams, riverine and riparian biodiversity, and ecology of exotic species, with particular attention given to the Rio Grande. In addition, he is an expert in groundwater–stream interactions and the effect of forest fire and flooding on riparian ecosystems.

Elizabeth I. Rogers is a research ecologist with White Water Associates, Inc., where she is involved in a variety of research, resource management. and education projects involving riparian areas. Dr. Rogers has been a principal scientist on the ecological studies of rivers required by the Federal Energy Regulatory Commission for relicensing of hydroelectric projects. In addition, she has conducted research on bird and mammal use of riparian areas of rivers, lakes, and ephemeral pools. Much of her research has been focused on creating recommendations for reducing impacts of forestry and agriculture in riparian areas. Dr. Rogers has taught both field and lecture courses on riparian functions and values to resource managers from state and federal agencies and the private sector, as well as real estate agents dealing with lakeshore properties. She received her Ph.D. in zoology (ornithology) from Michigan State University, her M.S. in environmental education from Southern Oregon State University, and her B.S. in biology from Central Michigan University.

Jack A. Stanford is the Bierman Professor of Ecology at the University of Montana and the director of the Flathead Lake Biological Station. He received his B.S. and M.S. from Colorado State University and his Ph.D. from the University of Utah. He is interested in the many natural factors and disturbances that interact to determine the distribution of species and productivity of food webs in aquatic and terrestrial environs, as well human influences that alter natural biogeochemical patterns. Dr. Stanford has extensive experience studying the ecology of mountain rivers (Flathead, Columbia, Missouri, Colorado) and their interaction with the shallow subsurface. In particular, he has studied how flow regulation by dams and diversions alters important ecosystem processes, like interstitial flow, gravel transport, temperature patterns, and channel–floodplain connectivity. Dr. Stanford was a member of the NRC Committee on Watershed Management.

Laura J. Ehlers is a senior staff officer for the Water Science and Technology Board of the National Research Council. Since joining the NRC in 1997, she has served as study director for eight committees, including the Committee to Review the New York City Watershed Management Strategy, the Committee on Riparian Zone Functioning and Strategies for Management, and the Committee on Bioavailability of Contaminants in Soils and Sediment. She received her B.S. from the California Institute of Technology, majoring in biology and engineering and applied science. She earned both an M.S.E. and a Ph.D. in environmental engineering at the Johns Hopkins University.